International Conference on Entrepreneurial Competencies in a Changing World (ECCW 2021)

SHS Web of Conferences Volume 135 (2022)

Ceske Budejovice, Czech Republic
24 November 2021

Editors:

Zdenek Caha
Michal Ruschak
Vendula Velková

ISBN: 978-1-7138-4390-0

Printed from e-media with permission by:

Curran Associates, Inc.
57 Morehouse Lane
Red Hook, NY 12571

Some format issues inherent in the e-media version may also appear in this print version.

For additional information, please contact EDP Sciences – Web of Conferences
at the address below.

EDP Sciences – Web of Conferences
17, Avenue du Hoggar
Parc d'Activité de Courtabœuf
BP 112
F-91944 Les Ulis Cedex A
France

Phone: +33 (0) 1 69 18 75 75
Fax: +33 (0) 1 69 28 84 91

contact-edps@webofconferences.org

Additional copies of this publication are available from:

Curran Associates, Inc.
57 Morehouse Lane
Red Hook, NY 12571 USA
Phone: 845-758-0400
Fax: 845-758-2633
Email: curran@proceedings.com
Web: www.proceedings.com

TABLE OF CONTENTS

INTERNATIONAL CONFERENCE ON ENTREPRENEURIAL COMPETENCIES IN A CHANGING WORLD (ECCW 2021)

24th November 2021
České Budějovice, Czech Republic

VŠTE

OTH REGENSBURG

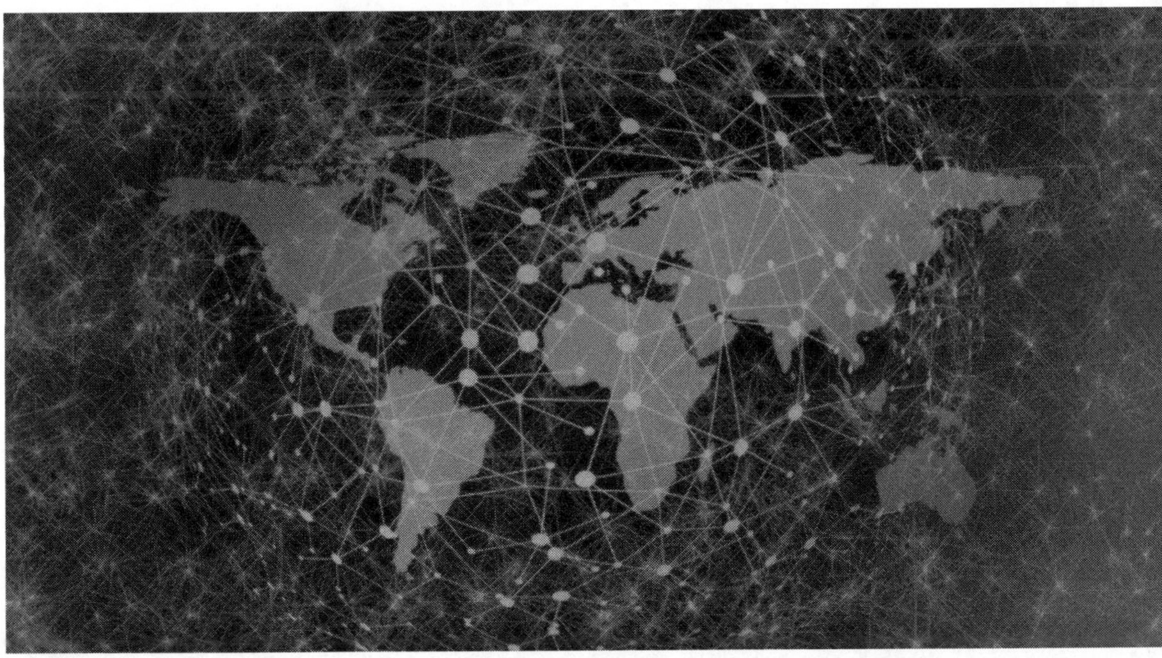

International Conference on Entrepreneurial Competencies in a Changing World (ECCW 2021)

http://www.podnikatelskekompetence.cz/

Editors: Zdeněk Caha, Michal Ruschak, Vendula Velková

About the conference ECCW 2021

Date of the conference: 24th November 2021

Due to the unfavorable situation associated with the covid-19 pandemic, the conference was organized online.

Target

The tareget of the conference is increasing and maintaining the competitiveness of enterprises not only in connection with the effects of the Covid-19 pandemic. Exchange of the latest knowledge to support the flexibility of companies and innovative practices, implemented in all primary and secondary functions of the company.

General topic

Strengthening the competitiveness of start-ups and existing companies in the post-Covid-19 period in the field of:

- Economics
- Strategic management and decision making
- Human Resources SMEs
- Innovations
- Legislative-legal area
- Marketing
- Business ethics and Corporate social responsibility

About organizers

Institute of Technology and Business in České Budějovice

Website: http://www.vstecb.cz/en/

Institute of Technology and Business (VŠTE) has about 4,500 students of technically and economically-oriented study programmes, two Master´s and six Bachelor´s study programmes with a wide range of specializations in civil engineering, mechanical engineering, transport and logistics and economics. Moreover, there are about 2,000 participants in lifelong learning, who officially do not have the status of "student, but it is one of the important activities of the school. This way it is possible to study two-year programmes of the European Council of Business Administration, MBA in Financial Management and BBA in Business Valuation. VŠTE is a higher education establishment with large project activities the main goal of which is to fulfil the long-term intentions of the Institute. The Institute is currently focused on systematic and permanent improvements in quality, deepening the cooperation with the application sphere, infrastructure development, transferring research activities into the application sphere, and increasing the employability of graduates. The Institute's project activities have significantly increased over the past two years.

Faculties

- Faculty of Corporate Strategy
- Faculty of Technology
- School of Expertness and Valuation

Ostbayerische Technische Hochschule Regensburg

Website: https://www.oth-regensburg.de/en.html

Ostbayerische Technische Hochschule Regensburg (OTH Regensburg) offers its students an excellent basis for a successful career. The technical university of applied sciences has around 11,000 students. 50 practice-oriented bachelor and master degree programmes in engineering, business studies, design, architecture, health and social studies offer high quality teaching and research. Modern facilities, such as 120 state-of- the-art laboratories and an award-winning library located on an attractive campus provide the basis for successful studies. OTH Regensburg's areas of expertise are energy & mobility, information & communication, life sciences & ethics, production & systems and building & infrastructure. These areas are tied together by expertise in sensor technology and digitalization. Cooperative applied research spans fields such as renewable energy, sensor technology, mechatronics, bioengineering, computer sciences and automotive engineering.

Faculties

- Applied Natural Sciences and Cultural Studies
- Architecture
- Civil Engineering
- Business Studies
- Electrical Engineering and Information Technology
- Computer Science and Mathematics
- Mechanical Engineering
- Social and Health Care Sciences

Guaranteed by

PaedDr. Mgr. Zdeněk Caha, Ph.D., MBA, MSc.
Prof. Dr. Sean Patrick Sassmannshausen
Prof. Ing. Marek Vochozka, MBA, Ph.D.

Scientific Committee

Prof. Ing. Marek Vochozka, MBA, Ph.D.
Institute of Technology and Business in České Budějovice (CZ)

PaedDr. Mgr. Zdeněk Caha, Ph.D., MBA, MSc.
Institute of Technology and Business in České Budějovice, Faculty of Corporate Strategy (CZ)

Prof. Dr. Sean Patrick Sassmannshausen
Ostbayerische Technische Hochschule Regensburg (DE)

Assoc. prof. PhDr. Ing. Ladislav Mura, PhD.
J. Selye University (SK)

Assoc. prof. Ing. Jiří Dušek, Ph.D.
The College of European and Regional Studies (CZ)

Prof. Ing. Alžbeta Kucharčíková, PhD.
University of Zilina (SK)

Ing. Marcel Kordoš, PhD.
Alexander Dubček University of Trenčín (SK)

Ing. Silvia Lorincová, PhD.
Technical University Zvolen (SK)

doc. Oleksandr Kushch, CSc.
Kremenchuk National University Mykhaylo Ostrogradskiy (UA)

Ing. Petra Pártlová, Ph.D.
Institute of Technology and Business in České Budějovice (CZ)

Organizing Committee

Ing. Michal Ruschak, MBA

PaedDr. Mgr. Zdeněk Caha, Ph.D., MBA, MSc.

Ing. Iva Klementová, Ph.D.

Ing. Vendula Velková

Ing. Michaela Procházková

Statement of Peer review

In submitting conference proceedings to *Web of Conferences*, the editors of the proceedings certify to the Publisher that

1. They adhere to its **Policy on Publishing Integrity** in order to safeguard good scientific practice in publishing.
2. All articles have been subjected to peer review administered by the proceedings editors.
3. Reviews have been conducted by expert referees, who have been requested to provide unbiased and constructive comments aimed, whenever possible, at improving the work.
4. Proceedings editors have taken all reasonable steps to ensure the quality of the materials they publish and their decision to accept or reject a paper for publication has been based only on the merits of the work and the relevance to the journal.

International Conference on Entrepreneurial Competencies in a Changing World (ECCW 2021), 24 November 2021, Ceske Budejovice, Czech Republic

Proceedings editor(s): Zdeněk Caha, Michal Ruschak, Vendula Velková

Date and editor's signatur

31. 1. 2022

Liability for Damage to Health Due to Limited Access to Health Care during Emergency of the Covid-19 Pandemic

Stanislav Bílek[1,*]

[1]Institute of Technology and Business, Faculty of Corporate Strategy, Okruzni 517/10, 37001 České Budějovice, Czech Republic

Abstract. Most successful companies have established effective marketing communication as a useful tool for selling their products to customers. The research explored the ways of communication of a selected internet subject, its methods of addressing customers and how effective the communication of the specific enterprise is not only during covid19 pandemic. We carried out a SWOT analysis, conducted a questionnaire survey and made detailed This paper discusses the responsibility / liability of the state and health service providers in relation to limited access by individuals to preventive as well as other outpatient or inpatient care as a result of the measures of the Government of the Czech Republic and the Ministry of Health of the Czech Republic during a declared state of emergency due to the coronavirus pandemic /referred to as SARS CoV-2/. The paper provides a legal analysis of the current legislation in health care services and the circumstances under which health care may be refused by its provider and discusses the possibilities of claiming compensation for damages incurred by affected persons as a result of limited access to health care. The survey is mainly focused on civil liability for non-pecuniary damage to health and human life and liability under the Crisis Management Act of the Czech Republic.
Keywords: emergency; general measures; liability for damage to health

1 Introduction

Limited access by individuals to preventive and outpatient health care during a state of emergency declared by the Government of the Czech Republic [1], which was subsequently extended by the Chamber of Deputies of the Parliament of the Czech Republic, entails one of the most serious impacts of the SARS CoV-2 pandemic, especially the effects on health of the affected persons. This happened repeatedly based on general measures adopted by the Ministry of Health of the Czech Republic [2], and health care providers were reducing preventive and outpatient care to follow these measures [3]. Compensation for damage to health, deterioration of health or death of a person causally related to limited access to health care may become the subject of many legal actions and disputes in the years to come where the liability of the state and health care providers for such damages will be addressed, taking into account the unprecedented state of emergency which has never before affected the function of public health to such extent.

A number of international studies across many disciplines were produced in the immediate aftermath of the SARS CoV-2 coronavirus pandemic outbreak, including papers examining the medical, sociological, legal and economic aspects of the pandemic. For example, the research of Firan and Fatine [4] who look at the economic consequences of the epidemic and point out that the choice of restrictive policies made it possible to reduce the negative medical impact of the pandemic. Another paper [5] emphasizes that the COVID-19 pandemic forced societies around the world to resort to social distancing in order to slow the spread of SARS-CoV-2.

But let us focus on the legal aspects of the pandemic that has occurred. The legal implications are also addressed by many international and national scientific studies and papers. Comazzeto [6] in his scholarly article discusses some aspects of the pandemic emergency, and analyzes the legal basis of the so-called lockdown with a focus on the definition of the "state of exception" in order to evaluate its function as the key concept in public law discourse. Barquero [7] reviews the legal requirements for court approval / ratification of health measures taken by public authorities, affecting the exercise of fundamental rights, including the interpretation doubts that have led to a high dispersion of criteria in judicial decisions.

The interplay between law and medicine can be seen in another author who addresses the system's response to the feared mass influx of patients in intensive care units as a result of the COVID-19 pandemic. Kern and Peters [8] analyze this and conclude, among other things, that it would be useful to be aware of and apply to the situation the basic knowledge of so-called disaster medicine even if this is not a disaster in the technical sense.

* Corresponding author: stanislav.bilek@mail.vstecb.cz

The issue of political, institutional and professional responsibility in the context of the SARS-COV-2 pandemic is currently widely discussed and involves several levels of investigation. One crucial aspect is the allocation of life-saving resources in a situation where there is lack of balance between the need and availability and the related issues of ethical and legal responsibility. In this context, Gambino et al. [9], for example, discuss the implications of the criteria used for the allocation of benefits in emergency and the issue of their legitimacy.

Apparently, the legal and ethical implications of the pandemic impact on medicine and health care have a broader scope across all countries of the world. It is obvious the situation that has arisen is quite unique, with unexpected economic, health and social impacts influenced by, among other things, targeted legislative interventions by the state in the effort to prevent system overload. In this context, the author of this paper is looking at the impact of limited access to health care for citizens as a result of the state measures, focusing on national law and health care providers in the Czech Republic. The analysis of the current legislation overlaps foreign legislation in parallel as the citizen's right to protection of life and health and to the provision of health care is guaranteed by international conventions for the protection of human rights. During emergency, the state measures primarily limited access to preventive health care, but also to other health care for citizens of the Czech Republic and foreigners.

The aim of this paper is analyze the legal aspects of responsibility / liability of the state (the Czech Republic) and health care providers in the Czech Republic due to limited access by citizens to health care during a declared state of emergency.

The author asked the following research questions to fulfil the aim of the paper: 1. Is the state responsible under the current legislation for damage to health caused by the postponement of health care due to the state emergency measures imposed during the state of emergency? 2. Is the health care provider responsible under the current legislation for damage to health caused by the postponement of health care due to the state emergency measures imposed during the state of emergency?

2 Methodology

The author uses a method of data collection consisting in content analysis of the current legislation, in particular constitutional laws, laws, government resolutions, and general measures as well as content analysis of case law and previously published scholarly articles related to the topic. The author's proposed measures are then built upon his private law practice.

The author analyzes (using legal interpretation methods) the content of the below documents to the extent necessary to fulfil the above aim. It is to be expected that national law has legal instruments to protect the affected persons in the normal course and function of the health care system. However, the crisis management legislation and the specific measures adopted by the state during emergency need to be particularly analyzed to determine the relationship (ratio) in emergency between the private citizen's health and life interest on the one hand and the public state interest in implementing measures to mitigate the global impact of the pandemic.

In terms of limited access to health care in the Czech Republic during the state of emergency, the following findings should primarily be taken into account.

The Charter of Fundamental Rights and Freedoms which forms the constitutional order of the Czech Republic reads (Article 31) that "*Everyone has the right to health protection. Citizens have the right, on the basis of public insurance, to free health care and to medical aids under the conditions laid down by law.*" According to the Charter of Fundamental Rights and Freedoms, Article 6, "*everyone has the right to life. Human life is worthy of protection even before birth. No one shall be deprived of life.*" [10].

The Constitutional Act No. 110/1998 Coll., on the Security of the Czech Republic, lays down the rules for limitations of the fundamental human rights. *The institution of emergency is explained in Article 5 which states that "the Government may declare a state of emergency in the event of natural disasters, environmental or industrial accidents, incidents or other dangers that threaten life, health or property values or the internal order and security to a significant extent. The Government shall immediately inform the Chamber of Deputies of the declaration of a state of emergency which may revoke the declaration." Article 6 of the same constitutional law provides further details of rules for declaring a state of emergency: "A state of emergency may be declared only with reasons for a specific territory and for a specific period of time. In parallel to the declaration of a state of emergency, the Government must define which rights provided for in a special law are restricted and to what extent, in accordance with the Charter of Fundamental Rights and Freedoms, and which duties are imposed and to what extent. The details shall be laid down by law. A state of emergency may be declared for a maximum period of 30 days. That period may only be extended subject to the prior consent of the Chamber of Deputies. The state of emergency shall end upon the expiry of the period for which it was declared unless the Government or the Chamber of Deputies should decide to lift it before the expiry of that period." [11].*

For the declaration of a state of emergency to be correct, among other things, the reasons therefore must be defined and it must be temporary and the territorial scope must be determined. Above all, however, the constitutional law imposes on the Government of the Czech Republic which rights stipulated in laws and the Charter of Fundamental Rights and

Freedoms will be restricted by such a measure. As is evident from the text of the government resolution on the declaration of a state of emergency, there is no list of restricted rights although the constitutional law requires this to be defined "in parallel" with the declaration of the state of emergency [11].

The details of a state of emergency are regulated by Act No.240/2000 Coll., on Crisis Management and on Amendments to Certain Acts (the Crisis Management Act). We will particularly look at the possibility of compensation for damages arising out of or in connection with the implemented crisis measures. This Act defines the competence and authority of public authorities and regional government authorities and the rights and duties of legal and natural persons in preparing for and managing emergency and crisis that are not related to the defence assurance of the Czech Republic against external attacks and in protecting the critical infrastructure, and liability for breach of those duties and obligations. Before we analyze the issue of the state's liability for damage, the terms of crisis management and crisis measures should be defined. The crisis management is defined by the Act Section 2 as *"The summary of management activities of crisis management authorities focused on analysis and evaluation of security risks and planning, organization, implementation and control of activities performed in connection with 1) preparation for crisis situations and their resolution, or 2) protection of critical infrastructure. A crisis measure denotes an organizational or technical measure intended to manage and mitigate a crisis situation and eliminate its consequences, including measures interfering with personal rights and duties."* [12]. For the purposes of this paper, it is therefore important to conclude that crisis measures are the package of all measures adopted by various competent public authorities in the context of an emergency. The form of crisis measures may be a government resolution, a law, a decree, a general measure, an administrative decision, or any actual measures adopted by public authorities [13]. The Crisis Act Section 5 provides an exhaustive list of specific measures that may be imposed by the Government. None is about limiting access to health care [12].

On 12 March 2020, the Government of the Czech Republic adopted a resolution declaring a state of emergency for the territory of the Czech Republic due to a threat to global health in connection with the demonstrated presence of coronavirus /referred to as SARS CoV-2/ in the territory of the Czech Republic for a period of 30 days starting 2.00 pm on 12 March 2020. The resolution was published in the Collection of Laws under No. 69/2020. It reads, inter alia, that: *"In accordance with Article 5(a) to (e) and Article 6 of Act No. 240/2000 Coll., on crisis management and on amendments to certain acts (the Crisis Management Act), as amended, the Government hereby orders and adopts crisis measures to manage the crisis with the specific implementation of the measures to be determined under a separate resolution of the Government. "*

On 9 April 2020, the Government extended the state of emergency until 30 April 2020 by resolution published in the Collection of Laws under No. 156/2020, with the prior consent of the Chamber of Deputies.

The Government of the Czech Republic repeatedly declared a state of emergency by its Resolution No. 957 of 30 September 2020, starting 5 October 2020, which was subsequently extended until 14 February 2021 with the approval of the Chamber of Deputies. Thereafter, the Chamber of Deputies did not approve the state of emergency, therefore the Government of the Czech Republic declared a new state of emergency by its Resolution No. 125 of 14 February 2021 for a period of 14 days from 15 February. The Constitutional Court of the Czech Republic, by its resolution no. ÚS 12/21 of 15 March 2021, rejected the petition for review of constitutionality of this action because of this declaration being a political act which is not subject to legal review. The Constitutional Court also described the declaration of a state of emergency as an unreviewable act of government in its previous decisions Pl. ÚS 5/21, Pl. ÚS 9/21 and Pl. ÚS 10/21. In the resolution Pl. ÚS 8/20 of 22 April 2020, the Constitutional Court explained that *"the declaration of a state of emergency by the Government is, in the present case, primarily an act of application of constitutional law; it constitutes an 'act of government' which has a normative impact, in principle not subject to review by the Constitutional Court and may only be 'reviewed' by the primary democratically elected political ('non-judicial') body, which is the Chamber of Deputies. The emergency declaration act could be annulled by the Constitutional Court if it should contravene the fundamental principles of the democratic rule of law and if it should constitute a change in the essential elements of the democratic rule of law. In the resolution by which the state of emergency was declared, the Government defined the rights to be restricted by reference to Sections 5(a) to (e) and 6 of the Crisis Management Act and to other more specific Government resolutions. Although such an implicit definition is general and vague, it cannot be regarded as an expansive violation or abuse of constitutional law. The above formal deficiency cannot be inferred to invalidate the decision to declare a state of emergency since an overly strict assessment of formal requirements would completely disregard the seriousness and exceptional nature of the situation which led to the declaration of a state of emergency."* [14].

On the other hand, the Municipal Court in Prague concluded that the declaration of a new state of emergency was unconstitutional and illegal, in its judgment No. 17 A 126/2020- 84 dealing with the distance learning case, providing the reasons that *"This is not a new state of emergency, but a de facto continuation of the state of emergency declared by Government Resolution No. 125, against the will of the Chamber of Deputies, which did not consent to the continuation of this state of emergency."* [15].

On 27 February 2021, Act No. 94/2021 Coll., on Emergency Measures during the COVID-19 Disease Epidemic, became effective and replaced this institution with the so-called pandemic emergency.

The following general measures adopted by the Ministry of Health of the Czech Republic are the first to be taken into account with regard to limited access to health care during the state of emergency. During the state of emergency from

12 March to 17 May 2020, the first general measure was adopted on 16 March 2020, Ref. MZDR 12066/2020-1/MIN/KAN, by which the Ministry of Health ordered the health care providers of acute inpatient care services to reduce the medical procedures within elective care to the extent strictly necessary, taking into account the state of health of the patients for whom such medical procedures were scheduled to ensure a sufficient capacity and resources (personnel, material and technical) for the treatment of patients with covid-19 disease, with effect from 17 March 2005 (repealed with effect from 18 May 2020 by the measure Ref. MZDR 12066/2020-2/MIN/KAN). This measure essentially restricted health care other than acute.

Similarly, during the so-called second wave of the pandemic, the Ministry of Health adopted the measure Ref. MZDR 46953/2020-1/MIN/KAN on 26 October 2020, by which acute inpatient care providers were ordered to immediately stop admitting new patients for the purpose of elective procedures. This order was repealed with effect from 24 November 2020 by the Ministry of Health's measure of 23 November 2020, Ref. MZDR 46953/2020-2/MIN/KAN. At the same time, however, this new measure ordered *"All providers of acute inpatient health care services with effect from 24 November 2020, and all providers of inpatient follow-up and long-term care, including intensive follow-up, to take measures in order to ensure the necessary and adequate personnel, material and technical resources for providing health services to patients with COVID-19 disease caused by the new SARS-CoV-2 coronavirus such that acute inpatient care providers who have previously been involved in the provision of inpatient care to patients with COVID-19 disease should maintain the necessary capacity of available acute care beds, including intensive care beds, for patients with COVID-19, according to the current epidemiological situation and predicted need for hospital care, and inpatient follow-up care providers should maintain the necessary capacity of available follow-up care beds to receive patients transferred from acute care"*. These measures were then extended until they were definitively repealed on 1 July 2021 by the emergency measure Ref. MHDR 46953/2020-8/MIN/KAN.

For the sake of completeness, note that the measures were adopted by the Ministry of Health as crisis measures under the Crisis Management Act, but with reference to Section 69(1)(d) and (i) and (2) of Act No. 258/2000 Coll., on the Protection of Public Health, as amended. And measures adopted under legislation other than the Crisis Management Act should also be regarded as crisis measures [14]. In accordance with the Act on Protection of Public Health, the Ministry of Health may order to, in particular, close day care or inpatient care facilities and allocate beds in health care facilities, and it may order or prohibit some other specific activities to eliminate an epidemic or the risk of an epidemic. Emergency measures are imposed to the extent strictly necessary in accordance with Section 69 of the Act [16].

A specific example of the implementation of such measures is the decision of the České Budějovice Hospital of 6 January 2021 to postpone all elective procedures / hospitalizations scheduled until 31 January 2021 where it is medically safe according to the emergency measure of the Ministry of Health of the Czech Republic "without significant impact on the health of patients". It was explicitly stated that *"The scheduling of new patients for these procedures will be stopped. The decision to postpone specific types of procedures will always be up to the head of the department. Patients concerned will be informed by the medical staff by phone, text message or email as in the spring and autumn. Patients will be rebooked for a new appointment when full operations resume. We ask those who have had a procedure cancelled to contact us if their health should deteriorate. This measure does not apply to the treatment of acute illnesses and diseases where the postponement could lead to damage to patients' health (in particular, cancer surgery, cardiovascular and thoracic surgery, or major orthopaedic procedures for symptomatic patients). For patients who have already been admitted to the hospital for these procedures, the scheduled procedure will be performed. Similar to the autumn, this is approximately a 50% reduction in care."* The partial limitation specifically affected elective care in the Surgery, Urology, Gynecology, Skin and Venereal Diseases, Neurosurgery and Cardiology Departments. Health care was similarly reduced by other providers regardless of their legal form and it was regularly reported in the media [3].

For analysis of the topic, it is also necessary to briefly describe the individual types of health care according to Sections 5 to 9 of Act No. 372/2011 Coll., on Health Services and Conditions of Their Provision (the Health Services Act), as amended. The types of health care are divided according to the urgency of their provision into urgent care, emergency care, essential care and elective care. This division should always be strictly considered in each individual case as the subdivision of each procedure refused or postponed by the provider will determine whether the provider proceeded in accordance with or contrary to the emergency measure. The forms of health care are then outpatient care, day care, inpatient care and health care provided in the patient's own social environment according to Section 6 of the Act [17].

In general, the analysis of legislation concerning the responsibility / liability of the state or health care providers for non-pecuniary damage must be based on Act No. 89/2012 Coll., the Civil Code, as amended, Act No. 240/2000 Coll., on Crisis Management and on Amendments to Certain Acts (Crisis Management Act), Act No. 82/1998 Coll., on Liability for Damage Caused by a Decision of a Public Authority or Maladministration, as amended,, Act No. 258/2000 Coll., on the Protection of Public Health, as amended, and Act No. 372/2011 Coll., on Health Services.

The legal regulation on compensation for non-pecuniary damage set forth in the Civil Code is only applicable to liability in relationships arising from private law, thus a healthcare provider's liability towards an injured individual. The state's liability / responsibility is not regulated by the Civil Code. Nevertheless, some of the general provisions of this law can also apply to liability in the relationships discussed below. Namely, Section 2951(2) (definition of adequate compensation), Section 2956 (compensation for injury on natural rights, including in respect of health, life and mental suffering), Section 2958 (scope of compensation for injury to health), and Section 2959 (scope of compensation in case of death or particularly serious bodily injury in relation to the victim's close persons) [18].

The Public Health Protection Act or the Health Services Act do not regulate liability in relationships. Nevertheless, Section 48 of the Health Services Act which regulates the conditions under which a provider may refuse to admit a patient for care, and Section 50 of the same Act, which regulates the right of a health professional not to provide a health service should be taken into account. Any violation of these rules constitutes an illegality which is one of the prerequisites for the provider's liability for non-pecuniary damage.

According to Section 48, among other things, *"The health service provider chosen by the patient may refuse to admit a patient if the admission of the patient should exceed the tolerable workload or if the admission of the patient is prevented by operational reasons, staffing or technical and material equipment of the health care facility; exceeding the tolerable workload means a situation where the provision of health services to this patient should result in reducing the level of quality and safety of health services provided to patients already admitted. The provider cannot refuse to admit or terminate the care of a patient if the patient requires emergency care, in case of childbirth or where health services are essential for the protection of public or occupational health, in an emergency / crisis situation, or in case of court-ordered protective treatment unless otherwise provided by other legislation."*

As provided in Section 50, among other things, *"A health care professional has the right not to provide health care services if the provision of such services should directly endanger their life or seriously endanger their health. A health professional may refuse to provide health services to a patient if the provision of such services was contrary to their conscience or religious beliefs. A health professional cannot refuse to provide health services to a patient if such refusal should endanger the patient's life or seriously endanger the patient's health and the provider is unable to arrange for the provision of health services by another health professional."*

In this context, note that according to Act No. 48/1997 Coll., on Public Health Insurance, as amended, the insured person has the right inter alia to *"timely and local availability of fully covered services provided by the contracted health care providers of the relevant health insurance company and to the provision of fully covered services to the extent and under the conditions set out in this Act"*. [19] This local and timely availability is guaranteed to the patient (the insured person) by their health insurance company.

The key legislation governing liability of the state and health care providers consists of the Crisis Management Act and the Act on Liability for Damage Caused by a Decision of a Public Authority or Maladministration.

Liability for damage caused by crisis measures is stipulated in Section 36 of the Crisis Management Act as follows: *"The state shall compensate loss or damage incurred by legal and natural persons in connection with crisis measures and exercises (Section 39 (4)) implemented under this Act. The state may only be exempted from this liability if it has been proved that the injured party caused the damage themselves. Monetary compensation shall be provided by the crisis management authority that imposed the crisis measure or exercise during which or as a result of which the damage or injury occurred. Compensation shall be claimed in writing by a legal or natural person, stating the reasons for the claim, from the competent crisis management authority within 6 months of becoming aware of the damage, but not later than 5 years after the damage occurred, otherwise the right shall lapse. The crisis management authority may, in circumstances eligible for special consideration, award compensation for damages after the expiry of the period for submitting an application or even without submitting an application, but not later than 5 years after the damage occurred. Compensation will not be granted to legal and natural persons who caused the damage event. The crisis management authority may claim reimbursement of the costs it has incurred in the compensation for damage from the person(s) who caused the accident, or any other event which led to emergency / crisis situation and necessitated the imposition of crisis measures."* [12]

For further interpretation, the author deems it important to recall the original content of the government proposal of the provisions in Section 36 (Section 53 and Section 54 of the government proposal. [Chamber of Deputies of the Parliament of the Czech Republic, Chamber of Deputies Document No. 447, 3rd term, 19992]. The latter determined the prerequisites of the state's liability for crisis measures as follows: *"Monetary compensation will be provided to legal and natural persons who have suffered damage in the course of a) activities of the authorities implementing crisis measures or rescue and liquidation work, b) the ordered provision of material resources, c) the activities referred to in Section 48(3)(f). Monetary compensation will also be granted to natural persons who have suffered damage or injury to health or property when they performed an imposed work duty, work assistance, personal assistance or voluntary assistance within an organised activity. In the event of death, compensation will be paid to the heirs."*

The second relevant Act No, 82/1998 Coll., as amended, above mentioned formulates the right to compensation for damages in Sections 7 & 8 as follows: *"The right to compensation for damages caused by an unlawful decision shall be vested in the parties to the proceedings where the decision by which the damage was caused was taken. The right to compensation for damages caused by an unlawful decision may be asserted unless otherwise provided for hereinbelow only if the final decision was annulled or reversed by the competent authority on the ground of illegality. The court deciding on compensation shall be bound by the decision of that authority. In cases other than those which deserve special consideration, a claim for compensation for damages caused by an unlawful decision may only be awarded if the injured party has exhausted, within the periods of time laid down by law, all the procedural remedies available to the injured party for the protection of their rights; such remedies include an ordinary appeal, an extraordinary appeal, except an application for a retrial, or any other legal remedies for the protection of the right, the exercise of which is connected with the initiation of judicial, administrative or any other legal proceedings, or an application for the stay of execution." The key point is Section 13 of the same Act which provides that "the state is liable for damages caused by maladministration. A breach of the obligation to perform an act or to render a decision within the time set by law shall also be maladministration. Where no time limit is laid down by law for the performance of an act or rendering of a decision, a breach of the obligation to perform an act or render a decision within a reasonable time shall also be regarded as maladministration. The right to compensation shall be vested in the person who has suffered damage as a result of maladministration. "* [20]

Since the Ministry of Health adopted emergency measures regulating the availability of health care under the Act on the Protection of Public Health, note that those are so-called general measures [13]. The Ministry of Health's measures concerning the limitations of health care were not repealed by a court decision, but only after they had been concluded as irrelevant in view of the epidemic development in June 2021.

Before proceeding to the actual results of the assessment of liability of the state and health care providers in the next part of this paper, the author gives an overview of some relevant expert opinions on this topic.

Novotná [21] concludes *"that the crisis measures of the Government of the Czech Republic were adopted in accordance with the law. Similarly, the general measures adopted by the Ministry of Health under the Act on the Protection of Public Health are considered lawful unless they interfere in a manner not permitted with the fundamental rights and freedoms of individuals. The state's liability as a result of crisis measures set out in the Crisis Act is, in principle, out of the question since the crisis measures are not the cause of the harm and injury suffered by patients as a result of the limitation or postponement of health care. If health care was limited or postponed by individual decision of a health care provider, it would be possible to hold the provider liable for the injury and harm consequently suffered since it would be caused by a violation of the Health Services Act."*

By contrast, Melzer [22] notes on the issue of compensation for damages and non-pecuniary harm under the Crisis Management Act: *"As previously mentioned, Section 36(1) of the Crisis Management Act only refers to compensation for damages, and not compensation for non-pecuniary harm. This text could imply that only pecuniary damages are compensated (Section 2894). On the other hand, however, the third paragraph expressly defines the method and extent of compensation for non-pecuniary damage to health arising from the performance of an imposed work duty, work assistance or assistance provided within an organised activity. This provision could either be special to paragraph 1 or it could indicate that non-pecuniary damage is to be compensated after all. The first interpretation, i.e. paragraph 3 considered as lex specialis in relation to paragraph 1 means that damage to health is only compensated in said case and other bodily injuries incurred by the victim as a result of the crisis measure are no longer compensable. The problem with the second option is that, at first sight, it leaves open the question of the type of non-pecuniary damage to be compensated. We consider that, given the considerable ambiguity in the relationship of the two provisions, the second option should be preferred. Human health and life are among the most important legal goods with the most intense protection. It would be absurd for the law to protect property interests while at the same time denying protection to these more intensely protected legal goods.*

A closer look will also clarify the scope of non-pecuniary damages to be protected. Section 2956 of the Civil Code must be applied. This provision must be interpreted as clearly establishing the right to compensation for non-pecuniary damages if the facts of compensation for damages are fulfilled, which also protects (at least some) natural rights protected by the provisions of Part I of the Civil Code, i.e. in particular the personality of a person. In the present case, Section 36 of the Criminal Code also protects the life and health of persons, and therefore the right to compensation for non-pecuniary damages arises out of Section 2956 of the Civil Code in the event of interference with these goods. In the case of death, Section 2959 of the Civil Code applies."

Bílek et al. [13] concludes in their analysis that *"Liability of the state for damages caused by crisis measures is objective, therefore the claim arises regardless of the fault, and the prerequisite for liability is not the illegality or incorrectness of an adopted crisis measure, but vice versa its faultlessness. It is therefore not appropriate to require the annulment of a measure taken. The actual damage and lost profits will be covered (for lost profits, see the decisions of the Supreme Court of Justice, Case No. 25 Cdo 3798/2007, Case No. 25 Cdo 1649/2007 and Case No. 25 Cdo 1422/2016) and the damage does not have to arise directly in connection with the declaration of a state of emergency, but "in connection" with the subsequent crisis measures adopted pursuant to law or in their summary. Both pecuniary and non-pecuniary damages will be compensated."*

Two Supreme Court judgments [23] and [24] from the pre-covid era are applicable to the state's liability under the Crisis Management Act, but they address flood situations where public authorities had to respond quickly, urgently and in a time-sensitive manner. The judgment of 11 April 2012, Case No. 28 Cdo 2519/2011 implies that *"The regulation of liability for damage arising out of or in connection with a crisis measure (including a measure taken in relation with a flood or to prevent its consequences) is a special regulation containing a special factual basis for the state's liability which is governed by the Crisis Management Act, not by Act No. 82/1998 Coll. or by a general regulation. It constitutes liability without regard to the fault (so-called strict liability) and unlike Act No. 82/1998 Coll., it does not require damage to be caused by an unlawful decision or maladministration. This special liability is based on the simultaneous fulfilment of three prerequisites, namely 1) the implementation of a crisis measure, 2) the occurrence of damage, and 3) a causal link between the crisis measure and the occurrence of damage. The state may be exempted from liability if it proves (the burden of proof lies on the state) that the victim caused the damage themselves. The text of Section 36(1) of the Crisis Management Act makes it clear that the state is the responsible and liable party (cf. the judgment of the Supreme Court of the Czech Republic of 17 June 2009, Case No. 25 Cdo 1649/2007 (Rc 10/2010))."*

The judgment of 22 October 2009, Case No. 28 Cdo 3798/2007, removes the state's responsibility in the event of an unavoidable event, stating that *"Compensation will not be granted in a situation where the state (crisis management authorities) acted upon or proceeded according to the current information."*

Although the opinion below does not address compensation for non-pecuniary damage, it should be stressed as relevant. It refers to a set of communications of the Ministry of the Interior of the Czech Republic from January and February of this year, rejecting the compensation claims of injured entrepreneurs for damages (lost profits) based on the provisions of Section 36 of the Crisis Management Act. The Ministry of the Interior justifies its position as follows: *"The above provision of the Crisis Management Act does not apply to crisis measures adopted in relation with the Covid-19 disease pandemic. In the Ministry's point of view, the provision of Section 36 of the Crisis Management Act applies to compensation for damages incurred by individuals solely as a result of the implementation of crisis measures that are specific in nature and directed against a specific person or a specific group of persons. On the contrary, this provision cannot be applied in the case of crisis measures that are sui generis law"* [25]. Furthermore, there is an argument in the same document about the institution of special sacrifice where an individual is only granted compensation for forced individual sacrifice for the benefit of the whole, and about the exclusion of the state's liability under Act No. 82/1998 Coll. for damages caused by unlawful legislation. The Ministry denies the objective liability of the state and argues that consequences of such a serious situation as a pandemic are unforeseeable. The Crisis Management Act only regulates compensation for material damage and loss arising out of specific regulations against a specific individual. It cites the pandemic itself, not the actions of the State, as the cause of damage [25].

As regards the liability of health care providers in relation to deferring care during a state of emergency, Dostál [26] states that *"the Ministry's order only applies to so-called 'elective care' under Section 5(1)(d). 'Elective care' by this definition is a residual category - it is care that is neither 'urgent' (Section 5(1)(a)), nor 'emergency' (Section 5(1)(b)), nor 'essential' (Section 5(1)(c)). If a patient requires care that fits into any of these three categories, they are not receiving elective care and the Ministry's order does not apply."*. He also notes that *"In addition to inpatient care, hospitals provide as a rule extensive outpatient care in hospital offices where patients come for check-ups, tests, or minor procedures. This care is not covered by the Ministry's order, so the hospital cannot reduce this care for reasons other than those usually permitted by law."*

Šolc [27] opposes these interpretations and argues that *"The term 'emergency inpatient care' in the Order does not refer to the form of care to be limited - indeed, this even would not make sense in terms of the purpose of the Order which does not apply to emergency care as explained above - but to the type of health care providers to whom the Order applies. It therefore defines not the substantive but the personnel scope of the order. Nor can it rationally follow from the quoted text that the Order should be limited to inpatient care for this type of health care providers. In real world, hospitals as providers of emergency inpatient care obviously offer a wide range of outpatient services. If indeed the Order did not apply to outpatient services, hospitals could only reduce access to health care to a very small extent. However, considering the actual operation of hospitals, this would be contrary to the very purpose of the Order, or it would not be possible to achieve it. The Order should therefore be rather interpreted as allowing emergency inpatient care providers to focus their resources as much as possible on the ongoing epidemic even if this should reduce access to non-inpatient care in their health facilities. Last but not least, remember that the outpatient care in hospitals can and is likely to spread infection. If a particular health care provider concentrates patients with COVID-19 in its inpatient ward, there is an increased risk of infection in their facility despite all hygiene and organizational measures. These are also reasons for limiting medical procedures in elective care. However, it is still the case that any limitations of outpatient or one day care must always be reviewed in terms of their proportionality and necessity for the purpose of the Order."*

As regards the state liability under Act No. 82/198 Coll. in view of the measures adopted by the Ministry of Health, Vojtek's conclusions [28] are particularly relevant *"In this situation where the definition of the person injured by an unlawful decision cannot be based on the status of a party to the proceedings, anyone who has suffered damage as a result of the decision is actively legitimate."* This makes anyone who has suffered damage as a result of a repealed general measure actively legitimate to claim compensation for damages provided the other conditions set out by law are met. Korbel [29] also states that *"As regards the possibility of claiming compensation for damages from a repealed general regulation, I am inclined to the conclusion that the law allows such an interpretation"*. Vojtek [28] defines *"A general measure is by its nature a decision sui generis"*.

3 Results

3.1 State Responsibility for Deferred Health Care

For research question 1 whether the state is liable under the current legislation for personal injury / damage to health caused by the postponement of health care due to emergency measures adopted by the state during a state of emergency, the author concludes that in his point of view, the state liability for personal injury / damage to health caused by the postponement of health care during a state of emergency can generally be based on two legal regulations. The first is the Crisis Management Act and the second option is to set up a claim by reference to Act No. 82/1998 Coll., as amended. These solutions could generally be used in parallel. As regards Act No. 82/1998 Coll., two legal instruments could be considered, namely a claim for compensation for damages caused by a decision of a public authority which has been annulled on the ground of illegality, or a claim for compensation for damages caused by maladministration, specifically in a situation where the state has failed to take an appropriate action or decision within a reasonable period of time.

As regards the Crisis Management Act, see the above analysis of the legal regulation of compensation for damages and the Ministry of the Interior's rejection of the compensation claims for material damages. Based on the information from the media, the author is aware of the first judgment of a civil court which rejected compensation claims under the Crisis Management Act on grounds similar to the Ministry of the Interior's interpretation.

The author holds that the state liability for non-pecuniary damage caused by deferred health care can be inferred from the Crisis Management Act, of course, subject to the conditions set out in the Act. The author is therefore critical of the negative conclusions as provided above, in particular for the reasons set out below.

General measures adopted by the Ministry of Health concerning the limitations of health care are not legal regulations; they are decisions sui generis as can be inferred from the constant case-law and literature cited in the previous part of the paper. If they were annulled in accordance with the procedure laid down in Act No. 82/1998 Coll., such a fact could even become the basis for a claim under that Act as if the decision was annulled in an individual case. It cannot therefore be accepted that compensation claims should be rejected on the basis that the state cannot be held responsible for the accuracy of the legislation.

Moreover, the author points out that the Crisis Management Act does not distinguish between general and individual measures for the definition of crisis measures. They can include legal regulations, decisions, and measures, or de facto actions, etc. The Crisis Management Act does not even distinguish which measure was relevant for the damage incurred, but the state's liability is explicitly based on any crisis measures under this Act. If the lawmakers had intended to limit the state's liability, they would have explicitly defined other conditions under which liability arises. This is how the government's draft Crisis Management Act was formulated as documented earlier in this paper. However, the lawmakers adopted a different wording of liability for damages, not limited in any substantive way. The author agrees with the exclusion of the state liability for adopted laws, so this argument could certainly be applied to the provisions set out in the Crisis Management Act. However, not to measures that are not legal regulations.

The author is also inclined to the conclusions of Melzer [22] and refers to his findings [13] that compensation for damages under the Crisis Management Act is based on the principle that it is compensation for damages caused by the due course of law, namely by a proper crisis measure. The prerequisite for liability is therefore not illegality, but rather a duly adopted measure of the state. The liability of the state is objective. The arguments rejecting liability as non-systemic and exceptional cannot therefore be accepted. It is not possible to circumvent an unambiguous textual interpretation by a purposive teleological interpretation.

If illegality is not a prerequisite for the state liability for damages under the Crisis Management Act, the instruments excluding illegality such as the performance of a legal obligation (the state acted properly and in accordance with the crisis legislation) or a last resort should not be applied. If it is argued that COVID-19 is a force majeure event, this institute is applicable to contractual liability. In principle, the epidemic as such is a force majeure event, so the author believes this institute is logically inapplicable. Nor can it be accepted in that respect that the pandemic as such, and not the measures adopted by the state, is the cause of the damage. The examination of a causal link between the measure taken and the harm / injury is a natural part of the court's application process, but claims cannot be rejected as a rule simply on the basis that the only adequate ground in the legal sense is the epidemic as such. Then the law would not address the right of recourse and it would essentially mean that the state is only responsible for situations that it caused itself.

At the same time, the author points out that the term "damages" used in the Crisis Management Act could lead to the conclusion that only material damage is subject to compensation, but this has already been refuted by Bílek et al. [13] and Melzer [22]. The Crisis Management Act originated at a time when the Civil Code did not differentiate between the terms of material and immaterial damages. Section 36 of the Crisis Management Act has remained unchanged since it was passed. It is therefore not reasonable to believe that the lawmakers had only material damages in mind. Values such as health and life deserve increased protection in accordance with the constitutional order and cannot be disregarded for rights relating to property values - indeed, quite the opposite.

The author is also aware of the argument of the so-called special victim institution where the state responsibility / liability is denied with reference to the fact that members of society, in general, benefit from the measures, and that compensation should therefore be provided only in the case of an individual's special extraordinary sacrifice for this general benefit. In the author's viewpoint, however, the measures would have to be applicable outside the crisis situation. The author agrees with the conclusions of Melzer [22]: "*It should be highlighted, however, that the Crisis Management Act does not regulate a general claim for sacrifice in the public interest (which is rather reflected in Section 35 of the Crisis Management Act), but its specific form. Hence, for example, the fulfilment of all the subconditions that were historically present (e.g. compensation does not apply to interventions in the form of a legal regulation), or are still associated with this general claim in related legal systems cannot be required.*" The author notes, and above all denies the conclusion that state measures are defensible to the extent that they can harm or are likely to cause harm to life and limb in any way. These natural human rights guaranteed by the constitution cannot be sacrificed to protect from other risks to the life or health of citizens. As above mentioned, the media report about threats to human health and life from diseases posing a greater risk than covid-19 in the Czech Republic. It is necessary to examine each individual case for a causal link between the measure and the injury suffered, not in general reject claims by applying a general rule of law teleologically reducing the unambiguous language of Section 36 of the Crisis Management Act. Finally, even the consequences to health and life caused by compulsory vaccination against covid-19 are compensated under the current legislation. Similarly, there is no reason why damage caused by other state regulations affecting the health and life of individuals should not be compensated, moreover if they are imposed not within the framework of setting general standards, but according to a sui generis decision of the Ministry of Health. The protection of public health as a public interest is not limited to one selected disease.

The Ministry of Health adopted its measures under the Public Health Protection Act, but it is still a crisis measure under the Crisis Management Act. In accordance with Section 97 of this Act, the costs associated with applying the measures imposed under this Act (limited access to health care, allocation of resources, etc.) are borne by the state.

For the entire duration of the pandemic (of course, if the actual present state of affairs in the Czech Republic can be referred to as such according to general standards), the state could have adopted the essential measures as legally binding. If general measures were used, it would be justified to impute liability under the Crisis Management Act.

Regarding the prerequisites for the state liability under the Crisis Act, the author refers to [13] and [22]. The key is also the judgment of the Supreme Court, Case No. 25 Cdo 1649/2007 of 17 June 2009 that summarizes these prerequisites and establishes the objective liability of the state for the measures taken. It must be emphasized that in each individual case it will be necessary to examine the causal link of the occurrence of the injury, objective predictability, and take into account the course and effects of the epidemic itself, and the behaviour of the individual in the particular case, i.e. how his or her behaviour contributed to the occurrence of the injury / harm [30].

The obligated person is the state and claims are brought against the state authority that imposed the measure. The state is represented by the Ministry of the Interior in civil proceedings.

As mentioned above, the second possible option for claiming compensation for damages is provided in Act No. 82/1998 Coll. Since the measures adopted by the Ministry of Health concerning health care limitations and the allocation of life-saving resources have not been repealed, compensation for damages cannot be successfully claimed under this Act since the annulment of an unlawful decision is one of the necessary preconditions for entitlement for compensation. However, the author refers to the institution of maladministration which, in theory, could be considered. This would be a technical basis for a claim that the Ministry of Health failed to take all appropriate actions to protect health and life. However, the author considers this legal procedure to be quite challenging that it cannot be reasonably recommended. The Ministry of Health adopted a whole series of measures during the epidemic, and it would therefore have to be professionally examined whether or not the flow of measures was sufficient, and correct, or timely, etc.

As a purely hypothetical construct, the reason for the annulment of the Ministry of Health measures may have been the lack of power to limit / reduce care other than one day and inpatient care. In fact, the health care providers also reduced access to outpatient care with reference to the Ministry's measure. The Ministry based its decisions on Act No. 258/2000 Coll., and in particular on the possibility to impose a 'ban on some other activities'. Whether the extent of the limited care would therefore stand up to judicial review is a matter of debate. However, if the health care providers interpreted the imposed measures as they should also postpone outpatient care, their liability should also be considered.

3.2 Liability of Health Care Providers

For the second research question, i.e. whether the liability of health care providers for damage to health resulting from the postponement of health care due to the emergency measures adopted by the state during a state of emergency is established under the current legislation, the author concludes that the liability of health care providers must be based on Act No. 372/2011 Coll., on Health Services, in conjunction with the Civil Code. Since the health care providers followed the measures of the Ministry of Health in limiting access to and deferring care, their liability cannot be determined. If a particular provider acted contrary to the Ministry's measure and limited access to health care contra legem artis beyond the defined scope, then it would be liable under the law. In principle, what the author cited earlier in this paper applies here.

The subject matter of a dispute may be the definition of the scope of deferred care due to the measures adopted by the government, i.e. whether the inpatient care providers can be ordered to defer outpatient care, or whether that scope was correctly interpreted by the providers. If their interpretation was wrong, they are liable for damages; if their interpretation was correct (which the author is inclined to think as is [...]), then the state's liability under the Crisis Management Act comes into play.

4 Discussion

In the discussion, the author highlights several key issues.

Having regard to the legal nature of a general measure, the author considered that there was a consensus that, for the purposes of Act No. 82/1998 Coll., it was a decision sui generis and if it should be annulled, the specific affected person could claim compensation for damages. If there is a relevant objection, in the context of claims by victims under the Crisis Management Act, that it is general law making, this poses a fundamental interpretation challenge. There is no logical reason why a general measure should have two different schemes whether or not it is repealed. In that respect, a uniform case-law system would be useful.

Another topic for discussion is the circumstances precluding the state liability under the Crisis Management Act. Using a linguistic and logical interpretation, the text of the Crisis Management Act leads to the conclusion that the state can objectively be held liable for material as well as non-material damages incurred by individuals in connection with the implemented measures even if they were lawful and of any legal nature. It is therefore only possible to identify circumstances precluding such liability by applying other general methods of interpretation, in particular teleological ones. If it is agreed that there need not be an element of illegality, do circumstances precluding illegality such as the performance of a legal duty or a last resort and others come into play? It would be appropriate to examine the question of damage or injury causation. Can the absence of foreseeability of damages or injuries be relied on in rejecting a claim, or is it sufficient to refer in general to the consequences of force majeure, i.e. the pandemic as such (moreover, in a situation where the state did not officially determine the parameters of the epidemic or pandemic during the state of emergency - the sickness was not determined and the positive rate of tests, or the number of people in hospital, etc. were used instead).

Discussions should also be held on legislative changes that would define more precisely which measures can be adopted during a state of emergency and in what legal form. Here the author means that if measures were imposed by generally binding legislation (laws, decrees, government orders), then many issues of interpretation discussed in this paper would be avoided.

5 Conclusion

The interpretation ambiguities as above documented in the assessment of the state responsibility / liability during a state of emergency, both on part of the professional and lay public, are not exactly the sign of legal certainty as a fundamental building block of the rule of law. In the author's opinion, Act No. 94/2021 Coll. on emergency measures during the COVID-19 disease epidemic, does not provide an adequate solution where, in fact, the same questions arise out of its application and interpretation as in the case of the Crisis Management Act.

The author believes that it would not be appropriate to apply the institution of a general measure to across the board measures imposed by, for example, the Ministry of Health. The author proposes that the Ministry of Health should be directly authorized by the Act on the Protection of Public Health and by the Crisis Management Act to adopt a decree (generally binding legal regulation) for the purposes of an emergency, or a form of government regulations may be considered in times of emergency. Where the situation requires the form of a law given the importance of a proposed amendment, this can only be recommended.

Furthermore, the Crisis Management Act and the Public Health Protection Act should clearly define what measures can be imposed. The author does not consider it appropriate for the Ministry of Health to adopt general measures based on the vague provisions of the Act on the Protection of Public Health in a crisis situation and under a declared state of emergency. This forms a dual legal regime. The Crisis Management Act does not address the substance of the measures to be imposed whileas the Pandemic Act does.

The state should make it clear under what conditions it shall be liable, or intends to be liable for harm / damages incurred during emergency / a crisis situation. The wording of the Crisis Management Act is so general that further legislative limits should be defined. The author does not advocate casuistic legislation, but the difference in the legal regulation of liability under the Crisis Management Act and the state's responsibility / liability under other legal regulations (even historical) is obvious.

The author proposes that the state should in any case be liable for injuries / damage to health and life incurred in connection with the measures imposed to limit access to health care during a state of emergency, even those resulting from generally binding legislation.

References

1. Government of the Czech Republic. *Resolution of the Government of the Czech Republic of 12 March 2020 Number 194* [online]. Available at: https://apps.odok.cz/attachment/-/down/IHOABMNHPBSV (2020)
2. Ministry of Health of the Czech Republic. *The general measure of the Ministry of Health of 16 March 2020, Ref. MZDR 12066/2020-1/MIN/KAN, of 26 October 2020, Ref. MZDR 46953/2020-1/MIN/KAN, of 23 November 2020, Ref. MZDR 46953/2020-2/MIN/KANN* [online]. Available at: https://koronavirus.mzcr.cz/doporuceni-poskytovatelum-zdravotni-pece/ (2020)
3. Hospital České Budějovice. *Nemocnice České Budějovice opět navyšuje rezervní kapacitu pro nemocné s covid-19* [online]. Available at: https://www.nemcb.cz/aktuality/nemocnice-ceske-budejovice-opet-navysuje-rezervni-kapacitu-pro-nemocne-s-covid-19/ (2021)
4. Z. Firano, F. A. Fatine. The COVID-19: Macroeconomics scenarii and role of containment in Morocco. *One Health.* (2020)
5. R. Aldaco, D. Hoehn, J. Laso, M. Margallo, J. Ruiz-Salmón, J. Cristobal, R. Kahhat, P. Villanueva-Rey, A. Bala, L. Batlle-Bayer, P. Fullana-i-Palmer, A. Irabien, I. Vasquez-Rowe. Food waste management during the COVID-19 outbreak: a holistic climate, economic and nutritional approach. *Science of the Total Environment.* (2020)
6. G. Comazzetto. The state of exception in the Italian legal system. Reflections on the impact of COVID-19 containment measures. *Biolaw Journal-Rivista Di Biodiritto.* **1**, 35-42 (2020)
7. P. C. Barquero. The judicial authorization or rectification of urgent sanitary measures in the COVID-19 pandemic. *Revista Juridica De Castilla Y Leon.* (2021)
8. B. R. Kern, M. Peters. Triage - Sifting, Sorting and Deciding in the COVID-19 Pandemic: Medical Legal Aspects. *Notarzt.* (2020)
9. F. Gambino, C. Petrini, L. Riva. Criteria for allocation of life-saving resources during the SARS-COV-2 pandemic: ethical implications and aspects of legal liability. *Annali Dell Istituto Superiore Di Sanita.* **57**(2), 113-120 (2021)
10. Economia. *Resolution of the Presidency of the Czech National Council of 16 December 1992 on the proclamation of the Charter of Fundamental Rights and Freedoms as part of the constitutional order of the Czech Republic. Constitutional Act No. 2/1993 Coll. as amended by Constitutional Act No. 162/1998 Coll* [online]. Available at: http://zakony.centrum.cz/listina-zakladnich-prav-a-svobod/
11. The Czech Republic. Constitutional Act No. 110 of 29 May 1998, Constitutional Act on the Security of the Czech Republic. In: *Collection of Laws of the Czech Republic.* Part 39, 5386-5387. ISSN 1211-1244 (1998)
12. The Czech Republic. Act No. 240 of 9 August 2000 on Crisis Management and on Amendments to Some Acts (Crisis Management Act). In: *Collection of Laws of the Czech Republic.* Part 73, 3475-3487. ISSN 1211-1244 (2000)
13. S. Bílek, Z. Caha, V. Velková. The state's responsibility towards entrepreneurs in the Czech Republic for the measures adopted in relation to the epidemic of coronavirus. *Innovative Economic Symposium 2020 – Stable Development in Unstable World (IES2020).* ISBN: 978-2-7598-9113-9 (2021)
14. The Czech Republic. *Resolution of the Constitutional Court of the Czech Republic Ref. Pl. ÚS 12/21 of 15 March 2021* [online]. Available at: https://www.usoud.cz/fileadmin/user_upload/Tiskova_mluvci/Publikovane_nalezy/2021/Usneseni/Pl._US_12_21_vcetne_disentu.pdf (2021)
15. The Czech Republic. *Judgment of the Municipal Court in Prague Case No. 17 A 126/2020- 84 of 23 February 2021* [online]. Available at: http://data.s3.atlascloud.cz/COVID/17A_126_2020_12.pdf (2021)
16. The Czech Republic. Act No. 258 of 14 July 2000 on the Protection of Public Health and on Amendments to Some Related Regulations In: *Collection of Laws of the Czech Republic.* Part 74, 3622-3662. ISSN 1211-1244 (2000)
17. The Czech Republic. Act No. 372/2011 of 6 November 2011, on Health Services and Conditions of Their Provision (Health Services Act). In: *Collection of Laws of the Czech Republic.* Part 131, 4730-4801. ISSN 1211-1244 (2011)
18. The Czech Republic. Act No. 89 of 3 February 2012, Civil Code. In: *Collection of Laws of the Czech Republic.* Part 33. 1026-1365. ISSN 1211-1244 (2012)

19. The Czech Republic. Act No. 48/1997 of 7 March 1997, on Public Health Insurance and on Amendments to Some Related Acts. In: *Collection of Laws of the Czech Republic*. Part 16. 1185-1264. ISSN 1211-1244 (1997)

20. The Czech Republic. Act No. 82 of 15 April 1998 on Liability for Damage Caused by a Decision of a Public Authority or Maladministration and on Amendment to Act No. 358/1992 Coll. of the Czech National Council, on Notaries and their Activities (Notarial Code). In: *Collection of Laws of the Czech Republic*. Part 31, 5058-5063. ISSN 1211-1244 (1998)

21. D. Novotná. Legal aspects of damage to health as a result of epidemiological measures limiting patients' access to health care. *Bulletin of Advocacy*. ISSN 1210-6348 (2021)

22. F. Melzer. Compensation for damage and injuries caused by emergency measures during a coronavirus pandemic. *Bulletin of Advocacy*. ISSN 1210-6348 (2020)

23. The Czech Republic. *Judgment of the Supreme Court of the Czech Republic Case No. 28 Cdo 2519/2011 of 11 April 2012* [online]. Available at: https://www.zakonyprolidi.cz/judikat/nscr/28-cdo-2519-2011 (2012)

24. The Czech Republic. *Judgment of the Supreme Court of the Czech Republic Case No. 28 Cdo 3798/2007 of 22 October 2009* [online]. Available at: https://iudictum.cz/200420/25-cdo-3798-2007 (2009)

25. Ministerstvo vnitra České republiky. *Poskytnutí informace – náhrady škody způsobené opatřeními států v souvislosti s pandemií nemoci COVID-19* [online]. Available at: https://www.mvcr.cz/clanek/poskytnuti-informace-nahrady-skody-zpusobene-opatrenimi-statu-v-souvislosti-s-pandemii-nemoci-covid-19.aspx (2021)

26. O. Dostál. *Práva pacientů v nouzovém stavu: Nemám COVID, potřebuji péči* [online]. Available at: https://advokatnidenik.cz/2020/03/19/prava-pacientu-v-nouzovem-stavu-nemam-covid-potrebuji-peci/ (2020)

27. M. Šolc, P. Šustek. *Nouzový stav: omezuje Ministerstvo zdravotnictví práva pacientů?* [online]. Available at: https://www.pravniprostor.cz/clanky/ostatni-pravo/nouzovy-stav-omezuje-ministerstvo-zdravotnictvi-prava-pacientu (2020)

28. P. Vojtek, *Liability for Damage in the Exercise of Public Authority*. Praha: C. H. Beck 3rd ed. ISBN: 978-80-7400-427-8 (2012)

29. F. Korbel, in: F. Ištvánek, P. Simon, F. Korbel. *Act on Liability for Damage Caused by a Decision of a Public Authority or Maladministration*. Praha: Wolters Kluwer. ISBN 9788075525215. (2017)

30. The Czech Republic. *Judgment of the Supreme Court, Case No. 25 Cdo 1649/2007 of 17 June 2009* [online]. Available at: https://www.zakonyprolidi.cz/judikat/nscr/25-cdo-1649-2007 (2009)

Predicting Students' Intentions to Create Green Start-ups: A Theory of Planned Behaviour Approach

Ahmed Chemseddine Bouarar[1,*], *Smail* Mouloudj[2], *Asma* Makhlouf [1], *Kamel* Mouloudj[1]

[1]University of Medea, Department of Commercial Sciences, 26000, Algeria, Spain
[2]University of Medea, Department of Commercial Sciences, 26000, Algeria & Visiting Researcher at University of Salamanca, Spain

Abstract. Start-ups play a fundamental role in the economy and have contributed enormously and effectively to solving economic and social issues. However, despite the significant environmental hazards the world is experiencing, the environmental dimension of its contribution is still significantly low, and the transition to green start-ups is still in its infancy especially in developing countries. The aim of the paper is to identify the factors that influence students' intentions to create green start-ups. Based on a convenience sample of 248 university students, a multiple regression analysis was performed to test our extended model. The empirical findings show that attitudes, subjective norms, perceived behavioural control (PBC), and environmental awareness positively influence the intention to create green start-ups. In addition, the findings demonstrate that students have a low level of environmental awareness. Based on the results, suggestions are provided on how to motivate university graduates to create green start-ups.
Keywords: eco-friendly behaviour; environmental awareness; green entrepreneurship; sustainability

1 Introduction

As environmental changes are occurring faster than expected, the world perceives a growing need to adopt more environmentally sustainable approaches [1], one of them being the green economy. In the green economy, both public and private sectors´ investments related to reducing carbon emission and pollution, increasing energy and resources efficiency, and preserving biodiversity and ecosystem services are the drivers for increasing income and employment [2]. The green economy also helps in overcoming environmental problems, depletion of natural resources, and ensuring the well-being of people at the lower position of the economic pyramid. Green entrepreneurship is considered as key component of the green economy. In the context of the sustainable development, the idea of green entrepreneurship is gaining a significant attention.

Shepherd and Patzelt [3] argue that green entrepreneurship helps preserve the ecosystem, improve the environmental quality and mitigate deforestation. Green entrepreneurship is playing an increasingly important role in environmental protection [4]. Promoting sustainable entrepreneurship among ambitious entrepreneurs could provide a sound solution to reduce the ecological footprint and help achieve the sustainable development goals [5]. However, entrepreneurship activities are conditioned by intention, people usually do not become entrepreneurs without some trigger and most importantly, intention. Krueger [6] defines entrepreneurial intention as "the commitment that an individual has to starting a new business and it can be considered a key antecedent of entrepreneurial behaviour." Hence, intention is the main factor that shapes one's intention to start a new business, and plays a significant role in the decision to start a new company [7]. The green entrepreneurship intention may be affected by several factors, such as perceived desirability, perceived feasibility, opportunity seeking, and reasonability [8]. Yi [4] argues that green entrepreneurial intentions have a direct positive effect on green entrepreneurship behaviour of university graduates in China.

In Algeria, the authorities are aware of the fundamental role that start-ups play in creating job opportunities and enhancing the growth. As a result, Algerian Start-up Challenge programme (ASC) was started in 2018. In 2020, the Algerian government established a ministry for start-ups, namely "Ministry of Micro-Enterprise, Start-ups and Economy of Knowledge" (MMESEK), whose aim was to remove any obstacles that may hinder start-up creation and functioning. ASC is the biggest Algerian program that generates opportunities for start-up owners and ecosystem actors to create innovation of tomorrow. The first edition focused on enhancing the emergence of start-ups in e-commerce and e-payment in Algeria; the 2nd edition (2020) organized in the form of a national competition aimed to bring together enthusiasts and actors of change in the world of entrepreneurship in Algeria; and the 3rd edition (2021) was organized in the form of specialized challenges by industry and focused on open innovation. In legislative terms, Executive Decree No. 20-254 was issued on September 15, 2020in the official Gazette. The decree governed the establishment of the National Committee that awards the mark of a "start-up enterprise", innovative project, and enterprise incubator. The main goals of this committee are to identify and encourage innovative projects as well as improving their ecosystem. Moreover, in

* Corresponding author: shemseddine26000@gmail.com

its efforts to support start-ups, the Algerian government has granted many tax incentives, including complete exemptions (and sometimes significant reduction) from some taxes and fees (such as the income tax, real estate tax, value-added tax, and professional activity fee), and customs rights of 5 to 10 years.

The decision to become a green entrepreneur may be regarded as voluntary and conscious. Therefore, it is highly recommendable to analyse how such a decision is made [9]. There is an increasingly huge volume of literature arguing that intentions play a fundamental role in the decision to start a new firm [7,9]. In the light of pressure from environmental degradation, entrepreneurs might be compelled to take into account environmental protection measures when they plan to start up an enterprise [10]. Going green has generated a huge enthusiasm in the society; however, green entrepreneurial intention is still evolving [8]. It is worth noting that much of the literature on green entrepreneurship lacks empirical research [11].

Since Ajzen developed the theory of planned behaviour (TPB) in 1991 [12], it has gained a major interest and has been enormously applied worldwide. Results confirm the applicability of the TPB to entrepreneurship, despite some contradictions of various studies [9]. Green entrepreneurship is an intentional and planned behaviour, a complex process consisting of several stages, where in the early stage, when a person becomes a green entrepreneur, a person develops a certain level of green entrepreneurial intention [4]. Sher et al. [5] and Waris et al. [13] applied the extended TPB to identify the drivers of sustainable start-up intentions among students in Pakistan. Kautonen et al. [14] confirmed the relevance and robustness of the TPB in the prediction of business start–up intentions and the following behaviour based on longitudinal survey data (2011 and 2012) from the adult population in Austria and Finland. In addition, some other studies use other theories to investigate green entrepreneurial intention. For example, Ramayah et al. [8] combined the entrepreneurial event theory (EET) and cultural values theory (CVT) as the theoretical framework to determine factors influencing the green entrepreneurial intention in Malaysia. However, in Algeria, no research has been carried out to understand the students' intentions to create green start-ups, which makes this study particularly important and valuable.

Therefore, this paper aims to answer the following research question: What is the influence of TPB constructs, and environmental awareness on intentions to create green start-ups? In this context, the objectives of this paper were to: (1) determine the factors that influence students' intentions to create green start-ups; and (2) develop and validate extended model that combines TPB with environmental awareness. Hence, a survey data from 248 students was used to test our extended model. The following part of this paper is structured as follows: The next section presents the theoretical framework, as well as the formulated hypotheses. The third section describes the methodology, including a sample definition and the measures of constructs. The fourth section presents the finding and discussion, which is followed by conclusion.

2 Literary research

2.1 Attitudes towards green start-ups

Attitude refers to "the degree to which a person has a favourable or unfavourable evaluation or appraisal of the behaviour in question" (p. 188) [12]. Attitude is the underlying variable of the direct and the indirect effects' confluence that stimulate the intention to behave or not to behave in a specific manner. In the context of entrepreneurship, the attitude towards start-ups was defined as "the degree to which the individual holds a positive or negative personal evaluation about being an entrepreneur" (p.570) [15]. The personal valuation reflects the individual's predisposition to own a business, which may help to form an entrepreneurial intention later; it could be converted to creating a start-up. Interestingly, it was found that U.S. and Turkish students show a low degree of entrepreneurial intentions, even though their attitude towards entrepreneurship is considered positive [7]. A study conducted by Saleem et al. [16] shows that attitude toward eco-entrepreneurship intention remains insignificant, which becomes strong and positive in the presence of two specific moderators, namely altruism and collectivist behaviour. In contrast to earlier findings, a positive attitude was found to be elicited by students' intention to create a start-up [17-19]. Likewise, Vamvaka et al. [20] who found that positive attitude and perceived self-efficacy are certainly the strongest predictors of entrepreneurial intention. Moreover, Peng et al. [10] found that attitude serves as a positive mediator between environmental values and sustainable entrepreneurial intention. Therefore, the following hypothesis is formulated:

H1: Attitudes positively influence intentions to create green start-ups.

2.2 Subjective norms

According to Bouarar and Mouloudj [21], subjective norm refers to "the level of social pressures exerted by a reference group that can influence a person's perception, feeling, preference, judgments, attitude, intention, and behaviour" (p.652). In the context of entrepreneurship, social norms represent entrepreneurs' perception of social pressure and significant references about certain entrepreneurial behaviour. Several studies confirm the existence of significant relationships between the intention to start a new business and the fact of knowing a person who owns a business, such as a parent and family members [17,22] or a role model. Previous research shows that subjective norm has a significant impact on entrepreneurial intention [23,24]. Peng et al. [10] found that subjective norm, as a mediator, has a positive impact on the link between environmental values and sustainable entrepreneurial intention. Therefore, the second hypothesis is formulated as follows:

H2: Subjective norms positively influence intentions to create green start-ups.

2.3 Perceived Behavioural Control (PBC)

PBC was introduced by Azjen [12] as a proximal antecedent of intention, extending the theory of reasoned action (TRA). PBC refers to "the person's perception of the ease or difficulty of performing the behaviour of interest" (p.183) [12]. PBC is related not only to the individuals' perception of their abilities to complete or execute an action, but also their perception to control it [9]. In the context of entrepreneurship, PBC reflects the entrepreneur's confidence in being capable to perform and control entrepreneurial activities. PBC was determined to have a significant impact on the entrepreneurial intention among undergraduate students [22]. Similarly, Yurtkoru [25] believes that PBC has a positive effect on Turkish university students' entrepreneurial intentions, although a slight one. Ashraf [23] clais that PBC has a positive influence on e-commerce bounded entrepreneurial intention. Therefore, the third hypothesis is formulated as follows:

H3: PBC positively influences the intentions to create green start-ups.

2.4 Environmental Awareness

Environmental problems are considered global challenges such as global warming, air pollution, and ozone layer depletion. Mouloudj and Bouarar [26] define environmental awareness as "the first level of environmental knowledge the individual owns about the causes and the effects of environmental damage" (p.875). According to entrepreneurial human capital (EHC) theory, individuals who acquire entrepreneurial knowledge presumably become entrepreneurs [27]. Li et al. [28] believe that education in green entrepreneurship promotes and boosts students' perception and behaviour towards green development. An intention to become a green entrepreneur is related to the commitment to buy and offer green products by using environmentally-friendly technology and materials to protect the environment against degradation and restrain or even reduce the depletion of natural resources. According to Peng et al. [10], environmental values have a significant and positive effect on sustainable entrepreneurial intention. It can be argued that environmental awareness is a key requirement to encourage environmentally-friendly behaviour; awareness is thus considered necessary along with other conditions for green entrepreneurship behaviour. The fourth hypothesis is thus formulated as follows:

H4: Environmental awareness positively influences the intentions to create green start-ups.

The summary of the hypotheses formulated within the constructs is shown in Fig.1.

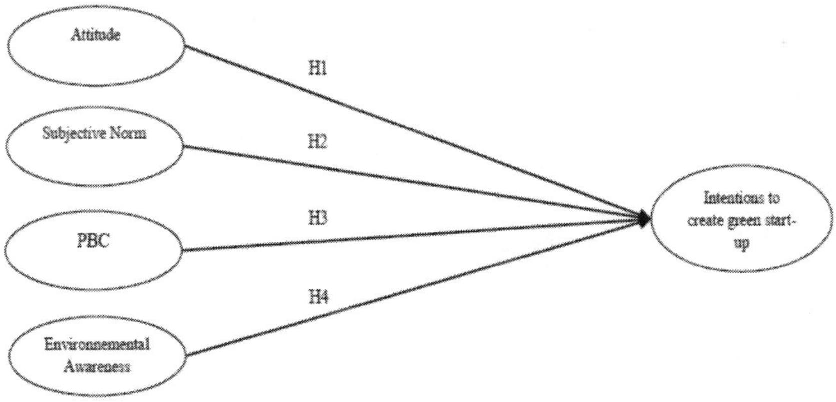

Figure 1. The research framework

Source: Authors.

3 Methodology

3.1 Sample and procedure

A structured questionnaire was designed for collecting data from a convenience sample. The population for the submitted study includes students of four Algerian universities (University of Algiers, University of Blida, University Setif, and University of Bouira). The selected students specialize in electronics, engineering, agriculture, biology, economic, and business. They were selected for the research because compared to other categories, the student category is more likely to be more knowledgeable and aware of the environmental hazards [26]. Moreover, researchers argue it was after graduation stage of development that most of entrepreneurs develop their intentions and engage in start-up activities [4,29]. Participation in the study was voluntary and completely anonymous. In total, 340 questionnaires were distributed, and 251 questionnaires were filled in. However, the analyses were performed on 248 questionnaires because 3 questionnaires were either incompletely or incorrectly filled in. The survey was conducted between 10 April 2021 and 20 June 2021. A total of 248 students were included in the study, with a response rate of 72.29%. Respondents' demographic analysis characteristics revealed that the sample included 155 (62.50%) men. The mean age among the students was 22.3 years, with the minimum age being 18 years and the maximum age 37 years. With regards to the education level of the students, the majority of participants are Bachelor students (51.21%), 36.69% participants are Master students, and 12.10% are doctoral students.

3.2 Measurement

The selected data collection method for this research is a questionnaire survey. The questionnaire used reliable and validated scales based on the aforementioned studies. The questionnaire used in this paper included three demographic questions (gender, age, and level of education) along with 16 items about five constructs (see Table 1). Three items adapted to those used by [9] are used to measure attitudes towards green start-ups. Three items developed by [21] are used to measure subjective norms. Three items developed by [9] are used to measure PBC. As for environmental awareness, four items adapted to those used by [26] are used. Finally, three items from [9] were used for measuring the intentions. A five-point Likert scale ranging from (1) strongly agree to (5) strongly disagree. The questionnaire was tested first and then modified to increase its reliability. The questionnaire was originally developed in English, and then translated into Arabic. The two versions of the questionnaire were distributed according to the students' wishes.

Table 1. Measurement of Constructs

Constructs	Measurement Item	Reference
Attitude	ATT1 Being a green entrepreneur implies more advantages than disadvantages to me.	[9]
	ATT2 A career of a green entrepreneur is attractive for me.	
	ATT3 If I had the opportunity and resources, I'd like to start a green firm.	
Subjective Norm	SN1 If I decided to create a green start-up, most people who are important to me will approve of that decision.	[21]
	SN2 My family expects me to create a green start-up.	
	SN3 My colleagues expect me to create a green start-up.	
Perceived Behavioural Control (PBC)	PBC1 To create a green start-up and keep it working would be easy for me.	[9]
	PBC2 I can control the creation process of a green start-up.	
	PBC3 If I tried to create a green start-up, I would be likely to succeed.	
Environmental Awareness	EA1 By creating a green start-up, I can influence environmental protection.	[26]
	EA2 I think I have the responsibility to protect the environment.	
	EA3 I think environmental problems will affect human life.	
	EA4 I think the environment is getting worse.	
Intentions	IN1 I had a preliminary idea for a green start-up to implement in the future.	[9]
	IN2 I'm willing to do anything to create a green start-up.	
	IN3 I´d like to start a green enterprise that would help solve environmental issues.	

Source: Authors.

4 Results

Cronbach´s alpha was employed to measure reliabilities among items of TPB constructs, environmental awareness, and intention (Table 2). The minimum coefficient was 0.774; a value that exceeds 0.6 means satisfactory internal consistency reliability [21,26]. Test of normality is an important postulate of regression analysis. The reliability of the regression result requires the normal distribution of the data. According to Byrne [30], the observed distribution is normal if both the skewness and kurtosis values are close to zero, with a measure of skewness statistics fluctuating around ±2 and kurtosis around ± 7. The results show that skewness ranges between 0.276 and 1.02, whereas kurtosis between -0.276 and +1.264, thus ensuring that the data used in the study is normally distributed.

Table 2. Reliability, multicollinearity, and collinearity test

Constructs	Cronbach's Alphas	Skewness	Kurtosis	Tolerance	VIF
ATT	0.774	1,025	1,264	,451	2,220
SN	0.944	,571	,302	,581	1,720
PBC	0.826	,500	-,179	,389	2,573
EA	0.808	,739	,495	,411	2,431
IN	0.882	,276	-,267	-	-

Source: Authors.

The presence of multicollinearity through tolerance and variance inflation factor (VIF) were assessed. When tolerance is lower than 0.20 and the VIF value is greater than five, there is an issue of multicollinearity. The results in Table 2 shows that the tolerance value is not lower than 0.2 and VIF does not exceed 5, which means that multicollinearity is not an issue in this analysis.

To test our research hypotheses, a multiple linear regression was employed. The hypotheses testing results are summarized and presented in Table 3.

Table 3. Multiple regression results for intention to create green start-ups

Model	B	t	Sig.	Decision
(constant)	.995	6.649	.000	-
ATT	.246	3.487	.001	H1 confirmed
SN	.161	2.628	.009	H2 confirmed
PBC	.215	2.836	.005	H3 confirmed
EA	.205	2.906	.004	H4 confirmed

Dependent Variable: Intention to create green start-up. Independent variables: Attitude (ATT), Subjective norm (SN), Perceived behavioural control (PBC), Environmental awareness (EA).
Notes: Model summary: R = 69.2%; R Square = 47.9%; Adjusted R Square = 46.9%; F = 55.820; $P = 0.000$ ($p<0.05$).

Source: Authors.

As shown in the table, attitudes have direct and positive influence on the intentions to create green start-ups ($t = 3.487$, $\beta = 0.246$, p=0.001), H1 is thus confirmed. However, subjective norms influence on the intentions to create green start-ups was found to be statistically significant ($t = 2.628$, $\beta = 0.161$, p=0.009), in this case, H2 was confirmed. The findings indicate that PBC has a direct and positive influence on the intentions to create green start-ups ($t = 2.836$, $\beta = 0.215$, p=0.005), and environmental awareness ($t = 2.906$, $\beta = 0.205$, p=0.004), Therefore, H3 and H4 are confirmed. The adjusted R square indicates the explanatory power of the predictor variable(s) on the respective construct. The attitude, subjective norm, PBC, and environmental awareness, explains about 47% of intentions to create green start-ups ($R^2 =$ 0.479).

5 Discussion

Our study indicates that attitude towards green start-ups is an important factor of influencing students' intentions to create green start-ups, which supports previous findings by [17-20]. This means that people form positive intention to engage in a creating green start-ups when their attitudes are positive; this also means that the stronger the attitudes are, the more likely it is to create green start-ups. However, intention may not lead to behaviour for several reasons, especially when fundraising is an issue. Undoubtedly, thousands of projects (green or not green) were not implemented due to lack of financial resources or administrative obstacles.

In the same vein, it was found that subjective norms have a strong positive influence on the intentions to create green start-ups. This suggests that for subjective norms, the perceptions of other important individuals (such as families, teachers, relatives, or peers) have a strong direct effect on the intention of creating green start-ups. This indicates that subjective norm reflects the perceived social pressure for conducting a particular behaviour, it is a factor of an individual's beliefs, represents social pressure put on entrepreneurs and is considered prior to any behaviour. Therefore, entrepreneurs may be influenced by information from environmental awareness groups who make claims about the positive effects of green entrepreneurship. Our finding is consistent with several studies indicating that subjective norms have a significant positive effect on sustainable entrepreneurial intentions [10,23,24].

Moreover, results show that PBC has a significant and positive influence on the intentions to create green start-ups. In this context, PBC reveals perception regarding access to resources and opportunities needed to carry out the behaviour, including access to financial resources, time, and other resources, and the person's self-confidence in the capacity to conduct the green entrepreneurship behaviour. This suggests that the high level of PBC should increase university students' intention to create green start-ups. In this sense, this finding was consistent with [22] and [25], according to whom the PBC positively influence students' entrepreneurial intentions. It is also supported by Ashraf [23], who found that PBC positively influence Facebook commerce entrepreneurial intention. In practice, it is expected that potential Algerian green entrepreneurs in the Algerian environment will encounter several obstacles, first financial obstacles that stem from the difficulty of convincing investors or sponsors of the feasibility of green projects, second, the institutional context in which they operate due to the monopoly power of incumbents in the energy industry, which hinders the provision of alternative energy sources. Thirdly, technological obstacles (lack of knowledge of green technology), administrative obstacles (lack of transparency when applying for government support). There might be other barriers such as legislation and law regulating the industry in which they operate.

Finally, the results indicate that environmental awareness has a strong direct influence on the intentions to create green start-ups. This means that students with high level of environment awareness and having good knowledge regarding the environment issues are more likely to engage in creating green project after graduation, in other words, students' intention is highly influenced by their environmental awareness level. Contrary to what was expected it has been found that students' environmental awareness is very low (mean=2.48), which means that they are not exposed to environmental issues and topics in their studies at university; in this regard, it is necessary to increase their awareness of this issue. Environment awareness may switch negative attitudes to positive ones tremendously, and can also influence the sense of environmental responsibility, which, in turn, may enhance people to create green projects.

6 Conclusion

Yi [4] considers green entrepreneurship as the solution to the environmental and social issues. In addition, research confirms that intentions play a significant role in the decision to start a new business [7]. This paper is aimed at developing and examining extended TPB model that evaluates how attitude, subjective norm, PBC, and environmental awareness influence the students' intentions to create green start-ups. To the best of our knowledge, the presented study is the first attempt to investigate this topic in Algeria. For the purposes of this paper, a questionnaire was distributed to a convenience sample of 340 students from six Algerian universities, and 248 properly filled in questionnaires were obtained. The findings show that most students were found to have low environmental awareness (Mean= 2.48). The findings also indicate that the three constructs of the TPB along with environmental awareness have a significant positive influence on the intentions to create green start-ups, and the four factors combined explain 47.9% of the variance in behavioural intention. All the hypotheses were confirmed; the findings from this paper thus contributes to the growing volume of research literature on green entrepreneurship intentions.

6.1 Managerial implications

The results indicate that attitude is the most significant factor of students' intention to create a green start-up; therefore, positive attitudes should be strengthened towards green entrepreneurship. In this vein, providing high quality information regarding the personal benefits of creating green projects is the main antecedents of a positive attitude, potential entrepreneur's beliefs regarding the profits that a green project could generate can significantly stimulate the intention to create a green enterprise in the future through promoting previous successful green project such as sharing videos of a successful green entrepreneurs on social media or airing videos on reliable and renowned TV channels (either local or international). The importance of launching green projects for increasing opportunities to enter international markets can also be highlighted. The results also indicate that social norms play a key role in shaping the intention to create green start-ups, which means that parents (as a reference group) should actively guide and enhance ambitious youth to create green products; opinion leaders such as university teachers, religious scholar) may effectively contribute as well. Furthermore, the results indicate that the intention to create a green start-up is highly influenced by the PBC; therefore, any obstacles that hinder the idea of creating a green start-up should be removed, mainly in the situation when green start-up projects in Algeria are still in the infancy stage and unclear. The obstacles that may hinder the idea of creating a green start-up to come into being include lack of knowledge regarding legal procedures, bureaucracy and complex administrative procedures, lack of knowledge regarding green financing sources, ambiguity on how to market green products, and lack of knowledge about green technology. Therefore, MMESEK ought to identify these obstacles and cooperate with the other ministries to find practical and useful solution. Finally, the results indicate that environmental awareness directly affects the intention to create green firms; therefore, different stakeholders (especially the MMESEK) need to encourage students to create start-ups as well as to increase their environmental awareness through TV conferences, organizing international conferences, and distributing leaflets. Universities can also play a pivotal role in increasing student environmental awareness through incorporating environmental issues in the educational curriculum and programs (such as green economy, green marketing, environment accounting, green entrepreneurship, etc.). Organizing national competition and awarding valuable awards for the best green innovative project can also be very fruitful to encourage young entrepreneurs to channel their innovative efforts and ideas towards innovative green projects.

6.2 Limitations and future research directions

This research is not without limitations. First, the sample size in this study is relatively small (248 students). Therefore, further studies should include a larger sample. Second, our study used convenient sampling, which limits the ability to generalize the results of the study. It is thus recommended to use random sampling to generalize results. Third, this study addressed students' intentions to create green start-ups; therefore, futher research into the current green entrepreneurs' behaviour is recommended to properly explore the transition from intentions into behaviour. Fourth, the current study incorporated environment awareness in the TPB; further research may include other variables such as entrepreneurial knowledge, perceived entrepreneurial risk-taking, individual traits, religious values, and any other variables that can enhance the explanatory robustness of the extended model.

References

1. R. Nordin, N.A. Hassan. The role of opportunities for green entrepreneurship towards investigating the practice of green entrepreneurship among SMEs in Malaysia. *Review of Integrative Business and Economics Research*. **8**, 99-116 (2019)
2. M. Vuola, M. Korkeakoski, N. Vähäkari, M.B. Dwyer, N.J. Hogarth, J. Kaivo-oja, J. Luukkanen, E. Chea, T. Thuon, K. Phonhalath. What is a green economy? Review of national-level green economy policies in Cambodia and Lao PDR. *Sustainability*. **12**(16), 6664 (2020)
3. D.A. Shepherd, H. Patzelt. The new field of sustainable entrepreneurship: Studying entrepreneurial action linking "what is to be sustained" with "what is to be developed". *Entrepreneurship: Theory and Practice*. **35**(1), 137–163 (2011)
4. G. Yi. From green entrepreneurial intentions to green entrepreneurial behaviors: The role of university entrepreneurial support and external institutional support. *International Entrepreneurship and Management Journal*. **17**, 963–979 (2020)
5. A. Sher, A. Abbas, S. Mazhar, G. Lin, Fostering sustainable ventures: Drivers of sustainable start-up intentions among aspiring university students in Pakistan. *Journal of Cleaner Production*. **262**, 121269 (2020)
6. N. F. Krueger. The impact of prior entrepreneurial exposure on perceptions of new venture feasibility and desirability. *Entrepreneurship Theory and Practice*. **18**(1), 5–21 (1993)
7. N. Ozaralli, N.K. Rivenburgh. Entrepreneurial intention: antecedents to entrepreneurial behavior in the U.S.A. and Turkey. *Journal of Global Entrepreneurship Research*. **6**(3), 1-32 (2016)
8. T. Ramayah, S. A. Rahman, S. K. Taghizadeh. Modelling green entrepreneurial intention among university students using the entrepreneurial event and cultural values theory. *International Journal of Entrepreneurial Venturing*. **11**(4), 394-412 (2019)
9. F. Liñán, Y. Chen. Development of cross-cultural application of a specific instrument to measure entrepreneurial intentions. *Entrepreneurship Theory and Practice*. **33**(3), 593-617 (2009)
10. H. Peng, B. Li, C., Zhou, B. M. Sadowski. How does the appeal of environmental values influence sustainable entrepreneurial intention? *International journal of environmental research and public health*. **18**(3), 1070 (2021)

11. J. Amankwah, H. Sesen. On the relation between green entrepreneurship intention and behavior. *Sustainability*. **13**(13), 7474 (2021)

12. I. Ajzen. The theory of planned behavior. *Organizational Behavior and Human Decision Processes*. **50**(2), 179–211 (1991)

13. I. Waris, M. Farooq, I. Hameed, A. Shahab. Promoting sustainable ventures among university students in Pakistan: an empirical study based on the theory of planned behavior. *On the Horizon*. **29**(1), 1-16 (2021)

14. T. Kautonen, M. van Gelderen, M. Fink, Robustness of the theory of planned behavior in predicting entrepreneurial intentions and actions. *Entrepreneurship Theory and Practice*. **39**(3), 655–674 (2015)

15. V. Souitaris, S. Zerbinati, A. Al-Laham. Do entrepreneurship programmes raise entrepreneurial intention of science and engineering students? The effect of learning, inspiration and resources. *Journal of Business Venturing*. **22**(4), 566–591 (2007)

16. F. Saleem, A. Adeel, R. Ali, S. Hyder. Intentions to adopt ecopreneurship: Moderating role of collectivism and altruism. *Entrepreneurship and Sustainability Issues*. **6**(2), 517-537 (2018)

17. K. Soria-Barreto, G. Honores-Marin, P. Gutiérrez-Zepeda, J. Gutiérrez-Rodríguez. Prior exposure and educational environment towards entrepreneurial intention. *Journal of Technology Management & Innovation*. **12**(2), 45–58 (2017)

18. B. Usman, Y. Yennita. Understanding the entrepreneurial intention among international students in Turkey. *Journal of Global Entrepreneurship Research*. **9**, 1-21 (2019)

19. W. Widayat, N. Ni'matuzahroh. *Entrepreneurial attitude and students business start-up intention: A partial least square modeling. Jurnal Manajemen dan Kewirausahaan*. **19**(1), 46-53 (2017)

20. V. Vamvaka, C. Stoforos, T. Palaskas, C. Botsaris. Attitude toward entrepreneurship, perceived behavioral control, and entrepreneurial intention: dimensionality, structural relationships, and gender differences. *Journal of Innovation and Entrepreneurship*. **9**, 1-26 (2020)

21. A.C. Bouarar, K. Mouloudj. Using the theory of planned behavior to explore employee's intentions to implement green practices. *Dirassat Journal Economic Issue*. **12**(1), 641-659 (2021)

22. S. N. A. Ambad, D. H. D. A. Damit. Determinants of entrepreneurial intention among undergraduate students in Malaysia. *Procedia Economics and Finance*. **37**, 108–114 (2016)

23. M. A. Ashraf, M. M. D. Alam, L. Alexa. Making decision with an alternative mind-set: Predicting entrepreneurial intention toward f-commerce in a cross-country context. *Journal of Retailing and Consumer Services*. **60**, 102475 (2021)

24. S. Sabah. Entrepreneurial Intention: Theory of planned behaviour and the moderation effect of start-up experience. In: M. Franco. *Entrepreneurship: Practice-Oriented Perspectives*. 87-101. Turkey: Ankara University (2016)

25. E. S. Yurtkoru, Z. K. Kuşcu, A. Doğanay. Exploring the antecedents of entrepreneurial intention on Turkish university students. *Procedia - Social and Behavioral Sciences*. **150**, 841–850 (2014)

26. K. Mouloudj, A.C. Bouarar. The impact of word of mouth on intention to purchase green products: An empirical study. *Revue Algérienne d'Economie* et *Gestion*. **15**(1) 871-890 (2021)

27. H. Ni, Y. Ye. Entrepreneurship education matters: Exploring secondary vocational school students' entrepreneurial intention in China. *Asia-Pacific Education Researcher*. **27**(5), 409–418 (2018)

28. H. Li, J. Chen, X. Xu, Z. Zhao. Research on the trend of green entrepreneurship education in colleges and universities. *China Market*. **52**, 71–74 (2016)

29. B. Fuller, L. Yan, S. Bajaba, L. E. Marler, J. Pratt. Examining how the personality, self-efficacy, and anticipatory cognitions of potential entrepreneurs shape their entrepreneurial intentions. *Personality & Individual Differences*. **125**, 120–125 (2018)

30. B.M. Byrne. *Structural Equation Modeling with AMOS: Basic Concepts Applications, and Programming*. London: Routledge (2016)

Energy-Saving Behaviour of Households: Methodological Aspects of Research

Rais Burganov[1,*], *Elma* Dolonona[1], *Lilia* Urasbahtina[1]

[1] Kazan State Power Engineering University, Department of Economics and Organization of Production, Krasnoselskaya Str. 51, Kazan, 420066, Russia

Abstract. The paper deals with the theoretical and methodological aspects of studying and modelling energy-saving behaviour of households in modern conditions. The main goal is to provide the scientific community in a concise form with these research results concerning the methodological support for the theoretical study of problems related to energy-saving behaviour of households. Extended version of the study on this topic will be published in a20003@mail.vstecb.cz different format. Scientific research methods are based on the use of abstraction and systematization of these energy-saving processes in households, the use of the principle "all other things being equal", and the study of official statistical materials. The theoretical basis was the teachings of representatives of institutionalism and the neoclassical direction of economic theory. The obtained results include a conclusion about the features of modelling energy saving of households, which should be taken into account when preparing practical recommendations in the field of energy consumption; on the criteria for modelling based on the grouping of households and the types of their behaviour in the field of energy saving. The fundamental variables in obtaining the effect of energy saving are the processes of organizing and self-organizing the process of energy consumption by households, which make it possible to apply effective energy-saving actions. Also, in the final part of the work, a system of variable factors influencing the energy consumption of households and used in modelling their behaviour is presented. The advantage of the study is that, for the first time in a systematic way, the theoretical and methodological aspects of studying the behaviour of households in energy saving are considered, which will allow obtaining a holistic view of trends in energy consumption. An in-depth study of the scientific foundations of the process of energy conservation in households can contribute to the formation of effective government and business programs and thereby rationalize the behaviour of households as consumers and suppliers of energy resources.
Keywords: household behaviour; energy saving; organizational behaviour; institutional infrastructure in energy saving

1 Introduction

All subjects of the national economy are interested in energy saving. However, in the main focus of the state bodies responsible for reducing the energy intensity of production are state institutions, branches of the national economy, and large economic units. The problems in the formation of energy-saving behaviour at the household level remain in the background. This is despite the fact that in the world and in individual countries there is an increase in energy consumption by households. However, the absence of a holistic and systematic approach to the study of the behaviour of households' energy saving does not allow making the correct strategic management decision, both at the state level and at the level of the households themselves. The need to study the behavioural aspects of household behaviour in energy conservation is a modern requirement. For example, the Global Commission on Urgent Action on Energy Efficiency following a conference at the International Energy Agency (IEA) recommended: "Use behavioural insights for more effective policies. People are at the centre of energy efficiency efforts, and understanding the behavioural sciences can help design smarter policies" [1]. However, in order to obtain real results in this area, it is necessary to determine the theoretical and methodological basis for studying the behaviour of consumers of energy of any kind. As hypotheses, one can propose the existence of significant differences between the positions of different theories regarding the study of the behaviour of households in energy saving (H1) and the multidimensionality of the processes of its manifestation (H2).

2 Methodology

In order to obtain scientific results, an arsenal of methodology of economic theory was used; namely, the methods of generalization and abstraction, systematization, the principle of "other things being equal", as well as the study of official statistical materials on the research topic. The theoretical basis was the teachings of representatives of institutionalism

* Corresponding author: burraabr@gmail.com

and the neoclassical direction of economic theory. As you know, at present, any action of market participants can be characterized in the form of models. In particular, situational modelling of household behaviour can be used. In this study, it is necessary to use a generalized model of household behaviour, which can be specified in the form of the following equation:

Energy saving = benefit + energy saving infrastructure + culture (1)

This kind of energy-saving model shows the need for a holistic solution to problems in the behaviour of the households, and requires taking into account all types of its manifestations.

In the scientific literature, you can find many publications devoted to the study of certain aspects of the behaviour of households in energy consumption.

Among the significant works of recent years, one can single out the studies of Boudet X. and others on the clustering of energy-saving behaviour in the family [2], Giovanni Frigo et al on the importance of individual energy consumption for human development and well-being [3], V. Strielkowski, in particular on the impact and scale of the penetration of renewable energy into households [4]. In Russian science, I. A. Bashmakov I. A., Bushuev V. V., and many others. In particular, one can note the article by I. A. Bashmakov on the need to improve the efficiency of energy use [5], and the fundamental work of V. V. Bushuev about the target vision of the energy sector of the future [6]. Certain aspects of the topic are considered in the writings of Burganov R. [7-8].

3 Results and Discussion

For the development of various models of energy-saving behaviour of households, all variables can be classified on different grounds; in particular, based on identifying key (significant) and non-key (insignificant) variables. Key (essential) variables predetermine the process of formation and development of the behaviour of the household in energy saving (Table 1).

Table 1. Model variables energy saving behaviour of households

Scientific foundations	**Behaviour variables**	
	Key	**Non-key**
Neoclassical theory	Necessity (need) for energy Benefits of energy consumption processes Rationality of action Elasticity Energy tariffs	Elasticity Energy tariffs
Institutional theory	Organizational, managerial, and institutional conditions for energy consumption The most important institution is a deal (contract)	Socio-psychological factors Emotional factors
Keynesianism	Part of ensuring reproduction Household behaviour can influence GDP growth, aggregate demand and supply.	Productivity from the real sector of the economy

Source: Authors.

As can be seen from the table, depending on the theoretical approaches, the key variables for modelling the energy-saving behaviour of households differ.

In general, the process of the formation of energy-saving behaviour of households depends on the implementation of many scientific ideas within the framework of the main scientific directions of economic theory.

But, in order to obtain effective, efficient, and systemic solutions, it is necessary to pay attention to new approaches to the energy-saving process. In particular, in energy-saving processes, one can single out the organizational mechanisms of the behaviour of households.

In fact, the study of any phenomenon from the point of view of its organization is the starting point of the analysis. Organization is a systems approach based on coordinating the actions of households as energy consumers with energy suppliers. Organizational forms include the purposeful activity of households in the formation of behaviour in energy saving. Organizational forms also include the distribution of responsibilities among members of the household, effective placement, and use of energy-receiving and energy-producing devices and equipment, etc. Households should use organizational forms in energy consumption, and self-organize in the field of reducing energy consumption of any kind.

Self-organization, i.e. the willingness of households to use technical and technological means (power receivers, control and measuring devices, etc.), to allocate astronomical time for control measures over the state of energy consumption, the implementation of contractual relations (for example, timely payment). Self-organization of households in energy saving can be defined as the orderly activities of households in a targeted impact on the optimization of energy consumption of any kind in order to meet their needs while creating favourable and comfortable living conditions. The process of self-organization is implemented more when living in private houses. According to Rosstat, more than 29% of Russians live in private houses, excluding summer houses and cottages. Self-organization in energy saving covers the organization of energy use for heating, air conditioning, hot water supply, cooking, lighting, household appliances, etc. The key elements of self-organization of households in energy saving as a system are the awareness of the state of energy consumption, the possibility of influencing the energy saving process, determining time periods for tracking the progress of energy saving, comparative analysis of energy saving data by periods of consumption, coordination of family members in behavioural actions, and the control of payment transactions for the energy used.

Also, the organizational form of the behaviour of households in energy conservation includes data management based on access to information about energy consumption using interactive services. The energy-saving behaviour of households based on data management, and the access to information on energy consumption using interactive services are important factors in the development of the energy market in the national economy.

In a mathematical model, the process of energy-saving behaviour of households can be described either by one or by many parametric representations. In a model with one parameter, one can take the growth of the population's income, which forms the energy-saving behaviour of energy users most significantly. In multi-parameter mathematical models, the influence of external (main and secondary) and internal (main and auxiliary) factors on the energy-saving behaviour of households is taken into account. Each variable can be positively or negatively reflected in the rational or irrational in the behaviour of households in energy saving (Table 2).

Table 2. Approximate classification of energy-saving factors in households

№	Factors	Characteristic of factor (s)	Examples
1	External	Influencing the energy saving process from outside the energy saving system	
	including:		
1.1	main	Defining goal and meaning of energy saving	The introduction of a tax on income received as a result of energy saving
1.2	minor	Influencing the actions and Interconnections of the elements of the energy-saving system,	The use of outdated house infrastructure of energy networks
1.3	... etc.		Environment The influence of the house committee
2	Internal	Influencing the energy saving process from within the energy saving system	
	including:		
2.1	main	Targets for Energy Saving of Family Members	Principle "When leaving, turn off the light"
2.2	auxiliary	Level of equipment with energy-saving devices	The presence of energy receivers in the household with the class "A +++"
2.3	... etc.		Energy literacy Energy-saving culture

Source: Authors.

In addition, to obtain a comprehensive analysis of the energy-saving behaviour of households, it is necessary to group them into homogeneous elements, and then draw an appropriate conclusion. The criterion for grouping (homogeneity) can be different circumstances, for example, household income, social status, living space per family, type of residential building, provision of an apartment and/or a house with general household energy metering devices, etc. (Table 3).

Table 3. Criteria for grouping households and types of their behaviour

№	Grouping criteria	Type of behaviour
1	Income:	Wasteful, ostentatious (excessive)
1.1.	High income	consumption
1.2.	Low income	Energy saving
2	Social status:	
2.1	The head of the family is the leader	Calculating
2.2	The head of the family is subordinate	Less calculating
2.3	The head of the family is non-working (pensioners, etc.)	Calculating
3	Living space per head of residents	
3.1	At the standard level	Calculating
3.2	Above the standard	Less calculating
4	Residing building type:	
4.1	Class A +	Calculating
4.2	Other classes	Less calculating
5	The provision of an apartment and/or a house with general house energy metering devices:	
5.1	Secured	Energy saving
5.2	Unsecured	Energy waste
6	Education of family members	
6.1	Higher education	More energy saving
6.2	Secondary education	Less energy efficient
7	Availability of energy competence	Energy saving

Source: Authors.

Of course, the selection of criteria for grouping households and their types of behaviour are incomplete and controversial. For example, according to social status, a large number of criteria can be distinguished.

The conditions and processes of saving energy are formalized in numerous legislative and regulatory documents of different levels. Only in Russia, at the federal level, over the past years, three state programs and three energy strategies have been created. However, the effectiveness and efficiency of implementations of the provisions of these documents is far from perfect. In particular, many of these provisions undergo adjustments and additions almost every year, as a result of which the level of confidence in these documents decreases. Timely payment for electricity use is perceived by many consumers as a norm of behaviour.

Of particular importance within the framework of institutional theory is the concept based on the use of a system of contracts and agreements. In particular, households in the field of energy supply can conclude such contracts as an agreement on the supply of electricity, on the provision of services for the development of a list of energy service activities; contract for the installation of energy service equipment; an agreement on the provision of services for the analysis and maintenance of energy service activities, etc. The process of concluding each agreement and energy contract has its own characteristics. Each type of agreement and energy contract includes costs incurred before its conclusion, during the process of conclusion, and after the conclusion.

In the studies of the founders of the course of institutionalism J. Commons, D. North, T. Veblen, W. Mitchell and others, as well as in the works of modern scientists, one can find many provisions suitable for analysing the behaviour of households in energy consumption and energy saving.

In modern studies, attention is paid to the consideration of new factors affecting the energy saving of households. For example, regarding property, more precisely, according to the economic theory of property rights. There are certain disagreements of interests regarding the consumption of electricity for the needs of households and for the maintenance of communal power grids. Common property is a significant consumer of electricity in apartment buildings; one of its aspects is discussed in the work of Moiz Masood Syed and his colleagues [9].

Establishing a rational relationship between family-wide and individual consumption of family members is a serious problem. Each family member can act in their own interests in energy consumption, which can cause a conflict about energy saving in the family circle.

Some aspects of the importance of individual energy consumption for human development and well-being are discussed in the work of Giovanni Frigo et al. [10].

Separate studies are devoted to the analysis of the influence of emotions on the possibility of reducing or increasing the volume of energy consumption. Saving energy depends on the mental and psychological state of each member of the household. In particular, in a joyful state, the consumer pays less attention to the need for energy saving. Revealing emotions through collective action and forms of social learning can lead to opportunities for transformative, socially embedded change — both reducing and improving energy use [11].

S of the impact of demographic evolution of households on their energy consumption should be also noted. This refers to the impact of the transition from one type of family to another (for example, from a single to a couple without children), which has not been studied previously [12]. The results of the study show that the evolution of households has a significant impact on their energy consumption patterns. However, the magnitude and direction of this effect are correspondingly weak and predominantly positive.

According to the institutional and technological concept, it is necessary for institutions to introduce the use of new energy-saving technologies and energy-receiving devices, for example class A +++, into the process of energy consumption of households. A number of works can be found, for example, the study of Brischke and his colleagues [13]; Zhang Y. and others [14].

A special role in energy saving is played by the moral and educational concept of institutionalism, according to which the behaviour of households depends on the functioning of such institutions as customs, traditions, and respect for energy and heat resources. The economic benefit expected from habit formation is a reduction in the costs associated with decision-making, observation, and/or trial and error (delusions) that would or could happen with less automatic behaviour [15].

Karlijn L. van den Broek provides an overview and organization of energy literacy, which focuses on understanding household electricity and gas use, providing a basis for classifying various conceptual and methodological approaches [16].

In general, modelling the process of saving energy in households based on the above factors is difficult and painstaking work and it is impossible to obtain real practical results without using modern methods of data analysis.

4 Conclusion

In conclusion, we can note the importance of a systematic theoretical study of the essence of the households behaviour in the field of energy conservation. Hypothesis H1 is confirmed. From the point of view of neoclassicists, that is, from the point of view of the "market", one can single out the desire of households to receive financial or other benefits based on the economic use of energy of any kind. The institutional characteristic of the energy-saving behaviour of households includes the creation of an appropriate infrastructure. When using different theoretical positions in the study of household behaviour, contradictory or even opposite results can be obtained. In particular, one can focus on obtaining only benefits due to energy savings, but the intention to invest in the installation of energy-saving technologies in the short term may not lead to its receipt.

According to the H2 hypothesis, it can be noted that the conditions and processes of energy saving are formalized in numerous legislative and regulatory documents of different levels, which indicates the multi-factor nature of obtaining an energy-saving effect in the behaviour of a household (Table 2). A comprehensive scientific direction is the study of the behaviour of households in the field of energy conservation based on the implementation of their organizational forms. Organizational forms include the distribution of responsibilities among family members, as well as the efficient placement of devices and equipment that receive and produce energy. Variables modelling household behaviour should group households according to different criteria. The criterion for grouping (homogeneity) can be various circumstances, for example, household income, social status, living space per family, type of residential building, provision of an apartment and/or a house with general house energy meters, etc.

In the context of the formation of a digital society and economy and the use of nature-like technologies in the energy-saving behaviour of households, transformational changes are objectively taking place, which should be carried out flexibly, taking into account the level of preparedness for innovations of consumers of various types of energy.

A key factor in changing household behaviour is the use of digital technologies in the process of energy consumption of all types. The state should create a system of incentive measures in the field of the formation of rational household behaviour, including postponing an effective organizational mechanism "supply-pay" in energy supply.

Acknowledgments

The reported study was funded by RFBR, project number 20-010-00099 [A1] "Theoretical and methodological approaches to the development of models of energy-saving behaviour of households in a nature-like economy».

References

1. L. Varadkar. Recommendations of the Global Commission for Urgent Action on Energy Efficiency [online]. Available at: https://iea.blob.core.windows.net/assets/d40d5638-1f45-42ac-b072-fe9e6417cc1e/Global-Commission-Recommendations.pdf (2020)

2. H. S. Boudet, J. A. Flora, K. C. Armel. Clustering household energy-saving behaviours by behavioural attribute. *Energy Policy*. **92**, 444-454 (2016).
3. G. Frigo, M. Baumann, R. Hillerbrand. Energy and the Good Life: Capabilities as the Foundation of the Right to Access Energy Services. *Journal of Human Development and Capabilities*. **22**(2), 218-248 (2021)
4. V. Strielkowski, E. Volkova, L. Pushkareva, D. Streimikiene. Innovative Policies for Energy Efficiency and the Use of Renewables in Households. *Energies*. **12**(7), 1392 doi:org/10.3390/en12071392 (2019)
5. V. Bashmakov. Improving energy efficiency and economic growth. Economic issues. **10**, 32-63 (2019)
6. V. Bushuev. Energy of Russia (selected articles, reports, presentations of 2014-2018). *Journal of Environmental Earth and Energy Study (JEEES)*. **5** (2018)
7. R. Burganov, E. Altynbaeva, L. Maimakova. Factor analysis of energy saving in households. *E3S Web of Conferences*. **209** doi:org/10.1051/e3sconf/202020903009 (2020)
8. R. Burganov, L. V. Maimakova, L. Urazbakhtina, L. A. Golitsyna. On the need to improve the methodology for calculating energy saving and energy efficiency of enterprises. *IOP Conference Series Materials Science and Engineering*. **791**(1), 012040 (2020)
9. M. M. Syed, G. M. Morrison, J. Darbyshir. Energy Allocation Strategies for Common Property. Load Connected to Shared Solar and Battery Storage. Systems in Strata Apartments. *Energies*. **13**(22), 6137 doi:org/10.3390/en13226137 (2020)
10. G. Frigo, M. Baumann, R. Hillerbrand. Energy and the Good Life: Capabilities as the Foundation of the Right to Access Energy Services. *Journal of Human Development and Capabilities*. **22**(2), 218–248 (2021)
11. M. Sahakian, B. Bertho. Exploring emotions and norms around Swiss household energy usage: When methods inform understandings of the social. *Energy Research & Social Science*. **45**, 81-90 doi:org/10.1016/j.erss.2018.06.017 (2018)
12. M. L. Chalal, M. Benachir., M. White, G. Shahtahmassebi, R. Shrahily. A Comprehensive Study on the Effect of Households' Evolution on Residential Energy Consumption Patterns. *Proceedings of 3rd International Sustainable Buildings Symposium (ISBS 2017)*. **6**, 485-500 doi:org/10.1007/978-3-319-63709-9_39 (2018)
13. L. A. Brischke, F. Lehmann, L. Leuser, S. Thomas, C. Baedeker. Energy sufficiency in private households enabled by adequate appliances. *Conference: ECEEE - Summer Study 2015*. 1571-1582 (2015)
14. Z. Xin-Gang, M. Xin, Z.Ying, L. Pei-Ling. Policy inducement effect in energy efficiency: An empirical analysis of China. *Energy*. **211** doi:org/10.1016/j.energy.2020.118726/ (2020)
15. D. Coldwel. The Economics of Habits. *Humboldt Journal of Social Relations*. **8**(2), 1-10 (1990)
16. K. van den Broek. Household energy literacy: A critical review and a conceptual typology *Energy Research & Social Science*. **57**(5) (2019)

Digital Transformation in HR

Jakub Chán[1,*], *Miluše* Balková[1]

[1]Institute of Technology and Business, Faculty of Corporate Strategy, Okruzni 517/10, 37001 Ceske Budejovice, Czech Republic

Abstract. The onset of the fourth industrial revolution and lack of human capital puts pressure on the development of own staff in terms of improving their qualifications. In the last four years, there has thus been introduced new methods in the management of staff development. The paper builds on the available texts in the area of theory and digital transformation of talent management and recruiting and puts them into the context of requirements arising from the principles of the fourth industrial revolution. Using matrix analysis, it examines the degree of compliance with the available implementation possibilities. The analysis performed leads to the conclusion that the requirements are best met by the method of exact description of partial tasks in the working process including its critical features and their subsequent assignment to workers according to the degree of compliance with their skills by means of combinatorics. This method thus also enables the management of worker development by assigning appropriate tasks.
Keywords: HR; human resources; Industry 4.0; industrial revolution; digitalization; digital transformation; skill management; matrix analysis

1 Introduction

With the onset of the Fourth Industrial Revolution, generally known as Industry 4.0, there are new requirements which literally force production companies to reconsider their current approaches to processes so that they are able to survive in the environment of ever-increasing competition whose success is based on the automation tools, AI or marketing using social networks. New challenges must respect both production and supporting processes. In the area of human resources management, these changes are reflected mainly in the change of the approach to staff development. In the times when the unemployment rate is extremely low, the labour market does not offer enough qualified workers. Companies thus need to obtain them by training low qualified or unqualified human capital using various growth and training programmes. In the last three years, there has been lack of even unqualified workers. They are often replaced by labour force from economically less developed countries which usually have high unemployment rate.

The focus of HR managers´ efforts to improve the qualification of human resources must thus dynamically shift given the current situation in the human capital market. Today we can thus see a shift from recruiting experts to working with human capital. Companies are aware of the fact that their potential success is generated mainly by appropriately allocated and qualified staff; there is thus a literal renaissance of skills monitoring, the so-called *talent management*. The driver of these transformation processes is represented by two main aspects in the long run – demographic and technological [1]. The demographic aspect refers mainly to the increasing average life expectancy. Combined with ever-increasing onset of automation technologies, it will bring an increase in unqualified labour force in the labour market. In the USA as the world´s strongest economy, unqualified labour force accounts for 47 % of all working people. About 60 % of working Americans perform tasks of which up to 30 % are subject to automation even now [1]. It is thus clear that organizations will have to succeed in monitoring, management and development of employee skills. It must be able to allocate a sufficient number of qualified workers in management positions and fill key medium-skilled positions with its own workers on the basis of their transformation from less qualified positions, while unqualified labour force will be gradually reduced as a result of the Fourth Industrial Revolution.

A question thus arises of how to cope with this emerging pressure and work effectively with available human capital, how to monitor the skills and competencies of employees, measure and develop them, and properly interpret the effects of growth or decline in skills. The specific benefit of this paper is the effort to find a comprehensive system that would facilitate and enable the mapping and development of employee skills using modern tools of database systems.

* Corresponding author: 27843@mail.vstecb.cz

2 Literary research

The 21st century is a period of innovative technologies and digitization. The transformation of HRM by means of digital tools puts an external pressure on changes in human behaviour, skills and competencies in organizations [2]. Technological development enables streamlining of production processes and brings changes in human labour, which might threaten employee wellbeing and challenge their existing skills and knowledge [3]. Claus [1] addressed the need for introducing talent management due to the unfavourable development in the human capital market. Using the application of scientific knowledge from other scientific disciplines closely related to HR, the author identified the need for introducing talent management in order to become competitive in the context of the fourth industrial revolution. Mothe and Nguyen-Thi [4] analysed the relationship of age diversity of employees and the degree of technological innovations in a company. On the basis of in-depth interviews with top managers, Whysall, Owtram and Brittain [5] conclude that the fast pace of technological changes created a gap between employee skills and dynamically changing requirements for their roles. This is confirmed by Dahlbom, Siikanen, Sajasalo and Jervenpää [6], who use qualitative interviews to describe the need for interconnecting data analysts and HR specialists, who usually do not have analytical skills. The application of HRA (Human resources analysis) shall also have a positive impact on reducing the subjectivity of evaluating employees, which may result in increased willingness to optimize work performance, as confirmed by Sharma, Sharma [7] using the analysis of available professional literature in the field of HR and HRA. Zhou, Liu, Chang and Wang [8] use a questionnaire survey implemented on a sample of 211 (mostly) manufacturing companies operating in China as the world´s largest labour market to describe the relationship between economic advance and digitization of HRM (Human resources management) and the improving performance of the company. The analysis of the obtained data indicates the existence of a positive correlation between a company´s performance and advanced digitized HRM system. Agarwal and Maurya [9] examine the relationship of quality and the power of corporate brand the way it is perceived by public in dependence on the degree of development of talent management in the organization. Using a questionnaire survey and subsequent regression and correlation analysis, they confirm the existence of a strong dependence between the degree of development of talent management and a positive perception of the organization by the public. Jones, Hutcheson and Camba [10] argue that the current Covid-19 pandemic can positively contribute to the development and digitization of systems (not only) in the field of HR. Companies were often forced to start investing in the digitization of their processes, mainly because of work from home.

It follows from the above findings that in the context of the Fourth Industrial Revolution, the world calls for the integration of HRA, talent management, and skills management in the everyday activities of organizations. However, the integration has been addressed in few publications only. Nunes, Pinto and Sousa [11,12] propose a development framework for employee technical skills by applying a structured training system. The effectiveness of the training system is then verified by mapping the OEE (Overall equipment effectiveness) indicator. Nunes, Pinto and Sousa [11,12] represent the method of improving competencies and skills of employees in production by means of continuous improvement of production process quality. By applying the method of QC Story (The process of control and improving the quality in production), subsequent analysis of key indicators, and questionnaire survey, they conclude that the focus on the process of quality improvement leads to the improvement of key process indicators as well as to the improvement of qualification, i.e. employee skills and knowledge in the long run. Anh and Lee [13] propose the application of the knowledge and skills of current employees as a model profile of a suitable applicant for a job. By mapping and subsequent analysis of excellent employees´ skills provides a certain model profile of an ideal potential job seeker. For creating the model and the compliance verification, analytical AI tools are used. It uses specific knowledge for the recruitment of employees, whose integration into the work process will be seamless; however, no future development of such employees is not provided by the system. Stadnicka, Arkhipov, Battaïa and Chandima [14] apply the model of knowledge and skills in the optimization of aircraft maintenance process as a critical process in terms of the quality and safety. By analysing the process steps in repairing a transport aircraft and subsequent assigning the necessary specific skills and knowledge from the field of technology and safety, they provide a comprehensive overview of requirements for a worker suitable for performing the specific critical task. For assigning the task to a specific employee, the so-called Hall's marriage theorem is used, i.e. combinatorial methods analysing the compliance of the requirements with the skills of a given worker. Kataoka et al. [15] focus on the application of skills management as a tool to improve the required skills of students taking their practical training. The application of the model is implemented in three steps. First, the expected skills are defined across all activities of the production process. Second, the verification methods and evaluation criteria are selected. The last step is the implementation of the model in the syllabus of specific training. The application of the training framework has a form of the so-called experiential learning. The course is repeated cyclically and its outputs are analysed until the individual reaches the required level of skills.

The available publications mostly reflect a specific, narrow field of the applicability of the system of mapping and development of employee skills. A comprehensive system applicable across processes and industries is not described.

3 Methodology

By studying available professional publications, we may conclude that there is a global need for the monitoring of employee skills and knowledge for the purpose of their management and development mainly in the case of low-skilled or unskilled positions. A matrix analysis is used to compare four selected systems of employee skills monitoring and their applicability for the manufacturing company. The selection criteria reflect the needs of the coming fourth industrial revolution and digital age. For the purpose of analysing available professional publication, the following research question was formulated: Which of the available methods best meets the requirements of the coming fourth industrial revolution?

The input alternatives for the evaluation using the matrix analysis are the systems by Nunes, Pinto and Sousa [11,12], i.e. the application of the training framework and subsequent verification of the effects on the production process by means of monitoring the OEE indicator, the systems by Stadnicka, Arkhipov, Battaïa and Chandima [14], i.e. the analysis of the partial process tasks and subsequent assigning of a suitable worker using the Hall´s marriage theorem, the method used by Ahn, Lee [13], i.e. the creation of an ideal job seeker´s profile by analysing the skills of excellent workers, and method used by Nunes, Pinto and Sousa [11,12], i.e. employee development by means of the application of QC story method – continuous improvement of the production process quality.

Available alternatives are compared in the following criteria:

a) multidisciplinary applicability

- the method is applicable across all production and non-production processes regardless of the field the organization operates in

b) forced costs

- *implementation of the method does not incur any costs; mainly the acquisition costs of software in which the method is implemented*

c) need for IT experts from practice (implementation and maintenance autonomy)

- implementation does not require hiring IT experts, the entity is able to manage the implementation without any help of IT experts → development of IT skills

d) permanent sustainability

- the methodology is repeatedly applicable; it reflects the current development in the field of IT as well as the automation of the production processes in the context of Industry 4.0

e) focus on personal growth

- the method is used for the development and management of an individual´s talent; it is applicable as motivation connected with the corporate remuneration system

f) comprehensibility at all organizational levels

- *all employees in the organization understand the basis of the methodology, understand what is being monitored, and for which purpose*

g) reduced evaluation subjectivity

- *the methodology is transparent enough so that it does not diminish employee trust and personal engagement*

The strength of the relationship between the variant and evaluation criterion is presented at four levels (*strong, medium, weak, and no relationship*), where no relationship represents the minimal or no match of the solution variant with the required criterion, while strong relationship refers to the absolute match. In the case of the evaluation criteria *forced costs* and *need for IT experts from practice,* the logic of assigning characteristic strengths of the features is logically reversed, i.e. the higher the need, the weaker the relationship and the lower degree of match with the required evaluation criterion. The strength of the relationship is then converted into a point scale (*0,5,10,15 points*). Summing of the points provides an accurate selection of the most suitable method showing the highest degree of match across the evaluation criteria.

4 Results

Table 1 shows a matrix analysis, i.e. the expression of the strength of the relationship between the selection criterion on the horizontal axis and the solution variant on the vertical axis. The strength of the relationship is then assigned based on the knowledge obtained from the specific professional papers on the scale of *no relationship – strong relationship.*

Table 1. Matrix analysis of available solutions

Symbol	Meaning
○	no relationship
△	weak relationship
◎	medium relationship
◎ (shaded)	strong relationship

Solution No.	Author	Published	Method	Multidisciplinary applicability	Forced costs	Need for IT experts	Permanent sustainability	Focus on personal growth	Comprehensibility for all levels of organization	Reduced evaluation subjectivity
1	**Pinto, Nunes, Sousa**	2020	application of training framework + verification by means of OEE indicator	△	△	◎	◎ (shaded)	○	△	○
2	**Stadnicka, Arkhipov, Battaïa, Ratnayake**	2017	analysis of process and its critical features – assigning operator using Hall´s marriage theorem	◎ (shaded)	△	◎	◎ (shaded)	◎	◎ (shaded)	△
3	**Lee, Anh**	2020	creation of an ideal job seeker´s profile + verification of the compliance using AI analytical tools	◎	○	○	△	◎ (shaded)	◎	△
4	**Souza, Nunes, Sousa**	2018	monitoring of the development using the QC story method	△	△	◎	◎ (shaded)	○	△	○

Source: Authors.

Table 2 shows the exact numerical expression based on the strength of the relationship (0-15 points). The last column of the table contains the overall sum of the points. The available methods under review show the compliance of the results of the criterion *permanent sustainability*, where three of them show a strong relationship; on the contrary, there is weak or no relationship in the case of the criterion *reduced evaluation subjectivity*. It can thus be concluded that the methods are applicable in the context of the fourth industrial revolution; however, it is necessary to develop the transparency leading to the reduced subjectivity of the evaluation of the employee, especially if the evaluation is connected with the remuneration system. The criterion of *forced costs* also shows match across all implementation schemes. This finding can be interpreted as a need for high input costs of the implementation of the method, mainly due to the dynamically changing requirements combined with the lack of HR staff skills, as confirmed by Dahlbom, Jervenpää, Sajasalo and Siikanen [6]. Particularly high implementation costs can be seen in the case of the implementation scheme No. 3 by Lee and Anh, which compares the candidate´s compliance with the profile of an ideal worker using AI analytical tools; autonomous implementation and administration from the side of the organization cannot be assumed there. On the contrary, the second implementation scheme shows the best results in the comparison with the requirements of the fourth industrial revolution, as at least medium strength of the relationship is achieved in the case of five out of seven criteria.

Table 2. Numerical expression of the strength of the relationship, criterion - variant

Solution No.	Author	Published	Method	Multidisciplinary applicability	Forced costs	Need for IT experts	Permanent sustainability	Focus on personal growth	Comprehensibility for all levels of	Reduced evaluation subjectivity	Σ
1	**Pinto, Nunes, Sousa**	2020	application of training framework + verification by means of OEE indicator	5	5	10	15	0	5	0	**40**
2	**Stadnicka, Arkhipov, Battaïa, Ratnayake**	2017	analysis of process and its critical features – assigning operator using Hall´s marriage theorem	15	5	10	15	10	15	5	**75**
3	**Lee, Anh**	2020	creation of an ideal job seeker´s profile + verification of the compliance using AI analytical tools	10	0	0	5	15	10	5	**45**
4	**Souza, Nunes, Sousa**	2018	monitoring of the development using the QC story method	5	5	10	15	0	5	0	**40**

Source: Authors.

5 Discussion

The total sum of the points of expressed strength of the relationship between the criterion and implementation variant indicates the most suitable method, which turns out to be the method No. 2, system developed by Stadnicka, Arkhipov, Battaïa and Chandima [14] based on the analysis of individual partial tasks and their subsequent assigning to individual workers according to their skills and knowledge, as it best meets the requirements arising from the coming fourth industrial revolution. The system is applicable across industries and processes. In its implementation and management, the participation of IT experts is not necessary. The system is sustainable and sufficiently flexible, focusing on the development of individuals and its transparency enables the reduction of the evaluation subjectivity. Also, the system of assigning specific tasks to employees on the basis of a detailed analysis of the task requirements and employee skills appears to be the most suitable available solution. The system enables the management of employee development by means of assigning tasks of various degrees of difficulty and critical features. The weakness is the overall transparency leading to increasing trust. This is represented by the need to apply the combinatorial method, the so-called Hall´s marriage theorem, to exactly assign a task to a suitable worker.

The imaginary last position is occupied by the systems used by Nunes, Pinto and Sousa [11,12]. These systems show the lowest degree of compliance especially with the criteria of focus on individual and the overall transparency of the system. Development is mapped by means of monitoring the indicators which are not directly related to the development of employees. Employee development is a secondary effect to the improvement of the production in the area of production quality or increasing the usability of machinery. The system is thus not fully transparent for employees and a question arises of "What is the share of individual employees in improving the indicators?" Therefore, this is not a suitable method of skills mapping with the potential connection to the organization remuneration system.

The system developed by Anh, Lee [13] shows the highest possible degree of compliance for the criterion *"focus on individuals"*, since the system creates the profile of an ideal employee on the basis of actual knowledge from practice and subsequently seeks for a suitable candidate with the highest degree of compliance with the profile. The main disadvantages can be seen in a very complex creation of the profile, which requires the participation of the IT experts from practice as well as high additional acquisition costs of software. Another disadvantage is its single-purpose use only in the stage of recruitment.

6 Conclusion

The objective of the paper was to determine how to meet the requirements of the HR system transformation, especially in the field of employee development in the context of the fourth industrial revolution and coming digital transformation. For this purpose, available professional texts concerning the given issue were analysed. From these texts, four specific implementation criteria were selected, which were put into the context of seven criteria meeting the basic ideas of the fourth industrial revolution and digital transformation, namely multidisciplinary applicability, minimal costs, reduction of the need for experts, permanent sustainability of the system, focus on the growth of individuals, transparency of the whole system, and efforts for the reduction of subjectivity in evaluating employees. The strength of the relationship between the implementation scheme and evaluation criterion was divided into four levels (strong, medium, weak, and no relationship), which were assigned relevant number of points on the scale of 0-15. Subsequently, matrix analysis of the strength of the relationship between the requirements arising from the fourth industrial revolution in HR and available solutions was performed. By summing the points assigned to the strength of the relationships, it can be concluded that the requirements of the fourth industrial revolution are best reflected and met by the implementation scheme No. 2, i.e. the authors Stadnicka, Arkhipov, Battaïa and Chandima [14], specifically, the system based on the exact knowledge of partial steps of a task including the definition of critical features and the subsequent assignment of the tasks to individual operators using the combinatorial method, the so-called Hall´s marriage theorem. However, the method of assigning a task to a specific employee may pose pitfalls in the case of its implementation in a small company due to its complexity and worse transparency of the system. After the exact analysis of partial steps of the working process, a question arises of how to find a suitable worker in a form understandable to the employees themselves. Therefore, there is some space for further research aimed at improving the transparency of the system across the organizational levels of the company in order to increase the trustworthiness of the system.

References

1. L. Claus. HR disruption - Time already to reinvent talent management. *BRQ Business Research Quarterly*. **22**(3), 207-215 (2019)
2. A. Sengupta, S. Lalwani, S. Goswami, P. Srivastava. Reinventing HR functions with SMAC technologies - an exploratory study. *Materials Today: Proceedings*. **46**(1) (2020)
3. A. Reiman, J. R. L. Kaivo-Oja, E.-P. Takala, T. Lauraeus, E. Parviainen. Human factors and ergonomics in manufacturing in the industry 4.0 context – A scoping review. *Technology in Society*. **65** (2021)

4. C. Mothe, T. U. Nguyen-Thi. Does age diversity boost technological innovation? Exploring the moderating role of HR practices. *European Management Journal*. (2021)

5. Z. Whysall, M.Owtram, S. Brittain. The new talent management challenges of Industry 4.0. *Journal of Management Development*. **38**(3) (2019)

6. P. Dahlbom, M. Jarvenpää, P. Sajasalo, N. Siikanen. Big data and HR analytics in the digital era. *Baltic Journal of Management*. **15**(1) (2019)

7. A. Sharma, T. Sharma. HR analytics and performance appraisal system: A conceptual framework for employee performance improvement. *Management Research Review*. **40**(6), 684-697 (2017)

8. M. Agarwal, K. K. Maurya. Organisational talent management and perceived employer branding. *International Journal of Organizational Analysis*. **26**(2) (2018)

9. Y. Zhou, X. Chang, G. Liu, L. Wang. The impact of HRM digitalization on firm performance: investigating three-way interactions. *Asia Pacific Journal of Human Resources*. **59**(1), 20–43 (2021)

10. J. D. Camba, S. Hutcheson, M. D. Jones, S. Hutcheson. Past, present, and future barriers to digital transformation in manufacturing: A review. *Journal of Manufacturing Systems*. **60**(9) (2021)

11. R. Sousa, S. Souza, E. Nunes. Developing organisational learning through QC story. *Total Quality Management & Business Excellence*. **31**(13-14), 1565-1587 (2020)

12. E. M. P. Nunes, L. Pinto, S. D. Sousa. A framework to improve training and development of workers' technical skills: effects on operational performance during company relocation. *Procedia Manufacturing*. **51**(6), 1806-1813 (2020)

13. CH. Ahn, D. Lee. Industrial human resource management optimization based on skills and characteristics. *Computers & Industrial Engineering*. **144** (2020)

14. D. Stadnicka, D. I. Arkhipov, O. Battaïa, R. M. Chandima. Skills management in the optimization of aircraft maintenance processes. *IFAC World Congres 2017*. **50**(1), 6912-6917 (2017)

15. T. Kataoka, K. Morikawa, K. Takahashi. Strategic Human Resource Management Simulation Considering Work Elements, Skills, Learning and Forgetting. *25th International Conference on Production Research Manufacturing Innovation: Cyber Physical Manufacturing*. **39**, 633-1640. (2019)

The Impact of the Coronavirus Pandemic on Small and Medium-sized Enterprises in Russian Federation

Christina Chernova[1], *Natalia* Neklyudova[1,*]

[1] Institute of Economics, Ural Branch of the Russian Academy of Sciences, 29, Moskovskaya St., 620014, Ekaterinburg, Russia

Abstract. The purpose of the article is to analyse the impact of COVID-19 on small and medium-sized businesses in the Russian Federation. Data from the research company "Evotor", the unified register of small and medium-sized businesses, as well as data from the Central Bank of Russia were used as material for the study. Based on the results, it can be stated that the pandemic has affected the economy in different ways. A number of small and medium-sized businesses and their revenue decreased in 2021 compared to 2019. But at the same time, the sector of domestic tourism has grown over the period under review. The largest increase in the average revenue of hotels in 2021 compared to 2020 was observed in the Kaliningrad region, Krasnodar Krai and Moscow.
Keywords: SMEs; COVID-19; domestic tourism; Russian regions

1 Introduction

Most countries in the world use the tools of fiscal and monetary policy to support the economies in the wake of the coronavirus pandemic. In 2020, drop in household income and uncertainty about the future im the new conditions of lockdown caused a sharp drop in aggregate demand. In Russia, this led to the decline in interest rate to the historically lowest level of 4.25% (from 27 July 2020 to 21 March 2021).

The most serious consequences for the Russian economy were a drop in demand for Russian exports, the decrease in oil prices, and the fall of domestic currency (rouble). The crisis is expected to have more serious consequences for the Russian economy since the state is currently facing the pressure of sanctions. In addition, since the beginning of 2020, the dynamics of GDP has been negative.

The reduction of interest rate to 4.25% was supposed to support aggregate demand and lending. The decrease in inflation expectations led to significant easing of monetary policy in the Russian Federation. However, the measure adopted could not bring significant results in the conditions of rouble depreciation by more than 20% against the US dollar and more expensive imports for businesses. In addition, the freezing of the economic activity, lockdown, has also negatively affected the effectiveness of this measure. In the short term, lower interest rates are unable to stimulate growth if most economic activities are banned.

Measures to reduce regulation, which were aimed at maintaining liquidity in the economy, were more pronounced. This can be explained by the fact that most economic activities were frozen, with many financial institutions facing problems [1].

Small and medium-sized enterprises in Russia seriously suffered from the pandemic. The negative effects have affected 69% of enterprises in the SME sector. At the peak-lockdown, 56.1% of enterprises were closed. According to surveys conducted by the SME Ombudsman, in 2020, a drop in demand affected 80% of enterprises, and to date, demand has not recovered for 52.6% of enterprises [2]. The turnover of small enterprises (microenterprises not included) decreased by 3.1 trillion roubles in 2020 [3]. The share of small enterprises with loan debts (including overdue ones) in the total number of small enterprises increased from 20 to 30% in 2020 [4].

The most severe blow of the pandemic occurred in April and May 2020, the stabilization of economic processes was outlined in June. Some researchers note that microenterprises and small enterprises had more opportunities for dismissing their employees; therefore, Russian regions with a high share of small enterprises in the economy are more susceptible to increased unemployment than the regions where the economy is based on large enterprises. The most vulnerable sectors were trade and services, transport, construction, and industry [5].

Based on the data from January to June 2020, Obraztsova et al. concluded that the average monthly decrease in the number of small business entities accounted for more than 2.5%, with the number of employees decreasing by an average

* Corresponding author: nnp81@mail.ru

of 3.2% per month. The authors of the article concluded that in general, medium-sized enterprises suffered less than small enterprises. In the first half of 2020, the reduction in the number of "young" firms set up in the preceding 3 months exceeded 19% due to the pandemic and quarantine measures. According to the calculations of the authors, this figure exceeded 30% in the Tula region, Krasnoyarsk Territory, St. Petersburg, and Moscow [6].

Epidemics in general can be an incentive to introduce new business models and technologies. For example, in 2003, an outbreak of pneumonia led to the introduction of online shopping by Chinese consumers, which accelerated the growth of a Chinese holding company operating in the field of online commerce, Alibaba [7]. The coronavirus pandemic has greatly affected the migration processes in society. Migrants quickly lost stable wages and their jobs due to the lockdown. The closure of borders reduced their mobility and migrants could not return home to their families.

The countries donor of labour force experienced a sharp drop in the stable flow of funds from Russia, which accounted for a significant percentage of their total GDP. In addition, hundreds of thousands of unemployed (possibly infected) migrants were forced to return to their homeland. In turn, Russia lost a source of cheap labour. In addition, the cost of moving between states has increased due to the lack of transport, the need to pay for covid tests, the use of various modes of transport for transfers [8].

The coronavirus pandemic and self-isolation of Russian citizens led to a collapse in demand for passenger taxi services by 20-50% or more in the spring of 2020. The economic problems that arose due to this collapse complicated even more the solution of the problems that had existed before the crisis in the taxi market of the Russian Federation [9].

Decrease in profit negatively affects the activities of any business. Due to the difficulty in complying with the rules of social distancing, in some cases, enterprises were forced to curtail their activities for a long period. The most affected sectors of the economy include the leisure sector, trade in non-food products, and hospitality sector. Forced suspension of activity affected mostly the construction sector and social services, some sectors of the manufacturing industry, fishing, forestry, and agriculture. The decline in commuting dramatically reduced the demand for public transport. However, at the same time, online sales increased, which led to a change in consumer preferences and increase in demand for delivery services. Consequently, the damage to those companies that were able to reorient themselves in the new conditions turned out to be less. In addition, over time, enterprises were able to evaluate the effectiveness of work from home, as it has allowed them to reduce the cost of maintaining offices [10].

Now the restaurant owners have to search for new formats that would take into account customer requests. On the one hand, guests are becoming more demanding; on the other hand, rather thrifty economical due to the consequences of the coronavirus pandemic [11].

The coronavirus pandemic has undoubtedly affected all banks, reducing their activities. However, since the second half of 2020, there has been an increase in assets among 59% of participants in the banking market [12]. The quality of banks' loan portfolios is of fundamental importance for interbank competition. In the medium term, the Bank of Russia assumes that there will be an increase in overdue loan debt and poor loan servicing in the banking sector. But there is a possibility of granting credit holidays or restructuring loans to borrowers affected by the effects of the coronavirus. Large banks, such as VTB and Sberbank provided significant support to branches of small and medium-sized enterprises affected during the introduction of self-isolation in 2020 in the form of interest-free irrevocable loans.

It should also be noted that in the middle of 2020, there was a decrease in the number of deposits, as depositors decided to purchase daily goods and, fearing a strong price increase in the future, made purchases for the future [13].

The latest support measures due to the coronavirus pandemic of small and medium-sized enterprises at the federal level include the following: According to the Resolution of the Government of the Russian Federation of September 8, 2021, № 1520 scheduled inspections of small enterprises will not be carried out in 2022 [14]. Medium-sized and small enterprises have the opportunity to receive a subsidy, provided that regional anti-covid measures apply to them [15]. The Ministry of Economic Development will provide a monthly subsidy to banks in order to compensate for the commission paid by enterprises for transfers under the fast payment system from July 1 to December 31, 2021. Banks will transfer subsidies to customer accounts [16].

2 Methodology

We analysed statistical data before and during the coronavirus pandemic in the Russian Federation. The data were obtained from the Evotor research company, the unified register of small and medium-sized enterprises, and from the Central Bank of Russia for the years 2019, 2020, and 2021. Evotor is an IT company, an ecosystem of technologies for automation of small, medium-sized and large enterprises. Since 2016, it has been a leader of technological transformation in small and microenterprises. This company is the largest Russian manufacturer of smart online cash registers: over 800,000 Evotor smart terminals have been activated in the country in the last 5 years. Every fourth entrepreneur in Russia uses Evotor. They provide more than 850 services for business management in the app store "Evotor.Market".

3 Results and Discussion

According to IT company Evotor in June 2021, hotels located in the Russian Federation increased their average revenue by 82% compared to 2020. Evotor used the information from the receipts of 750,000 Evotor terminals in the Hotel segment for June 2019, 2020, and 2021 to determine the average revenue per terminal [17].

Table 1 shows that the largest increase in the average revenue of hotels in 2021 compared to 2020 was observed in the Kaliningrad region, Krasnodar region, and Moscow.

Table 1. Dynamics of average revenue of hotels in 2021 compared to 2020 and 2019, percentage

Russian regions	Growth compared to June 2020	Growth compared to June 2019
Sverdlovsk region	76	4
Republic of Tatarstan	71	31
Republic of Karelia	68	18
Irkutsk region	37	19
St. Petersburg	130	13
Moscow	200	22
Kaliningrad region	220	63
Krasnodar region	200	36

Source: Author based on the data of Evotor [17]

The increase in the average revenue of Russian hotels can be explained by the fact that in June 2021, the demand of Russians for holidays within the country increased due to the limited opportunities for recreation abroad. For example, tourists from Russia have been deprived of the opportunity to go on vacation to Turkey since 15 April 2021. Flights to this country were not resumed until 22 June 22 2021.

The analysis of the unified register of small and medium-sized enterprises shows that in all federal districts of Russia, the number of small and medium-sized enterprises has decreased in 2021 compared to 2019, except for the Central Federal District, where the value increased by 0.44 % compared to 2019. In 2021, the increase in the number of small and medium-sized compared to 2020 was 3.9% in the Central Federal District, 0.52% in the North-Western Federal District, 1.92% in the North Caucasus Federal District, 0.44% in the Volga Federal District, 0.08% in the Far Eastern Federal District. However, this indicator decreased by 0.26% in the Southern Federal District, by 0.17% in the Ural Federal District, and by 0.2% in the Siberian Federal District by 0.2% (Figure 1).

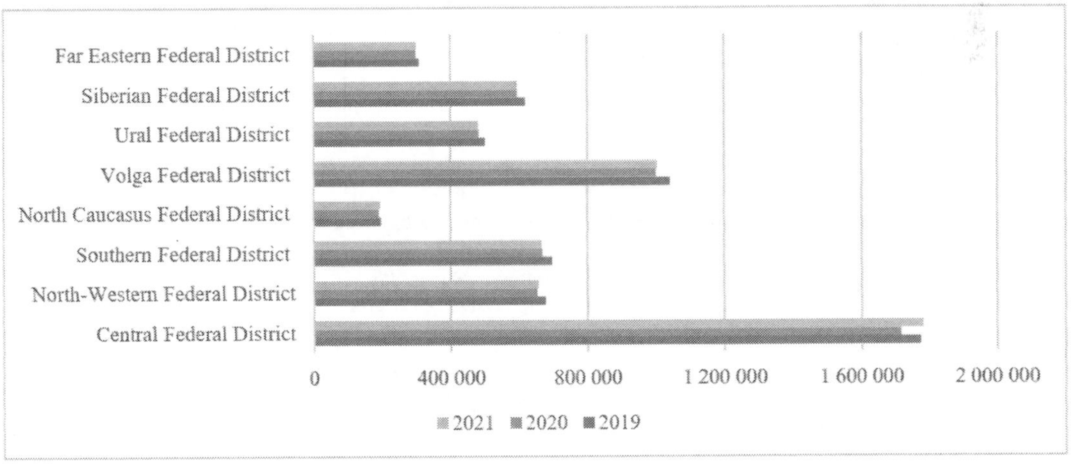

Figure 1. Number of small and medium-sized enterprises in Russia by federal districts

Source: Author based on the data from unified national register small and medium-sized business entities [18]

In addition, we concluded that the number of medium-sized enterprises in the Russian Federation grew during the pandemic (by 3.4% in 2020 compared to 2019 and by 5.5% in 2021 compared to 2019). This does not apply to microenterprises and small enterprises, as the number of microenterprises decreased by 3.7% in 2020 compared to 2019, and this indicator in Russia increased by 1.5% in 2021 compared to 2020. The number of microenterprises decreased by 3.4% in 2020 compared to 2019, and this indicator decreased by 1.8% in 2021 compared to 2020 (Table 2).

Table 2. Number of firms in Russia before and during the pandemic by forms of entrepreneurship

	2019	2020	2021
microenterprises	5,570,745	5,367,233	5,445,653
small enterprises	225,556	217,951	213,990
medium-sized enterprises	16,666	17,239	17,576

Source: Author based on the data from unified national register of small and medium-sized enterprises [18]

The number of banks in Russia decreased by 23% as of 1 September 2021 compared to 1 January 2019. This indicator decreased monthly during the monitored period (Figure 2).

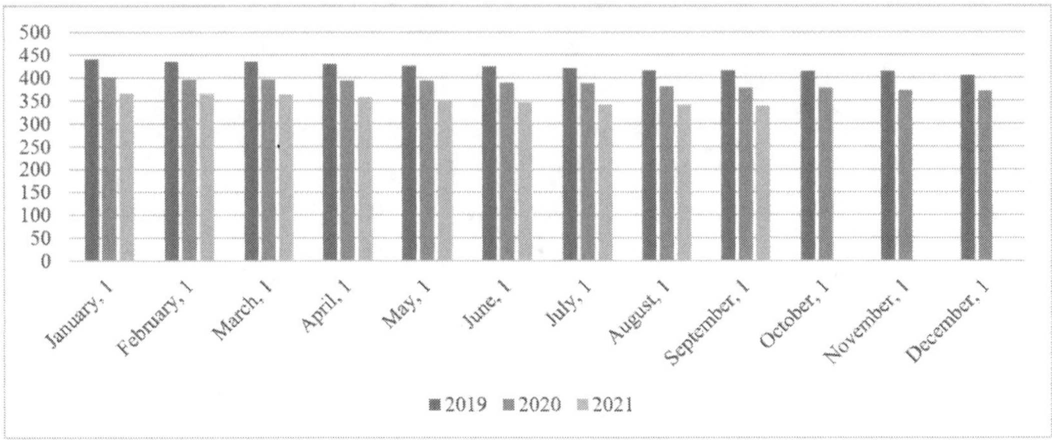

Figure 2. Quantitative characteristics of Banking Sector of the Russian Federation in 2019-2021

Source: Author based on data from the Central Bank of the Russian Federation [19]

4 Conclusion

The Covid-19 pandemic has affected sectors of the economy in a different way. Some industries have been seriously affected, while others suffered less. The tourism industry is considered to be one of the most affected sectors of the economy due to the introduction of restrictive measures to stop the spread of the coronavirus disease. Domestic tourism has become a trend this year in Russia. People who are tired of a prolonged regime of self-isolation prefer to make short trips at weekends. In the current situation, the Russians prefer travelling by car to flying. Tourist places have become more popular, trips to which were planned, but long postponed. Instead of hotels, travellers prefer cheaper alternatives, e.g. apartments where they can cook without having to visit cafes or restaurants. Accommodation is booked mainly using aggregators or hotel websites. Thus, while most small and medium-sized enterprises have been seriously affected by COVID-19, the pandemic works in favour of domestic tourism. Russian government, for its part, has also provided substantial support to domestic tourism, in particular in the form of returning part of the funds spent on trips to Russia, the so-called tourist cashback. Thus, tour operators and accommodation facilities - hotels, hotels, hostels - were provided some support. Currently, we are expecting the winter season. As usual, ski resorts are very popular at this time. If ski resorts are open, it would be advisable to resume the tourist cashback programme so that the Russians start to prefer domestic destinations for their holidays again.

The article is prepared under the plan of the Institute of Economics of the Ural Branch of RAS for 2021–2023.

References

1. Y. K. Zaytsev. Monetary and fiscal policy measures during the COVID-19 economic crisis in Russia. *Finance: Theory and Practice.* **24**(6), 6-18 (2020)
2. Institute for the Economics of Growth P. A. Stolypin. *Special report of SME Ombudsman to the President of the Russian Federation - 2021* [online]. Available at: http://doklad.ombudsmanbiz.ru/2021/7.pdf (2021)
3. Rosstat. *Institutional transformations in the economy* [online]. Available at: https://rosstat.gov.ru/folder/14036?print=1 (2021)

4. Central Bank of the Russian Federation. *Loans to small- and medium-sized enterprises grow by 22.6% in 2020* [online]. Available at: https://cbr.ru/press/event/?id=9619 (2021)

5. L. Yemelyanova, A. V. Lyalina. The labour market of Russia's Kaliningrad exclave amid COVID-19. *Baltic Region*. **12**(4), 61-82 (2020)

6. O. I. Obraztsova, A. Y. Chepurenko. SME Policy in the Russian Federation: Renewal After the Pandemic? *Public Administration Issues*. **3**, 71–95 (2020)

7. P. A. Esin. World market development scenario in the context of the coronavirus crisis. *Izvestiya VUZ. Applied Nonlinear Dynamics*. **28** (2), 158–167 (2020)

8. S.V. Ryazantsev, I. N. Molodikova, A.D. Bragin. The effect of COVID-19 on labour migration in the CIS. *Baltic Region*. **12**(4), 10-38 (2020)

9. Y. Ezrokh. State Regulation of the Russian Passenger Taxi Market: Accumulated Problems and Solutions. *Economic Policy*. **15**(4), 138-163 (2020) doi:org/10.18288/1994-5124-2020-4-138-163

10. I. Miles, V. Belousova, N. Chichkanov, Z. Krayushkina. The Impact of the Coronacrisis on KIBS Sector. *Foresight and STI Governance*. **15**(1), 6-18 (2021) doi:org/10.17323/2500-2597.2021.1.6.18

11. I. N. Fedeneva, I. Y. Sewruikov, O. V. Leushina, R.A. Yakshigulov. Public Catering in the Structure of Tourist Services (on the Example of Novosibirsk). *Human. Sport. Medicine*. **21**(1), 166–170 (2021)

12. RBC. *Russian banks have shown a record growth in assets for six years* [online]. Available at: https://www.rbc.ru/finances/25/12/2020/5fe482489a79476ad2ed7afd (2020)

13. O. Y. Donetskova. Interbank competition in the Russian banking market. *Finance: Theory and Practice*. **25**(1),143-156 (2021) doi:org/10.26794/2587-5671-2021-25-1-143-156

14. GARANT-SERVICE. The Resolution of the Government of the Russian Federation of September 8, 2021 № 1520 *«On the specifics of conducting scheduled control (supervisory) activities in 2022, scheduled inspections in relation to small businesses and on amending certain acts of the Government of the Russian Federation»* [online]. Available at: https://www.garant.ru/products/ipo/prime/doc/402677354/ (2021)

15. GARANT-SERVICE. The Resolution of the Government of the Russian Federation of September 7, 2021 № 1513 *«On approval of the Rules for Granting Subsidies from the federal budget in 2021 to small and medium-sized businesses and socially oriented non-profit organizations operating in municipalities most affected by the deterioration of the situation as a result of the spread of a new coronavirus infection»*[online]. Available at: http://www.garant.ru/products/ipo/prime/doc/402697350/ (2021)

16. Ministry of Economic Development of the Russian Federation. *Reshetnikov: 53 banks compensate businesses for 100% commission for quick payments*. Available at: https://www.economy.gov.ru/material/news/reshetnikov_53_banka_kompensiruyut_biznesu_100_komissii_za_bystrye_platezhi.html (2021)

17. EVOTOR. *The turnover of Russian hotels in June exceeded the pre-pandemic level.* [online]. Available at: https://evotor.ru/research/oborot-rossijskih-gostinits-v-iyune-prevysil-dopandemijnyj-uroven/ (2021)

18. Federal tax service of Russia. *Unified national register small and medium-sized business entities* [online]. Available at: https://rmsp.nalog.ru/index.html (2021)

19. Central Bank of the Russian Federation. *Quantitative characteristics of Banking Sector of the Russian Federation* [online]. Available at: https://cbr.ru/statistics/bank_sector/lic/ (2021)

Analysis of the Czech Advertising Market and its Main Players

Michal Konečný [1,*], *Róbert* Világi[2], *Yaroslava* Kostiuk[1,3], *František* Pollák[1]

[1]Institute of Technology and Business, Faculty of Corporate Strategy, Okruzni 517/10, 37001 Ceske Budejovice, Czech Republic
[2]University of Economics in Bratislava, Faculty of Business Management, Dolnozemska cesta 1, 852 35 Bratislava, Slovakia
[3]University of Zilina, Faculty of Operation and Economics of Transport and Communications, Univerzitna 1, 010 26 Zilina, Slovakia

Abstract. At a time of global crisis, also at a time of accelerated digitalisation of businesses, sensible investment in communication with their customers is becoming very topical. The paper focuses on the development of investments in Internet advertising in the period 2012-2020. The method used for the research was a comparative analysis, which showed that the amount of money invested in internet advertising increased every year and that, unlike other types of advertising, no year was recorded for internet advertising during the period under review, when the value of investments in a given year would be lower than the value in the previous year. The results show that online advertising is taking an increasing share of total advertising investment. Another finding is that television advertising maintains a dominant position in the advertising market and is not directly affected by the development of investment in Internet advertising. However, the growth of investment in Internet advertising has a direct impact on the volume of investment in print advertising and OOH (Out of Home) advertising. The given types of media showed the largest decrease in terms of market share and negatively copied the development of investments in Internet advertising. In the case of radio advertising, there was also a decrease, but its dependence on Internet advertising has not been proven. However, the question arises as to the development of investments in the coming years. It is possible to anticipate the continuing strengthening of the role of Internet advertising and it will also be interesting to see whether Internet advertising will jeopardize the position of TV advertising. The results of the research will help both researchers and companies in the process of their decision to invest in selected types of advertising.
Keywords: advertising market; investments; internet advertising; strategic management

1 Introduction

Today, the digital world is an integral part of corporate culture, and not only corporate data, but also data about customers and customer behavior is an integral part of the company. A significant competitive advantage for the SME sector is the availability of cutting-edge technologies and technological progress, but not every company uses the offered technologies of the digital world. Today's manager must follow current trends and customer behavior, because today he already has access to technology without significant investments. According to figures from the Czech Statistical Office [1], 82% of Czech households already had access to the Internet in 2020, and 54% of the human population shop online. It can be seen that Czech entrepreneurs have also become aware of this trend, and in recent years they have placed more emphasis on activities in this communication channel. The Internet has thus become an irreplaceable tool for the acquisition of new customers of domestic companies. The aim of this paper is to find out how advertisers were willing to invest in Internet advertising in the period 2012-2020 and how the phenomenon affected the amount of investment in other forms of advertising. Television advertising, print, radio and OOH - Out of Home advertising, which represents the types of advertising affecting the advertising audience away from home, will be considered as other forms of advertising.

1.1 Current State of the Adressed Problem at Home and Abroad

Misuraca, Scepi and Spano [2] state that the advent of Web 2.0 has changed everyday life in many ways, such as a system involving the purchase of products and the use of services, so that the revolution has inevitably affected businesses by making customers increasingly demanding. Vila, Cardoso and de Araujo [3] argue that communication is a key tool for informing consumers about products and services. They argue that the current consumer trends that come with new technologies and the Internet must be considered when creating a communication strategy with consumers, thus creating the concept of 2.0 communication. Opresnik [4] states that in the context of the growing popularity of social networks, blogs and podcasts, customers can spread their views on products or services to potentially millions of audiences, and

* Corresponding author: 27826@mail.vstecb.cz

that with the expansion of people's Internet access, anyone interested in addressing product problems and services provided by companies to deal with them.

Pollak et al. [5] state that the COVID-19 pandemic redefined the practices in all business sectors, as in the first half of 2020, companies faced an upredictable and unexpected situation which turned into both threats and opportunities. As a result, brick-and-mortar branches of most stores were closed and a significnt part of business transactions and activities has moved to the online environment. Khurramov and Fayzieva [6] the term new media refers to traditional and different ways of providing information to consumers based on the latest digital technologies. Khajehasani, Abolizadeh and Dehyadegari [7] say that, as technological progress grows, so does the number of Internet users, and Internet communication has become the routine of the 21st century. The Internet as a means of providing business products and services is already a common tool and even necessary for some businesses. Soviar et al. [8] state that the digital environment is playing an increasingly important role in the customer's decision-making. Pollák and Markovič [9] claim that it is also a place for sharing the customer's opinions and experiences with a specific company, where a dissatisfied customer can pose a reputational risk to the company.

Kumar and Mittal [10] state that, because interactions take place anywhere and anytime, they go far beyond the traditional customer connection. Brown and Hayes [11] state that the consumer is exposed to more than a thousand advertisements before going to bed. Zhang et al. [12] state that in the case of financial market trading, online advertising is based on real-time bidding and uses an auction mechanism to automate media purchases. Therefore, how optimally the bids are submitted is crucial for increasing profit and return on investment. The high risk of user behavior and the uncertainty of costs caused by auction competition can lead to significant risk in estimating campaign performance. Gasilov et al. [13] state that the effectiveness of advertising is determined not only by the number of people who come into visual contact with the advertising media, but also by the degree to which their characteristics match the parameters of the target audience of the advertisement. Consumer motives are important as they help us understand what influences them when making purchasing decisions.

The aim of the research will be to find out how advertisers were willing to invest in Internet advertising in the years 2012-2020 and how the phenomenon affected the amount of investment in other forms of advertising in the Czech Republic through the following established research questions: What is the development in the Czech Republic investing in internet advertising? What is the relationship between investing in online advertising and other types of advertising?

2 Methodology

The aim of this paper is to find out how advertisers were willing to invest in Internet advertising in 2012-2020 and how the phenomenon affected the amount of investment in other forms of advertising in the Czech Republic. The data were obtained from questionnaire surveys and were based on the annual financial statements of the surveyed subjects, published annually by the Association for Internet Development (SPIR). In order to maintain the protection of sensitive information, the ppm factum agency was entrusted with the processing of these investigations for the years 2012-2015. For the years 2016-2020, the MEDIAN agency was entrusted with their processing [14]. The research dataset will include Internet operators whose media dominate the Czech market and thus affect the majority of Internet users. The performance of other types of media types will be determined based on data by Nielsen Admosphere agency.

The research methodology will include the following procedure:

1) We analyze the obtained data using a comparative method.
2) Then we process the data in Google Sheets.
3) The cost values for the calendar years will be assigned in rows to the corresponding columns, which will represent specific years, and the percentage share of investments in individual types of ads in the total advertising investments in a given year will also be expressed. For each one year, a new column will be created, in which the amount of increase or decrease in investment will be determined by deducting the monetary value of the zero year from the costs for the year under review. The year-on-year change will be expressed in both monetary and percentage values.
4) Subsequently, the values found for the years 2012-2020 will be compared with each other.

The obtained results will be demonstrated by means of a graphical representation in the following chapter.

3 Results and Discussion

Based on the proposed methodology of input data processing, the amount and percentage of investments made in the advertisement and in its individual media types was determined. Figure 1 graphically shows the observed values of advertising investments.

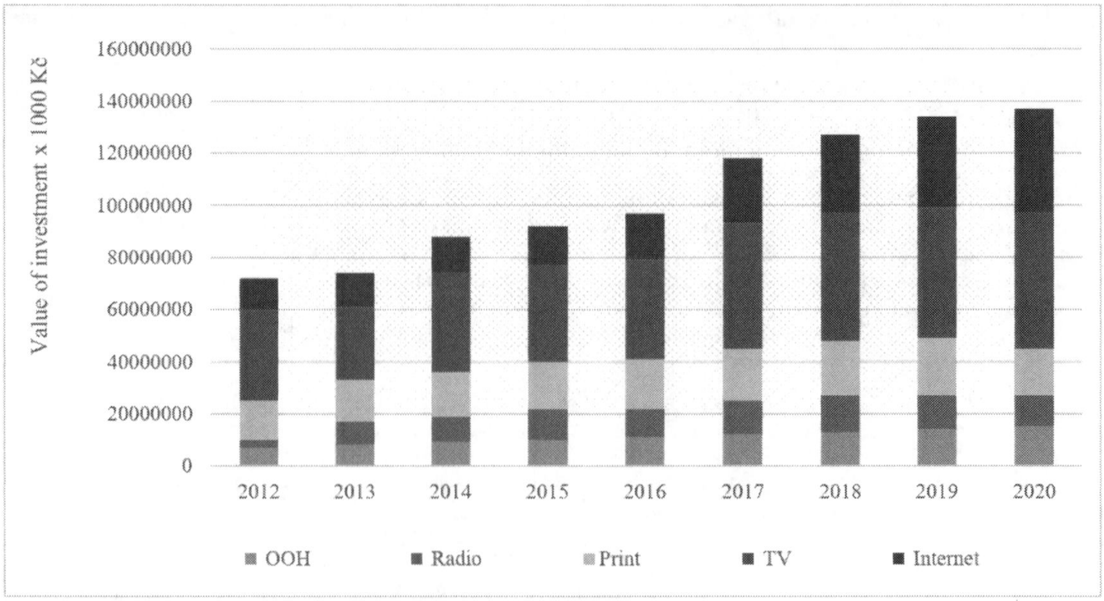

Figure 1. Development of investments in advertising in the Czech environment

Source: Author.

Investments made in the Czech advertising market have doubled over the period under review and show constant growth every calendar year. The amount within the total absolute investments was CZK 63.99 billion in 2012, but in 2020 it already reached the amount of CZK 130.55 billion. Within 2013, total investment in advertising increased by CZK 5.52 billion.

Developments in online advertising investment have been growing for a long time. Year-on-year absolute growth within individual years was CZK 2.16 billion in 2013, CZK 1.28 bn. in 2014, CZK 0.46 billion in 2015, CZK 4.68 billion in 2016, CZK 3.58 billion in 2017, CZK 5.28 bn. in 2018, CZK 5.78 bn. in 2019 and CZK 5.14 billion in 2020. As for the largest year-on-year financial increase, it was recorded in 2019. The smallest absolute increase in investments was recorded in 2015.

In the long run, the most invested advertising medium - television advertising - is showing steady absolute growth similar to the case of Internet advertising. However, 2013 was an exception, as there was a one-off decrease in investments by CZK 2.32 billion (-7.69%). In 2016, the highest absolute and relative growth was recorded, when investments increased by CZK 6.88 billion (19.39%).

Currently, the third most important advertising channel after TV and the Internet - the press - also recorded its best result in 2016, when investments increased by CZK 1.16 billion (6.47%) but shows a long-term stagnation in contrast to the two most successful types of advertising. The years in which the press showed a negative increase in invested funds were 2014 (-0.78%), 2015 (-0.07%) and 2020 (-13.06%).

In 2013, radio advertising recorded a record growth, when investments in it increased by 447.15%. In 2018, advertisers spent the most funds, namely CZK 8.04 billion. (10.37%). Absolute year-on-year growth over the last two years showed a decrease in investments in 2019 by -1.35% and in 2020 by -8.02.

OOH advertising was the last monitored medium in which investments fell by -19.44% in 2020. In previous years, however, no extreme was recorded in the year-on-year absolute development of investments.

Subsequently, an analysis of the share of media types in the total investment in advertising was performed, graphically shown in Figure 2.

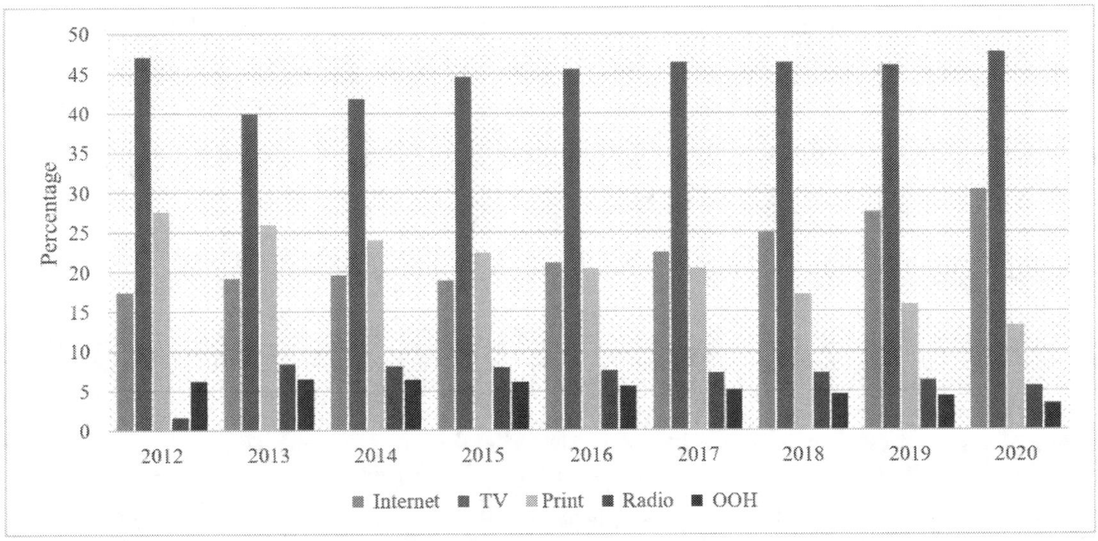

Figure 2. The share of media types in total investment in advertising

Source: Author.

The development of the share of investments in individual media types in the total investments in advertising was as follows: the most invested type in 2012 was television advertising, representing 47.06% of all investments. This was followed by the press with 27.59%, followed by the Internet with 17.43%, OOH with 6.25% and finally radio with 1.67%. The share of TV advertising in 2013 fell to 39.99% and the share of print to 25.95%. In the given year (2013), on the other hand, an increase in Internet advertising to 19.14%, radio advertising to 8.41% and OOH to 6.51% was recorded. A slight increase in TV advertising (41.83%) and Internet advertising (19.57%) at the expense of print advertising (24.01%), radio (8.13%) and OOH (6.46%) was recorded in 2014. TV advertising in 2015 strengthened and took up 44.54% and the share in all other media was reduced, namely the share of Internet advertising was reduced to 18.88%, the share of print advertising to 22.44%, the share of radio advertising in 8.01% and the OOH content to 6.13%. TV advertising (45.37%) and Internet advertising (21.12%) increased in 2016, while advertising in the press (20.39%), radio (7.55%) and OOH (5.58%) decreased. TV advertising (46.34%) and Internet advertising (22.46%) are growing again in 2017, the share of print advertising (20.39%), radio (7.28%) and OOH (5.11%) decreases. The only medium that recorded an increase in market share in 2018 is the Internet (24.95%), the share of TV advertising (46.31%), print advertising (17.13%), radio (7.28%) and OOH (4.59%) decreased. This trend continues in 2019, when the share of Internet advertising increased to 27.50% and, conversely, the share of TV advertising (45.97%), print advertising (15.84%), radio (6.35%) and OOH (4.35%) decreased. Growth was recorded in 2020 in the case of Internet (30.26%) and TV advertising (47.61%). Advertising in the press (13.19%), radio (5.59%) and OOH (3.35%) decreased. In Graph 2, the findings are shown graphically.

Based on the obtained results, the determined research questions can be answered:

What is the development of investments in Internet advertising in the Czech Republic?

Investments in Internet advertising show steady growth, with their market share increasing from 17.43% in 2012 to 30.26% in 2020. No negative growth in its investments was recorded in any calendar year, which is a unique finding, as for all other types of advertising media, at least one such year has been recorded. The results show that online advertising is growing, and advertisers are willing to invest more and more money in it. The increase in the number of Internet users can explain the stable growth of investments, which makes the environment even more attractive for advertisers.

What is the relationship between investing in online advertising and other types of advertising?

No direct dependence was found between Internet and television advertising. As for the share of TV advertising in the total advertising investment, even during the dramatic increase in investment in Internet advertising, it remains at similar values for the entire period under review and maintains its position as a leader in the advertising market. However, it is different in the case of print advertising, which saw a dramatic decline in market share from 27.59% in 2012 to 13.19% in 2020. It can be stated that this is due to its direct dependence on the willingness of investors to invest in online advertising. In the case of OOH advertising, there has been a significant decline, with its market share falling from 6.25% in 2012 to 3.35% in 2020. In this case, it can also be stated that there is a direct dependence on online advertising. Radio advertising is also showing a declining trend, but it is not as dramatic as in the case of print advertising and OOH advertising. It is therefore not possible to prove a relationship between investment in internet and investment in radio advertising.

At the expense of online advertising, print advertising and OOH advertising are the most neglected advertising media by advertisers. In the case of print advertising, this development can be explained by the growing number of mobile users and also by the declining willingness of people to read print media, but these media began to create content in the online environment, where they offer paid advertising. However, the value of a given advertisement is already included in the investment in internet advertising. Thus, a decrease in market share in this case does not necessarily mean a decrease in the revenues that companies earn from paid advertising. However, similar possibilities are not offered to OOH advertising operators and therefore the result of the given contribution can serve as a basis for subsequent research into the usability of OOH advertising in the coming years.

4 Conclusion

Technology has become an integral part of the market economy. It is not discussed in the scientific world, not only is the question of their need being addressed, but their scope for use in SMEs is also being examined. The aim of the paper was to analyze the development of investment in online advertising in the Czech Republic during 2012-2020 and to determine whether there is a relationship between the development of investment in Internet advertising and other types of advertising, such as TV, print, radio and OOH advertising. The results show that investments in online advertising are still growing and each year occupy a larger market share in the total investment in advertising.

Potential for further research: There is an assumption that the growth of investment in online advertising will continue in the coming years due to constant technological progress and constant growth in the technical literacy of users. However, the question remains as to what effect this increase will have on television advertising and its hitherto dominant position, and whether the given trend will not lead to a revolution in the advertising market and also to the extinction of certain media types.

Application in practice: The results of the research will help both researchers and companies in the process of their decision to invest in selected types of advertising.

Acknowledgments

This contribution is a partial output of the work on the project VEGA MŠ SR no. 1/0140/21.

References

1. Czech Statistical Office. *Podíl nakupujících na internetu se meziročně zvýšil z 39 na 54 %* [online]. Available at: https://www.czso.cz/csu/czso/podil-nakupujicich-na-internetu-se-mezirocne-zvysil-z-39-na-54- (2020)
2. M. Misuraca, G. Scepi, M. Spano. A network-based concept extraction for managing customer requests in a social media care context. *International Journal of Information Management*. **51**(101956) (2019)
3. N. A. Vila, L. Cardoso, A. F. de Araujo. Communication 2.0 in the Spanish Tourism Sector: An Analysis Of The Official Tourism Promotion Websites. *International Journal of Marketing, Communication and New Media*. **5**(5), 5–25 (2019)
4. M. O. Opresnik. Effective Social Media Marketing Planning – How to Develop a Digital Marketing Plan. *Social Computing and Social Media – LNCS*. **10913**, 333–341 (2018)
5. F. Pollák, R. Vavrek, J. Váchal, P. Markovič, M. Konečný. Analysis of Digital Customer Communities in terms of their interactions during the first wave of the COVID-19 pandemic. *Management and Marketing*. **16**(2), 134–151 (2021)
6. O. Khurramov, S. A. Fayzieva. The main convenience of internet marketing from traditional marketing. *Academy*. **1**(52), 30-32 (2020)
7. S. Khajehasani, A. Abolizadeh, L. Dehyadegari. The role of management and strategy in the development of E-marketing. *Recent Advances in Computer Science and Communications*. **13**(4), 641-649 (2020)
8. J. Soviar, M. Holubčík, J. Vodák, M. Rechtorík, F. Pollák. The Presentation of Automotive Brands in the On-Line Environment – The Perspective of KIA, Peugeot, Toyota and VW in the Slovak Republic. *Sustainability*. **11**, 2132 (2019)
9. F. Pollák, P. Markovič. Size of Business Unit as a Factor Influencing Adoption of Digital Marketing: Empirical Analysis of SMEs Operating in the Central European Market. *Administrative Sciences*. **11**(3), 71 (2021)
10. V. Kumar, S. Mittal. Mobile marketing campaigns: Practices, challenges and opportunities. *International Journal of Business Innovation and Research*. **21**(4), 523–539 (2020)
11. D. Brown, N. Hayes. *Influencer marketing*. Routledge: Taylor & Francis Group (2008)

12. H. Zhang, W. Zhang, Y. Rong, K. Ren, W. Li, J., Wang. Managing risk of bidding in display advertising. *Proceedings of the 10th ACM International Conference on Web Search and Data Mining* (WSDM 2017). 581-590 (2017)

13. V. Gasilov, M. Karpovich, M. Preobrazhensky, K. Borodin, T. Dao. Optimization of advertising media placement on transport infrastructural sites on the basis of big data technologies. *IOP Conference Series: Materials Science and Engineering*. **890** (2020)

14. Průzkum interních výkonů SPIR. *Zadavatelé v loňském roce investovali do internetové reklamy 39,5 miliard korun* [online]. Available at: http://www.inzertnivykony.cz/ (2020)

Rewarding Employees through Specific Wage Scales – Analysing the Remuneration Scheme Using Fixed Wage Elements in a Selected Company

Dana Lakronová[1,*]

[1]Institute of Technology and Business, Faculty of Corporate Strategy, Okružní 517/10, 37001 České Budějovice, Czech Republic

Abstract. The presented article examines rewarding employees on different wage scales by exploring a remuneration scheme using fixed wage elements of a specific enterprise. The submitted study evaluates the remuneration scheme in České Dráhy Plc. and its influence on the company. The achieved results of formulated hypotheses will be analysed through a structured interview and observation methods.
Keywords: wage scheme; remuneration; wage scale; fixed elements; work evaluation

1 Introduction

The human factor constitutes the principal part of the production process. Employees must therefore be properly motivated to perform at their best. Adequate remuneration of personnel is a crucial aspect throughout companies, constituting one of the critical elements of motivation. The remuneration scheme is always tailored to the requirements of a particular company to satisfy the needs not only of their employees, but also the company as a whole.

Specific remuneration schemes provide workers with financial or non-financial reward for their contribution to the company, also focusing on strengthening their motivation and inspiring loyalty. Enterprises may choose from various methods and factors of remuneration. The wage scheme must be effective, depending on a variety of factors including adequate wage structures. Creation of a high quality and elaborate remuneration system should thereby be the major focus of the department of human resources. The human factor encompasses a unique investment capital requiring effective management. Nowadays, companies must strive for winning the loyalty of their workers, as quality and well-paid employees ensure competitiveness of the enterprise. In my research, I focused on České dráhy Plc. The company was involved in large number of railway accidents that had happened during the summer. In the submitted article, I aim to inspect wages of selected posts in the company. I chose one position and compared it with other transport organizations to explore the impact of workers' salaries on railway transport safety.

The article is divided into a theoretical and application part. The theoretical-methodological part involves professional literature expounding upon terms concerning wage schemes, such as wages, wage scales and remuneration system. It further refers to methods used for gathering the material in the application part. The application part deals with describing a specific enterprise, its wage scheme and impact on the functioning and prosperity of the enterprise, including its influence on the staff motivation.

Hypothesis 1: Remuneration scheme is highly impactful on the staff performance.

Hypothesis 2: The tariff remuneration scheme scales with personnel's increased qualification and their bonuses.

2 Literary research

The topic of employee remuneration has widely been discussed in professional literature, including national and foreign studies. This article focuses on defining the specific terminology, referring to basic terms such as employee remuneration, wages, wage elements etc.

The highly instrumental publications involve "Motivating and remunerating employees" [1] or older works of Nakonečný [2], who clearly explains the importance of remuneration as an efficient motivational tool for employees to enhance their performance. Financial reward constitutes a key factor of a long-term working relationship, allowing employers to stimulate their staff to higher performance by giving bonuses. However, it is presumed that the harder and demanding the work is, the higher the requirements for skills are and the higher the performance of the workers should be. Enterprises should also try to preserve wage justice, where separate activities are classified according to their difficulty. Employees will thereby better understand the difference between their wages and salaries of their colleagues who perform more or less demanding work [3].

* Corresponding author: 18361@mail.vstecb.cz

Wages are, however, only one aspect contributing to work motivation and do not necessarily have to be the crucial one. Armstrong [4] argues that money is a positive stimulus not only because of a sheer need for it, but also implying explicit recognition of employees. The reward must clearly reflect the effort or degree of worker's responsibility. This is a different view on the issue of remuneration, Armstrong [5]. Wöhe and Kislingerová [3] suggest that the business economy determines employees' wages according to their working performance. Wages as a way of remuneration of one production factor accounts for a business quantity affecting incurred costs. Workers must be remunerated under employment contracts independent from the current situation of the enterprise. Urban [6] in his publications "Management and Leadership Psychology" and "Corporate Human Resource Management: Dimension of Human Resource Management" [7] broadens the knowledge of human resource management and business management. Mukherjee [8] in his work "Principles of Management and Organizational Behaviour" outlines principles of human resource management and organizational behaviour of employers.

Of significance might also be an article from scientific databases by Wakabayashi [9] "Fixed Salary or Incentive Contract? The Effect of Stickiness of Compensation Contracts". The study explores contractual theory, collective agreements and fixed wage elements. Bajzikova [10] in her article "The Minimum Wage in Compensation Systems in EU and the Slovak Republic" analyses remuneration schemes, perks and drawbacks of minimum wages, assessing their role in remuneration systems. The study further concerns remuneration in terms of bonuses and other job perks depending on workers' performance. Yoon [11] in his book "Mechanisms underlying creative performance: Employee perceptions of intrinsic and extrinsic rewards for creativity" describes the principle of viewing workers on the grounds of intrinsic and extrinsic bonuses for their creativity and performance. Bi-xi [12] in his article "Wage system innovation in small and middle manufacture enterprise" focuses not only on small and medium-sized manufacturers, but also suggests convenient methods of improving remuneration schemes tailored to various manufacturing companies. Hanes [13] in his study "The rise and fall of the sliding scale, or why wages are no longer indexed to product prices" outlines a gradual decline of sliding scales and their practical experience.

3 Methodology

The following chapter briefly describes the enterprise concerned, systematically exploring methods of collecting the relevant data for the research in question.

The first part expounds upon basic terms relating to employee remuneration, specifically focusing on expressions such as wages, wage scale, and different types of remuneration, also comparing opinions of various experts.

The practical part fully describes the selected company, provides basic information and analyses the particular system of the company. It further deals with the decisive criteria for remunerating individual jobs, the stability and employees' motivation including corporate costs and expenses. This part also compares the remuneration of engine drivers in two different companies.

The employed approaches involve observation methods and structured interview with the head of the human resource department (whose job content is: a manager of transport and logistics and deputy chairman of railwaymen trade union) comprising five questions. The interview targets specific criteria of remunerating workers, the remuneration scheme in the whole company, comparison of the contractual salary with the tariff salary, the impact on the performance and behaviour of employees and their motivation.

The final part of this article discusses the findings, either confirming or rejecting the formulated hypotheses. The conclusion summarizes all recorded findings as well as it suggests appropriate measures for overhauling the particular wage scheme. For the aforementioned purposes, I chose two positions to analyse in terms of remuneration, involving an engine driver and the head accountant. I will thoroughly discuss this issue in my diploma thesis.

For my research, I chose a Public limited company České dráhy established on 1st January 2003 under Act 77/2002 Sb. as a successor legal subject of the original national organization České dráhy. The company tightly integrates passenger transport and haulage with rail transport routes on the regional and national railways owned by the state.

Upon transforming České dráhy organization, three successor establishments emerged: Public limited company České dráhy, providing services in passenger transport and haulage and ensuring serviceability of the rail transport route, Railway Administration National Organization, disposing of the state property, making the transport route available to railway carriers and ensuring modernization of the transportation infrastructure, and the Railway Inspection as an organizational unit for resolving extraordinary events.

České dráhy constitutes a railway enterprise ranking amongst the biggest employers in Czechia. The group of companies České dráhy involves České dráhy Plc. as a passenger railway transport undertaking and ČD Cargo Plc. as a haulier. Organizational units of ČD (České dráhy) comprise the managing directorate, detached branches of the business, organizational units and executive units. ČD includes several subsidiaries such as:

- ČD Travel Ltd.,
- Dopravní vzdělávací institut Plc.,
- Výzkumný Ústav Železniční Plc.,
- ČD-Telematika Plc.,

- ČD-Informační systémy Plc.,
- ČD Reality Plc.,
- ČD Restaurant, Plc.,
- RailReal Plc.,
- Žižkov Station Development Plc.,
- Smíchov Station Development Plc.,
- DPOV Plc.

4 Results

The following part considers all results achieved from observing and interviewing.

4.1 ČD Remuneration code for ČD workers

All employees, including ČD workers, are entitled to wages for the performed work, involving tariff wages and other related wages components provided according to the remuneration code. In compliance with the code, employees are further entitled to compensation for the on-call duty and wage reimbursement. ČD Company presents a wise choice for workers due to its coherent monetary policy. Basic wage and related wages components paid by the enterprise give security and social stability to our employees. České dráhy ranks third on the Best Employer of 2020 Chart including companies exceeding 5,000 employees.

The ČD remuneration scheme is based upon hourly pay. The wages reflect the monthly tariff or the wages negotiated in the contract, complying with working hours or calculated as a product of the hourly tariff of the declared tariff rate and hours worked. Most workers of the company are remunerated under specific tariff rates. Senior managers are paid pursuant to the contractual salary.

České dráhy Plc. collaborated with the Ministry of Labour and Social Affairs of the Czech Republic on the list of jobs and related tariff rates. The tariff rates are classified according to the complexity, difficulty and responsibility, labour intensity and significance of the job performed for ČD. It is imperative to realize that particular characteristics may not allow for all the possibilities and at times they just describe the performed work in general.

Employees are classified into specific tariff rates within one employment according to the qualification difficulty of the predominantly performed work. If certain requirements for specific tariff rates are identical, the decisive factor for ranking into a higher tariff rate is fulfilling the requirement that separates texts of higher rates from the lower ones. In certain positions, the amount of wages is set just according to an individual agreement. The application of the tariff rates and contractual wages in separate organizational units is governed by prescribed systematization upon meeting particular pre-conditions. The employees are remunerated under The Employee Remuneration Code of ČD suggested in Table 1. The terms relating to the educational background are explained in the list below the table.

Table 1. Tariff rates of České dráhy Plc.

Tariff rate	Monthly tariff	Indicative qualification prerequisites (educational background + standing)
1	14,600	A, B, C (+ apprenticeship or training)
2	14,980	A, B, C, D (apprenticeship or training)
3	16,420	D, E
4	18,250	E, J, H
5	19,990	J + 1 year of standing, H, K
6	21,900	H + 2 years of standing, K + 1 year of standing, M, L
7	23,700	H + 3 years of standing, K + 2 years of standing, M + 1 year of standing, L + 1 year of standing
8	25,410	H + 4 years of standing, K + 3 years of standing, M + 2 years of standing, L + 2 years of standing
9	27,220	M + 3 years of standing, L + 3 years of standing, N + 1 year of standing, R
10	29,000	M + 4 years of standing, L + 4 years of standing, N + 2 years of standing, R + 1 year of standing
11	30,650	M + 6 years of standing, L + 6 years of standing, N + 2 years of standing, R + 2 years of standing, T + 1 year of standing

12	32,560	N + 4 years of standing, R + 3 years of standing, T + 1 year of standing
13	33,380	R + 4 years of standing, T + 3 years of standing
14	35,580	R + 6 years of standing, T + 5 years of standing
15	39,830	R + 8 years of standing, T + 7 years of standing V + 5 years of standing
16	43,950	T + 8 years of standing, V + 7 years of standing

Source: Author based on the data of Collective agreement of ČD 2020.

The list of qualification prerequisites and educational background

A *Without education (incomplete primary education)*

B *Without education (completed primary education)*

C *The basic education (one-year and two-year vocational school)*

D *Lower secondary education (three-year preparation at a secondary vocational school)*

E *Lower secondary vocational education (vocational education schemes, vocational schools)*

J *Secondary or secondary vocational education without the general secondary education exam or vocational certificate*

H *Secondary vocational education with the vocational certificate (a graduate of non-graduation educational schemes)*

K *Completed secondary education*

M *Completed secondary vocational education with the general secondary education exam (without apprenticeship)*

L *Completed secondary vocational education with apprenticeship and general secondary education exam*

N *College education (colleges, conservatories, post-secondary specialized education)*

R *Bachelor education (University education completed by bachelors exams)*

T *Master education (master, engineer and medical schemes)*

V *Doctoral education (doctoral study programme, postgraduate studies, scientific education)*

Discussing employee remuneration also involves negotiating between the employer and trade unions on the budget of wage costs.

The negotiated wages involve a so-called fixed wages component (guaranteed wages component), which the employee may always take for granted. It equals no less than two thirds of the overall wages. The rest underlies performing other pre-negotiated conditions evaluated by the senior worker. Remuneration according to tariff rates provides workers with financial security. All employees, paid either under a specific tariff rate, or pursuant to the contractual salary, are motivated to higher work performance by bonuses added to the basic wages. In the event that an employee fails to perform an employee's duty, the bonuses are reduced. In this way, performance bonuses are considered as motivational tools. Most operating employees are also liable to indemnify the employer if the damage is caused by employees' work performance. In such a case, the employee is liable to indemnity not exceeding three monthly salaries.

With an exception of specific cases where contractual wages apply, all ČD employees are subject to tariff wages. It is thereby presumed that all professions share the same base remuneration. Certain specifics may be found in so-called operating professions, e.g. administrative and operating positions. It generally applies that all employees' bonuses provided by our company are predominantly governed by The Labour Code. To a certain extent, the amount of bonuses is subject to negotiation between the employer and trade unions. From my point of view, these bonuses may then be considered as a motivational tool. Educational background is imperative for assigning the worker a particular job and the respective tariff rate. Above all, the educational background is highly impactful on the bonuses determined by the length of service. In particular, higher education provides the employee with two extra years of standing.

Engine driver – operating position

- The job involves expertise in controlling, service and maintenance of driving vehicles of passenger transport trains, higher quality trains and international trains within railway services.
- The position falls into 10th, 11th and 12th tariff rate according to the service classification regarding respective train categories. The engine driver is further entitled to negotiate on a contractual salary. In such a case the engine

driver, besides other things, controls, services and maintains driving vehicles of trains. The engine driver shall also submit a certificate of proving his/her language knowledge and official authorization for a position of an engine driver on international railway lines.

- Time allocation amounts to 36 hours a week.
- The main bonuses to wages involve personal evaluation, bonuses determined by the length of service, bonuses for working overnight, bonuses for working on Saturday and Sunday, bonuses for working on holiday and bonuses for the work release programme.
- This position includes very specific wage bonuses:
    - bonuses for the on-call duty
    - bonuses for avoiding profit leaks
    - compensation for extraordinary events, remuneration when working under unfavourable weather conditions, including the consequences
    - remuneration for taking part in a training course, examination, studying on the job and retraining
- Operating employees are entitled to "compensation for separate shifts" - a profession where a worker spends working hours outside his/her domicile, including without limitations employees in transport, e.g. a chief conductor. He/she is paid 60% of his/her hourly wage up to 2 hours, of which the amount increases to 75% when the time of the separate shift exceeds 2 hours.

Head accountant - administrative position

- The job involves specialized and professional administration of employees' accounts, providing and inspecting annual and monthly financial statements including tax checkouts. The post further includes thorough knowledge of the payroll software and legislation pertaining to wage accounting and providing checkout as required by the competent administrative bodies.
- The position falls into 12th tariff rate with the time allocation 40 hours a week.
- The position is exclusively administrative; workers are not entitled to issue a statement on working overtime.
    - Administrative workers are entitled to take the accrued days off.
- The salary of the administrative worker comprises: the tariff, bonuses determined by the length of service, personal evaluation or reimbursement of traveling costs.
    - The worker is entitled to the reimbursement of travelling costs if he/she is commissioned to work outside his/her domicile, e.g. he/she is appointed as a substitute for a sick employee.

4.2 Comparing wages of engine drivers in České dráhy Plc. and RegioJet

Specific requirements of both companies involve an evaluation of worker's state of health pursuant to Regulation No. 101/95 Sb. MD being without criminal records and responsibility. České dráhy offers training course if the candidate does not have an engine driver license. The candidate must also have his/her secondary education completed and submit his/her vocational certificate of transport or related education. RegioJet requires competence to drive a railway vehicle and expertise in technical areas of railway transport.

Table 2. Comparing salaries and perks

	České dráhy	RegioJet
Net wages	31,000 CZK	34,000 CZK
Bonus	13,000 CZK	12,000 CZK
Meal vouchers	110 CZK	No
Employee discount on the train ride	Yes	Yes
Days off	5 weeks	5 weeks

Source: Author.

The suggested comparison shows that both companies offer a large number of job perks. České dráhy has the biggest competitive advantage in its long-established tradition and corporate stability. On the other hand, RegioJet suffers from being a complete novice on the market. Although the enterprise has recently been trying to entice potential candidates by pushing up wages of engine drivers, the real-life situation might be quite different.

5 Discussion

I have reasonable cause to believe that railroad accidents predominantly involve fault of a human factor - either of a railway employee, or drivers of road vehicles. All the same, railway transport is considered one of the safest ways of transportation. Railway lines are secured by various safety and communication devices, scaling with the traffic volume on the respective lines. Available funds allocated to the greater safety and their potential profit-making are imperative for preventing tragic accidents. The level of railway safety is always thoroughly examined when it comes to railroad crashes, although we cannot simply say that safety factors are involved in all railway accidents. Even the lowest degree of safety makes the railway traffic completely secure. It leads us to the question of the human factors involved in the process of railway traffic control. As an example, I can give a railway line controlled only by telephone communication, and a line secured by automated systems using IT technologies and modern safety devices.

There has been a profound lack of certified engine drivers on the market, affecting all carriers - České dráhy, RegioJet and Leo Express. The companies try to maintain these driving specialists by a pay rise. In the event of ČD, the situation is not that critical, as the enterprise does not face a rapid labour turnover. The organization also benefits from a long-term elaborate and effective preparatory programme for trainees. The training of new engine drivers is financially and time consuming, with the duration of no less than 9 months. The course, however, usually lasts longer, since not all carriers devote the same amount of time to the training. Other organizations try to attract new workers on a pay increase. RegioJet has raised its basic wages by 12 - 14%, whereas Leo Express has pushed up its salaries by 5 %.

6 Conclusion

In the article, I analysed and evaluated the current remuneration scheme in České dráhy Plc. assessed its effectiveness and the impact on the stability and security of the company, coming up with constructive suggestions within the aforementioned scheme.

In the theoretical part, I focused on a general description of the remuneration scheme, bonuses for the performed work and overall elaboration of the system. In the practical part, I introduced České dráhy Plc. exploring its remuneration system and testing formulated hypotheses. I also found an answer to the issue of rail transport safety, considering recent accidents and their possible link to the company remuneration policy. As to suggested remedial measures, the corporation itself has involved in its employment contracts for operating employees that the working position occupied in the company may not be held in any other competing company at the same time. These measures have not been imposed until recently. České dráhy Plc. ranks third amongst the best employers in the Czech Republic. Its employees are thereby very well looked after and are offered jobs and excellent salary packages in a stable environment, the information which was found out by inspecting the collective agreement. Even though the enterprise has firmly set its tariff rates, certain positions can negotiate a contractual salary outside the fixed tariff rates.

The chamber of deputies is currently considering a bill to amend the Railway Act, which stipulates uniform laws of the railway transport throughout the European Union, on the national level introducing the monitoring of the prescribed time spent by driving a railway vehicle on a national or regional line. The amendment aims at ensuring safety of the railway transport on the EU market. The modification of the prescribed time spent by driving railway vehicles on a national or regional line enables better controlling, sticking to the rules governing the shift length and the time spent by driving the railway vehicle in international transportation. General rules and regulations governing the operating and technical linkage between the national and regional lines and lines of other members of the EU were imposed.

Based upon the employed methods, the formulated hypotheses were successfully confirmed.

References

1. J. Urban. *Motivace a odměňování pracovníků*. Praha: Grada. ISBN 978-80-271-0227-3 (2017)
2. M. Nakonečný. *Motivace pracovního jednání a jeho řízení*. Praha: Management Press (1992)
3. G. A. Wöhe, E. Kislingerová. *Úvod do podnikového hospodářství*. Praha: C.H. Beck, ISBN 978-80-7179-897-2 (2007)
4. M. Armstrong. *Řízení lidských zdrojů: nejnovější trendy a postupy*. Praha: Grada, ISBN 978-80-247-1407-3 (2007)
5. M. Armstrong. *Odměňování pracovníků*. Praha: Grada, Expert (Grada). ISBN 978-80-247-2890-2 (2009)
6. J. Urban. *Psychologie řízení a vedení. Praha: Ústav práva a právní vědy*. ISBN 978-80-87974-15-5 (2017)
7. J. Urban. *Řízení lidí v organizaci: Personální rozměr managementu*. Praha: Wolters Kluwer (2013)
8. K. Mukjerjee. *Principles of Management and Organizational Behaviour*. New Delhi: Tata McGraw-Hill Education Private Limited (2009)

9. T. Wakabayashi. Fixed salary or incentive contract? The effect of stickiness of compensation contracts. *Asia-Pacific Journal of Accounting and Economics*. (2019)

10. L. Bajzikova. The minimum wage in compensation systems in EU and the Slovak Republic. *Proceedings of the 31st International Business Information Management Association Conference, IBIMA 2018: Innovation Management and Education Excellence through Vision 2020*. 1550 (2018)

11. H. J. Yoon, S. Y. Sung, J. N. Choi. Mechanisms underlying creative performance: Employee perceptions of intrinsic and extrinsic rewards for creativity. *Social Behavior and Personality*. **43**(7), 1161-1180 (2015)

12. Z. Bi-Xi, X. Xiang-Tian. Wage system innovation in small and middle manufacture enterprise. *2007 IEEE International Conference on Control and Automation, ICCA*. 752 (2007)

13. C. Hanes. The rise and fall of the sliding scale, or why wages are no longer indexed to product prices", *Explorations in Economic History*. **47**(1), pp. 49-67 (2010)

14. České dráhy. *Collaborative Agreement* (2020)

Adaptation of Small and Medium-sized Enterprises During the COVID-19 Pandemic

Natalia Neklyudova[1,*]

[1] Institute of Economics, Ural Branch of the Russian Academy of Sciences, 29, Moskovskaya St., 620014, Ekaterinburg, Russia

Abstract. The article analyses static data that show the deterioration of the situation of SMEs around the world, both in developed and developing countries. Measures taken by governments to support SMEs and overcome the crisis were considered in the analysis. In most cases, they turned out to be insufficient to help reach the pre-pandemic economic level. The article also confirms that the self-employed and female entrepreneurs have suffered most. Industries such as tourism, construction industry, wholesale and retail trade, real estate, and services have suffered major loss.
Keywords: SMEs; pandemic; crisis; COVID-19

1 Introduction

Last year, many enterprises around the world were forced to close their operation in relation to to the COVID-19 pandemic, and those that could not cover the accumulated costs due to falling sales will no longer open. Often, large enterprises have a greater capacity to withstand COVID-19-like shocks, for example, if they have been able to make higher profit, gain easier access to finance, adapt to home delivery services, or diversify their businesses. Shocks in China, the EU, and the United States has had the greatest impact on the world trade, and as a result, developing countries have been affected as well.

This study shows how COVID-19 has affected small and medium-sized enterprises, based on recently published studies of World Bank enterprises in 13 countries. The study shows that enterprises of all size categories have suffered serious losses in many respects. Sales of small and medium-sized enterprises are declining more and their cash is depleting faster than those of large enterprises operating in the same sector and country. This is especially true for enterprises that depend on imports and exports, as well as companies operating in the service sector [1].

Specific attention is paid to how the pandemic has affected small and medium-sized enterprises (SMEs) in Russia. The indicators of SMEs in Russia are traditionally lower than in other major or actively developing economies of the world. The SME sector is not a growth driver in Russia [2]. Recently, there has been research on crisis management in small enterprises, including pandemic response [3-7], but there are few sources of information concerning low- and middle-income countries.

At the end of July 2020, the World Bank published new studies analysing enterprises from 13 different countries that had measured the impact of the virus on business, ranging from Italy and the Russian Federation to Albania and Zimbabwe. Despite their limitations, these surveys, along with the many other sources used in this study, show that the impact of the crisis on small enterprises is different compared to large enterprises [8].

There is thus a need to improve the anti-crisis management practices and mechanisms of SMEs, including addressing crises in the wave of the economic impact of the pandemic.

2 Methodology

The research methodology is based on the articles published by modern scientists on problems of crisis management in the period of the COVID-19 pandemic. Methods of statistical and economic analysis, expert assessments, methods of bankruptcy diagnosis, and forecasting were used. For the purpose of the study, statistical and analytical data from the World Bank, OECD, RBC, and the NAFI Analytical Centermare used. The impact of COVID-19 on SMEs in the federal districts of the Russian Federation is analysed using the data from the Federal Tax Service of the Russian Federation on the number of enterprises and employees in the SME sector in November of 2017-2021.

* Corresponding author: neklyudova.np@uiec.ru

3 Results

Some sectors and regions with a disproportionately high share of small and medium-sized enterprises (SMEs) have been significantly affected by COVID – 19 restrictions. Although all enterprises and economic sectors have been directly or indirectly affected by government closure measures, some areas with a disproportionately high share of SMEs have been most seriously affected, at least initially.

This applies mainly to the tourism sector. It is estimated that international tourism will decrease by about 80% in 2020 [9]. Despite the fact that domestic tourism has developed relatively better, it remains significantly below the pre-Covid level. For example, the estimated decrease in domestic tourism in Spain and the United Kingdom, is 45-50% in 2020 compared to 2019. Cultural events related to the closure of museums, theatres and cinemas have also been severely affected. A significant recovery of international tourist flows is not expected until 2021, and the achievement of the pre-crisis level is not expected until 2023 [9].

Besides the tourism sector, the economic sectors most directly affected by isolation measures, at least initially, include production of vehicles for the transport sector, construction, wholesale and retail trade, air transport, accommodation and catering services, real estate, professional services, and other personal services (e.g. hairdressers) (OECD, 2020). These sectors alone account for 40% of the total average employment in OECD countries [9].

SMEs ensure the majority of employment in the most affected sectors: on average, 75% in OECD countries and almost 90% in Greece and Italy. Microenterprises with fewer than 10 employees, those probably most endangered by cash shortages, employ about 30% of employees in these sectors and up to 60% in Greece and Italy [10].

Before the COVID-19 crisis, the financial situation of SMEs was generally favourable. Long-term interest rates were very low compared to historical standards, and monetary policy became increasingly flexible to compensate for relatively weak economic activity. Credit conditions were improved for SMEs with a low level of loan refusals and a gradual transition of loan portfolios to longer maturities.

In addition, digital tools contributed to the emergence of new online banks, some of which specialize in providing services to self-employed persons and owners of small businesses. In addition to bond issuing, asset-based financing also became an alternative source of financing for a growing number of SMEs. For example, in recent years, there has been a development of financial methods, such as leasing and hire purchase, as well as factoring, which allows SMEs to convert part of their assets into money to raise funds in the short term. In addition, the volume of venture capital investments for SMEs is growing, partly due to the government promoting the development of equity financing for SMEs [1].

Along with these positive trends, it is important to note that a third of all SMEs in the European Union continue to rely exclusively on domestic sources of income for their day-to-day operations and investments, which partly explains the slow credit growth in recent years. In addition, financing models still differ significantly between small and large SMEs, especially in Europe, with the former using more self-financing options and short-term loans, and fewer government subsidies and asset-based financing [11]. In addition, there are signs that the current crisis has a strong impact on equity, trade financing, and alternative financing, which is declining from the recent progress.

At this stage, small enterprises are still experiencing more serious financial difficulties compared to large ones. As a result, SMEs are more dependent on their internal financial resources and cash flows than large enterprises, both in the area of investment and in covering their current expenses, such as the remuneration of their employees.

These problems make SMEs more vulnerable to economic downturns and falling incomes than large enterprises are. This vulnerability is even more obvious for microenterprises that can rely only on bank financing or a limited number of alternative sources of financing.

However, the crisis caused serious problems for start-ups that had been established before its outbreak. Numerous surveys confirm that these young enterprises have been seriously affected by the crisis. More than 40% of new enterprises slipped into red zone (having only three-month cash or less to maintain their operations). Almost 3 out of 4 start-ups surveyed have experienced a decline in revenue and liquidity problems, and 41% of them needed to raise capital within the next three months [12].

There is growing evidence that the consequences of the COVID-19 pandemic for the self-employed and entrepreneurs were disproportionately negative. This negative impact was more significant than in the case of large enterprises, and in general in the case of employees. For example, an international study by Eurofound shows that the probability of becoming unemployed during the pandemic was higher among the self-employed (13%) than among employees (8%) [13]. This is consistent with the studies conducted in OECD countries, including e.g. in Australia, Germany, the United Kingdom and the United States [1].

In addition, there is evidence in OECD countries that subgroups of entrepreneurs, such as women and minorities, have been more affected, which partly reflects problems related to the access to finance, the economic sectors in which they usually work, and the increase in women's household responsibilities during the crisis.

The number of women entrepreneurs in the United States decreased by 10% between February and June 2020, while the number of male entrepreneurs decreased by 7% only [14]. Similarly, data from Germany show that female entrepreneurs are more than 30% more likely to lose income than male entrepreneurs are. There is also data from Canada and the UK that women entrepreneurs on average have reduced the number of hours worked about 1.5 times more than male entrepreneurs have. Many of these gender differences can be explained by differences in the sectors in which male and female entrepreneurs work. In addition, women are more likely to take on more household and care responsibilities during a pandemic [9], which limits their time for business activities.

The smaller the size of the enterprise and the earlier it is established, the more likely it is to refuse state support. Out of 32 OECD countries, 33% of SMEs established 1-2 years ago received state support, compared to 39% of enterprises established 2-3 years ago and 45% of SMEs with at least five-year existence. Newly established enterprises were even less likely to receive support. Only 15% of SMEs that started their activities in 2020 have received state support [15].

SMEs have continued to struggle during the pandemic and the recovery phase: in 32 OECD countries, 42% - 96% of the surveyed SMEs in December 2020 said that additional support would be needed in the future. Enterprises which operate in countries with stricter deterrent measures are more likely to need further government support. For example, in New Zealand, a country with relatively mild deterrent measures, only 58% of SMEs are expecting support [12]. By comparison, in Chile, the stringency index is twice as high as in New Zealand, which indicates stronger deterrent measures.

The indicators of SMEs in Russia are traditionally lower than in other major or actively developing economies of the world (see Table 1).

Table 1. Indicators of the SME sector in some countries

	Share in GDP (%)	**Share of employed persons (%)**
Russia	20.8	27.0
USA	56.0	52.0
Germany	43.0	57.0
China	60.0	83.0
Japan	55.0	72.0
Malaysia	39.0	48.0
Vietnam	40.0	50.0
Uzbekistan	56.9	78.0
Israel	62.3	68.7
South Africa	42.0	47.0

Source: Special report of SME Ombudsman to the President of the Russia Federation [2].

According to various estimates (IMF, FAS of Russia, analytical centres), the SME sector is not a driver of growth in Russia. The share of the public sector, including state-owned companies, in the Russian economy is 40-70%, while the share of the SME sector is 2.5 times lower. According to SME Ombudsman, the segment of medium-sized enterprises is the most undeveloped [2].

The sector of small and medium-sized enterprises in Russia has been stagnating for 10 years (see Table 2). Moreover, there are no processes of transformation of small enterprises into medium-sized ones.

Table 2. Indicators of the SME sector in Russia

	2010	2015	2016	2017	2018	2019	2020
Share of SME (%)	19.9	19.9	21.6	21.9	20.2	20.8	19.8
Share of employed persons in SME (%)	27	28	26	27	27	26	27

Source: Special report of SME Ombudsman to the President of the Russia Federation [2].

The analysis of the Federal Tax Service on the number of SME and the number of persons working in the sector indicates that the number of enterprises itself has not decreased, it has even slightly increased, while the number of people employed decreased significantly, in the Far Eastern Federal District very drastically (see Fig.1).

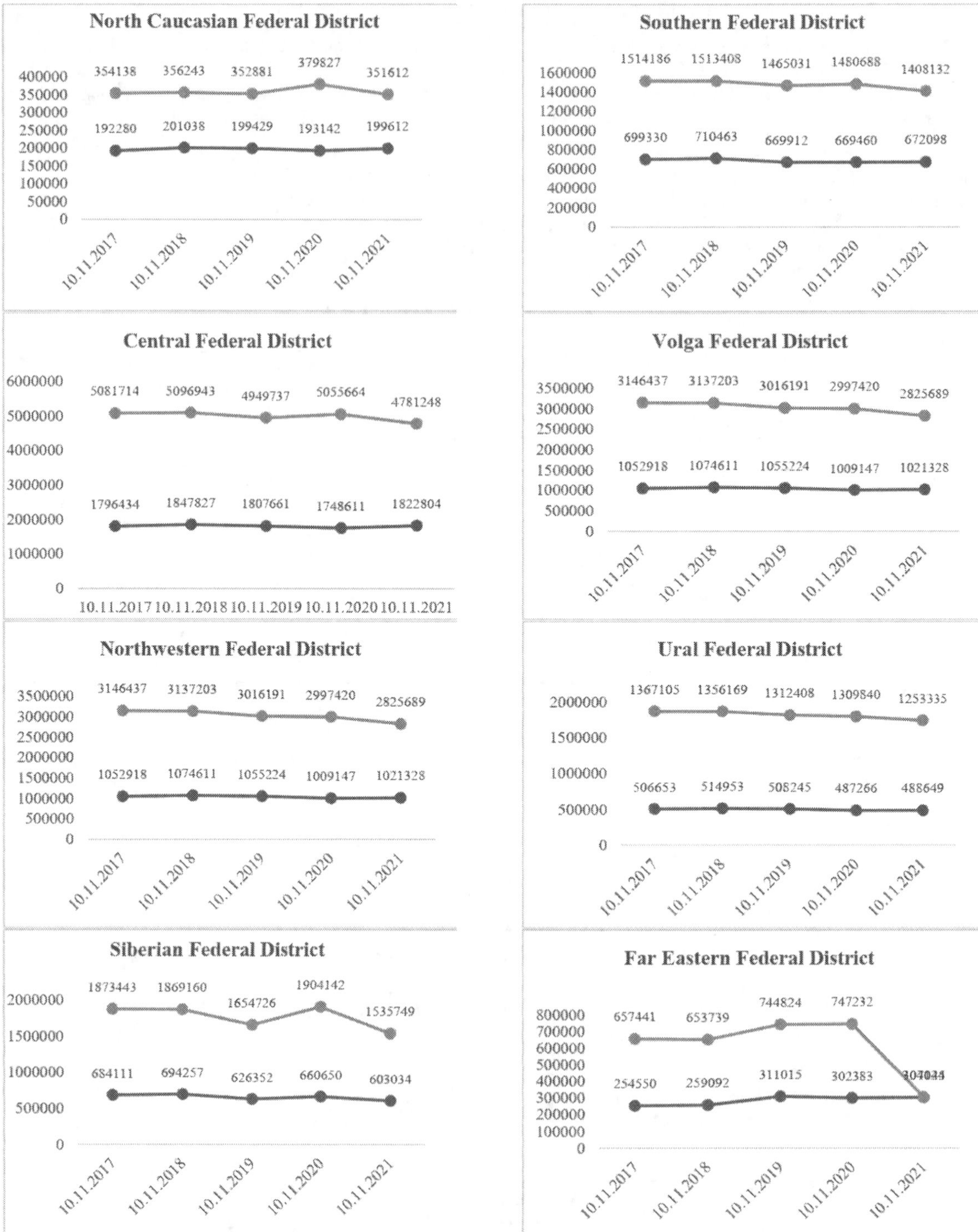

Figure 1. Number of SME (upper line) and employed in SME (lower line) in Russian federal districts

Source: Author based on the data of Federal tax service of Russia [16]

This indicates that the main measures of state support should be aimed at maintaining employment. Statistical data on the impact of coronavirus on SMEs in districts is still limited. However, the current level of knowledge points to serious failures in the most affected regions. More than 36% of enterprises in the Russian Federation consider tax rates to be the biggest obstacle, compared to ca. 10% of enterprises in other countries. As for the access to finance, this was mentioned as the biggest obstacle b\ 15% of enterprises. More than 8% of enterprises mention corruption as the biggest obstacle to their business activities [17].

According to Rosstat, 96% of SMEs are microenterprises with a 37% share of the sector's turnover and a 29% share of employment, while small enterprises ensure 49% employment and account for 63% of the turnover. However, this information does not take into account informal workers and self-employed person, which represents ca. 21.3% [18].

When non-working days were announced in the Russian Federation with the preservation of wages, this has led to the suspension in the activities of SMEs, especially those that were unable to switch to the online format. For example, in some travel agencies, after the introduction of anti-epidemic restrictions, revenue has decreased by 90%. According to experts, the cost of one weekday of inactivity in a coffee shop in the centre of Moscow is 100-150 thousand roubles, while in the case of a clothing store, it ranges between 40 and 100 thousand roubles, and 60 - 80 million roubles in the case of a shopping centre [17].

Consumers postpone the purpose of non-essential goods and services, such as clothing, tourism, durable goods, etc. The drop in domestic demand directly hit the domestic suppliers. However, in the case of global value chains, consumer spending in high-income destination markets affects the demand for their exports.

At the same time, Mail.ru Group estimates that as of 8 July 2020, a third of medium-sized and small businesses did not receive payment for their services from customers during the pandemic [19]. According to the Federal Tax Service, 305.6 thousand individual entrepreneurs and 208 thousand legal entities ceased their activities as of 1 July 2020 [20].

4 Discussion and Conclusion

This study focuses on the situation of small and medium-sized enterprises (SMEs) with fewer than 100 employees in the wave of the pandemic. The study discusses the differences in the impact of the pandemic on enterprises of different size categories as well as their responses and how development finance can best support them in low-and middle-income countries [15]. Focusing on small firms makes sense for many reasons.

SMEs are the largest provider of employment and supply of goods and services in low-income countries; therefore, disruptions in this segment have serious social and social consequences for the poor and rural population. Moreover, long-term closure of small enterprises leads to the loss of intangible capital, skills, and innovative potential, which may result in a deep economic downturn in countries even long after the pandemic.

Thus, some of the development goals, whether aimed at supporting the incomes of the poor during an economic downturn, providing goods and services, or creating an appropriate framework for future recovery and development of industry, are directly related to the situation of SMEs during a crisis. The COVID-19 crisis has been the most serious global recession since the end of World War II. Physical distancing, lockdown and restrictive measures introduced worldwide to stop the spread of the pandemic have led to unprecedented shocks to the business environment both in OECD and non-OECD countries. This recession was even more sudden than the financial crisis of 2008-2009. GDP declined significantly in all major OECD countries in the first two quarters of 2020.

In different countries, the stricter the isolation measures, the stronger the initial decline in economic growth has been. Facing a significant health crisis, many governments reacted by adopting various measures of social distancing (isolation) to prevent the spread of the virus, which has also significantly reduced economic activity.

As restrictions related to the economic activity were eased in the summer, GDP partly recovered, but remained below the pre-crisis level. While the total GDP in the OECD region was 11.6% lower in Q2 of 2020 than Q2 of 2019, the gap narrowed to 3.8% in Q3 of 2020. Nevertheless, due to the emergence of new options in recent months, the pace of recovery slowed down, and in Q4 of 2020, the GDP level remained 3.4% lower than in the previous year [12].

At the time of implementing this study, the prospects for overcoming the SME crisis continued to improve, as evidenced by recent economic forecasts, but they remain uncertain and uneven across countries. The brighter prospects are mainly related to the gradual introduction of effective vaccination, support for macroeconomic policies, especially in the United States (USA) and the EU, and signs that the economy is coping better with measures to stop the spread of the virus. Global economic activity has now returned to its pre-pandemic level, but it will remain weaker at the end of 2022 than before the pandemic. There are also noticeable differences in the impact of the crisis and the pace of recovery in different countries. The recovery is still affected by the risks of new outbreaks of the virus with the appearance of its variants in different places and the problems associated with vaccine rollout to a sufficient extent.

To survive the pandemic, SMEs need grants, not loans. Most simply do not apply to banks for loans, despite record low-interest rates. However, not all governments have a budget for grant support. Eight out of 10 enterprises of all size categories expect some form of government help in higher-income countries, such as Cyprus [20]. In low-income countries, such as Georgia or Moldova, small public funds are allocated to larger enterprise. In Africa, even fewer people expect government support; most governments in low-income countries have not taken any measures to support SMEs.

The combination of limited budget and already high corporate debt creates the greatest risks for countries with above-average income.

The main measures that can contribute to economic growth include the reduction of tax burden on enterprises; limiting the growth in tariffs for infrastructure services and raw materials monopolies and reduction of administrative pressure on business. In the short term, governments are recommended firstly to ease restrictions to attracting employees from neighbouring countries; secondly, to compensate SME expenses for PCR tests and personal protective aids. In the case of lockdown, governments should pay a 100% for employee downtime at the expense of the state directly, without necessary participation of the employer.

The COVID-19 pandemic is a unique crisis, affecting enterprises in many ways: supply (reduced labour supply because employees stay at home, lack of resources, disrupted supply chains); demand (reduced demand from lost and returning consumers, precautions, cautious investors); uncertainty (inability to count on a future income stream that justifies, for example, the replacement of employees or equipment); and unavailability of finance.

The article was prepared under the plan of Institute of Economics of the Ural Branch of RAS for 2021–2023.

References

1. S. Adžić, J. Al-Mansour. Business analysis in the times of COVID-19: Empirical testing of the contemporary academic findings. *Management Science Letters*. **11**(1), 1-10 (2021)
2. Institute for the Economics of Growth P. A. Stolypin. Special report of SME Ombudsman to the President of the Russian Federation - 2021 [online]. Available at: http://doklad.ombudsmanbiz.ru/2021/7.pdf (2021)
3. M. Mandviwalla, R. Flanagan. Small business digital transformation in the context of the pandemic. *European Journal of Information Systems*. 1-17 (2021) doi:org/10.1080/0960085X.2021.1891004
4. Kuckertz, L. Brändle, A. Gaudig, S. Hinderer, C. Arturo, M. Reyes, A. Prochotta, K. M.Steinbrink, E. S. C. Berger. Startups in times of crisis–A rapid response to the COVID-19 pandemic. *Journal of Business Venturing Insights*. **13,** e00169 (2020) doi:org/10.1016/j.jbvi.2020.e00169
5. J. C. Alves, T. C. Lok, Y. Luo, W. Hao. Crisis Management for Small Business during the COVID-19 outbreak: Survival, resilience and renewal strategies of firms in Macau. *Research Square*. 1-29 (2020)
6. I. J. Akpan, D. Soopramanien, D. H. Kwak. Cutting-edge technologies for small business and innovation in the era of COVID-19 global health pandemic. *Journal of Small Business & Entrepreneurship*. 1-11 (2020)
7. V. Ratten. Coronavirus (covid-19) and entrepreneurship: changing life and work landscape. *Journal of Small Business & Entrepreneurship*. **32**(5), 503-516 (2020)
8. I. Adian, D. Doumbia, N. Gregory, A. Ragoussis, A.Reddy, J. Timmis. Small and medium enterprises in the pandemic [online]. Available at: https://openknowledge.worldbank.org/bitstream/handle/10986/34552/Small-and-Medium-Enterprises-in-the-Pandemic-Impact-Responses-and-the-Role-of-Development-Finance.pdf?sequence=1 (2020)
9. OECD. *Coronavirus (COVID-19): SME Policy Responses* [online]. Available at: http://www.oecd.org/coronavirus/policy-responses/coronavirus-covid-19-sme-policy-responses-04440101/ (2020)
10. J. Juergensen, J. Guimón, R. Narula. European SMEs amidst the COVID-19 crisis: assessing impact and policy responses. *Journal of Industrial and Business Economics*. **47**, 499-510 (2020)
11. A. W. Bartik, M. Bertrand, Z. B. Cullen, E. L. Glaeser, M. Luca, C. T. Stanton. *How are small businesses adjusting to COVID-19? Early evidence from a survey* (No. w26989). National Bureau of Economic Research [online]-Available at: https://www.nber.org/system/files/working_papers/w26989/w26989.pdf (2020)
12. J. Sheth. Business of business is more than business: Managing during the Covid crisis. *Industrial Marketing Management*. **88**, 261-264 (2020)
13. Eurofound. *European Working Conditions Survey* [online]. Available at: https://www.eurofound.europa.eu/surveys/2020/european-working-conditions-survey-2020 (2020)
14. T. Ritter, C. L. Pedersen. Analyzing the impact of the coronavirus crisis on business models. *Industrial Marketing Management*. **88**, 214-224 (2020)
15. H. He, L. Harris. The impact of Covid-19 pandemic on corporate social responsibility and marketing philosophy. *Journal of Business Research*. **116**, 176-182 (2020)
16. Federal tax service of Russia. *Unified Register of Small and medium-sized businesses* [online]. Available at: https://ofd.nalog.ru/ (2021)

17. RBC. *Research on the impact of the COVID-19 pandemic on Russian business* [online]. Available at: https://sapmybiz.rbc.ru/RBK_Issledovanie_vliyaniya_pandemii_COVID_19_ na_rossiyskiy_biznes.pdf (2020)

18. NAFI Research Centre. *The impact of the coronavirus epidemic on business and the need for state support* [online]. Available at: https://nafi.ru/projects/predprinimatelstvo/rossiyskiy-biznes-i-koronavirus-chast-1-predprinimateli-o-vliyanii-epidemii-na-ikh-biznes-i-o-potreb/ (2020)

19. Mail.ru Group. *Research by Mail.ru for business* [online]. Available at:https://biz.mail.ru/blog/contractors-survey/ (2020)

20. Federal tax service of Russia. *Statistics on state registration of legal entities and sole proprietors in the Russian Federation as a whole* [online]. Available at: https://www.nalog.ru/rn77/related_activities/statistics_and_analytics/regstats/ (2020)

21. J. B. Carnevale, I. Hatak. Employee adjustment and well-being in the era of COVID-19: Implications for human resource management. *Journal of Business Research*. **116**, 183-187 (2020)

Factors in the Development of Entrepreneurial Competencies

Anastasiya Pesha[1,*]

[1]Ural State University of economics, Department of labour economics and HR management, 8 Marta 62, 620144 Yekaterinburg, Russia

Abstract. Entrepreneurship is one of the main factors of economic growth, which ensures sustainable development of countries. Educational institutions and corporate universities develop programmes to develop entrepreneurial competencies of students and employees. Companies are introducing new functional divisions whose activities are related to revealing internal entrepreneurial competencies. This paper provides a systematic review of research in order to analyse the factors affecting the development of student and employee entrepreneurial competencies. Analysis of global, national, intra-organizational and personal factors in the development of entrepreneurial competencies makes it possible to form the most flexible and risk-resistant system of teaching entrepreneurship in universities and organizations.
Keywords: competencies; entrepreneurial competencies; intrapreneurship; innovation; factors; sustainable development; competitiveness

1 Introduction

Use The economic situation, large-scale technological changes, globalization trends and the competitive environment, observed as the phenomena of the digital economy, are reflected in the entrepreneurial activity of the population. Entrepreneurs contribute to increasing the economic well-being of people and innovative social development [1]. The development of entrepreneurship has a positive effect on reducing the unemployment rate by creating new jobs, thus ensuring the economic development as well as the social stability of society [2]. To ensure stability and development, companies are constantly working to improve their business models, products and services provided. One of the tools for this improvement is the development of intrapreneurship. As a result of the activities of internal entrepreneurs, new products and services are introduced, new corporate enterprises are initiated, new markets and business development horizons are opening [3]. Hypotheses about a positive correlation between the effectiveness of intrapreneurship, the development of entrepreneurial competencies, and business growth are obtained empirically [4].

What is currently happening with the indicators of entrepreneurial activity? The COVID-19 pandemic has affected the economies of countries and entrepreneurs′ activities. Some entrepreneurs were forced to close their businesses, revise their activities in new conditions, and for some people, the current situation opened up new opportunities for the implementation of business ideas. According to the Global Entrepreneurship Monitor 2020-2021 (Global Entrepreneurship Monitor, GEM), compiled by Elam A.B., Brush C.G., Greene P.G., Baumer B., Dean M. and Heavlow R., in 2020, indicators of early-stage entrepreneurial activity fell by more than 25% in 9 out of 35 surveyed countries (4 of them in Europe [5, p. 43]). At the same time, during the pandemic, the above indicator of entrepreneurial activity increased by more than 25% in 4 countries - Oman, Egypt, Panama and Colombia [5]. It should be noted that the indicators of entrepreneurial activity in various countries presented in the study are ambiguous; therefore, it is impossible to say that the COVID-19 p has significantly affected them. This pandemic was an extreme event for which countries had not been prepared.

The scientific community becomes increasingly interested in the development of entrepreneurial competencies, the development of intrapreneurship, arguing that the need to study entrepreneurial competencies from the point of view of their "potential" (opportunities for their application in the field of entrepreneurship), and not only "capital" (the availability of special education and entrepreneurial experience) is human [4, 6]. A large number of studies by scientists in this area are aimed at studying the possibilities of higher education in the formation of entrepreneurial competencies [1, 7]. The factors of entrepreneurial competencies′ development were studied by Palos-Sanchez P.R., Baena-Luna P., Pena A.C. [8]. Factor analysis of domestic entrepreneurship was performed by scholars such as Bouchard. V, Basso O., Menzel H. C., Aaltio I., Ulijn J. M. and others [9, 10].

The objective of this paper is the analysis of global, national, intra-organizational and personal factors that influence the formation of entrepreneurial competencies and the development of intrapreneurship.

* Corresponding author: myrabota2011@gmail.com

Among the research questions formulated in this paper, there is highlighted the analysis of sources representing different approaches to the interpretation of intrapreneurship; research on the main differences between entrepreneurship and intrapreneurship; classifications of entrepreneurial competencies, as well as factors of different levels of influence on the development of entrepreneurial competencies of people and intrapreneurship in companies.

2 Methodology

Literature review was chosen as the main research method, as it best meets the objective and research questions. The literature review included the following stages: formulation of research questions, review and selection of relevant studies taking into account the selected definitions, analysis of the results.

Relevant papers were selected from the international abstract databases Web of Science and Google Scholar, search definition "entrepreneurial competence". The search included papers published from the year 2010 to the present, except for some papers most popular in the scientific community which deal with the topic of research with the highest citation rates. The year 2010 was selected since at this time, a sharp increase in the interest of the scientific community in the problems of entrepreneurship factors and the development of entrepreneurial competencies was recorded (the number of publications per year in Google Scholar> 200, more than 60% of publications deal with the factors of entrepreneurship development / entrepreneurial competencies). Selection criteria for publications, in addition to the key search definition, were as follows:

1. Studying the factors / conditions for the development of entrepreneurial competencies;
2. Included analysis of entrepreneurship and intrapreneurship characteristics.
3. The empirical component as the basis for the research results presented in the papers.

On the basis of the selected criteria, more than 50 papers that meet these criteria were studied; the most significant ones are included in References.

3 Results and Discussion

3.1 Features of intrapreneurship and the main approaches to its understanding

There are several definitions of the concepts of "intrapreneur" and "intrapreneurship". Some scholars define intrapreneurship as a synonym for the concept of "corporate entrepreneurship" [11], while others describe the existence of certain differences between these two phenomena, pointing out that in corporate entrepreneurship, the initiation of entrepreneurial activity starts from the top down, while in the case of intrapreneurship, it is from the bottom up [3]. There are many scientists who believe that intrapreneurship is a sub-area of entrepreneurship [12]; others define it as "innovative practice within an organization, with the help of which employees take on new types of business activities and use various opportunities" [13, p. 813]. Nevertheless, all definitions of "intrapreneurship" have certain features in common: the manifestation of initiative and the introduction of innovations (entrepreneurial behaviour), the implementation of ideas within the framework of an existing company and for its benefit. We agree most with the concept of intrapreneurship formulated by Vargas-Halabí T., Mora-Esquivel R., Siles B., who claim that "intrapreneurship is a process in which an individual or group of individuals, within the framework of an existing organization, identify, pursue, and encourage innovative opportunities and create a new organization, renewing the organization or introducing product and process innovations" [14, p. 89]. Based on the existing interpretations, the most complete definition of the concept of an internal entrepreneur was formulated, which we adhere to in our research: An intrapreneur is an employee of an organization with an entrepreneurial mindset [15], capable of creating, identifying and using new opportunities to turn an idea into a profitable finished product [16] by actively taking risks and innovations [17]. In addition to entrepreneurial thinking, the following competencies are necessary for an intrapreneur: seizing opportunities; innovation; stimulating ideas; planning; risk management; resource management; maintaining relationships; building interaction with others [14].

Analysis of existing approaches shows there are two levels of manifestation of intrapreneurship - organizational and personal. The organizational level is associated with the creation of conditions and entrepreneurial orientation, while the personal level is connected with the entrepreneurial behaviour and competencies of employees when they are involved in intrapreneurship [3]. From our point of view, both of these levels deserve equal attention. The creation of intra-organizational conditions is one of the main factors in identifying entrepreneurial competencies of a specialist. At the same time, it is necessary to influence personal attitudes, intention and development of entrepreneurial competencies not only within the organization, by creating the conditions indicated above, but even in the process of professional training and personal development of future specialists at universities. The factors of development of entrepreneurial competencies are discussed in more detail in Chapter 3.2.

Summing up the results of this chapter, it should be noted that scientists distinguish a number of differences between intrapreneurship and entrepreneurship. Having analysed various points of view, the key ones were formulated as follows:

- entrepreneurial efforts and activities are carried out within a company with an established policy and culture;
- entrepreneurial activities are carried out as a result of interconnected entrepreneurial activities of several participants;
- entrepreneurial activities can be implemented at different organizational levels;
- the organization assumes all risks associated with the implementation of intrapreneurs´ initiatives;
- intrapreneur´s flexibility in decision-making and achieving a business goal is limited;
- the decision-making process in intrapreneurship is lengthy, conditioned by the approval of the company's management;
- the ability of the intrapreneur to use the resources of the organization to develop the idea;
- intrapreneurs do not think about the financial component of their initiative, they are focused on drawing up innovative and feasible plans;
- intrapreneurs´ activities have organizational security;
- in case of failure to implement entrepreneurial activity, intrapreneurs still receive remuneration from the company.

We believe that when implementing the policy for the development of intrapreneurship in the company or for the formation of entrepreneurial competencies of university students, the above differences shall be taken into account, since motivation, self-confidence and, finally, the implementation of various kinds of business initiatives of employees and alumni can depend on this.

3.2 Key factors in the development of entrepreneurial competencies

Intrapreneurship is very multifaceted and is influenced by numerous factors, some of which are of an objective socio-economic nature. The degree of influence of this type of factors can vary from country to country due to specific conditions of economic development and cultural characteristics. These factors are presented in this article in two groups x - global factors and national factors. of the study by Kim Y. T., & Heo C. M. analyses the significant influence of the factor of social values on the formation of entrepreneurial competencies and the implementation of entrepreneurial intentions [18].

In addition to global and national factors of influence, we identified the meso-level - the level of education and educational environment, which has a significant impact on the development of not only the corresponding range of competencies, but also on the identification of entrepreneurial intentions. The impact of teaching entrepreneurship competencies at universities on the implementation of entrepreneurial intentions has been addressed in many studies published in the last 2 years. As an example, there can be mentioned the article by Lv Y. et al., who conducted a survey including more than 5000 undergraduates in China. The results show that the development of entrepreneurial competencies plays an intermediary role for the practice of entrepreneurship and entrepreneurial intention [1].

The globalization of economic processes, strong competition and high demand for innovation complicates the organizational environment, changing the approach of companies to human resources and the use of their talents. Companies that are focused on attracting employees to innovative internal corporate behaviour must create an organizational climate that would motivate the employees to entrepreneurial initiatives. The internal entrepreneurial climate is characterized by "a positive correlation between management support, remuneration system and employee autonomy" and affects the innovativeness of employees and the transformation of ideas into real goods and services of companies [19]. The concept of "entrepreneurial climate" includes the style of leadership and support from management. The table below thus does not present these factors separately, as highlighted by a number of authors [10]. In addition to the entrepreneurial climate, the key intra-organizational factors for the development of entrepreneurial competencies represent new ways of working (NWW), internal competitions of ideas for the development of innovative culture in the company, which allow employees to gain recognition and be aware of their importance for the company, and the creation of conditions for personnel initiatives, as research shows, has significant impact on the development of domestic entrepreneurship [6]. The authors also pay great attention to the development of communications and relationship of trus within the team as a factor in the innovative behaviour of employees [20].

A review of factors in the development of entrepreneurial competencies will be incomplete if the key group of them - personal factors – are not considered. This group of factors has the greatest impact on the development of competencies and on the formation of entrepreneurial intentions [2]. First of all, it is necessary to mention motivation as a driving factor of personal development. In addition to motivation, studies conducted show the high importance of extraversion, openness and awareness for the development of entrepreneurial competencies [21]. Moreover, Schmitt-Rodermund E. empirically investigated and confirmed the existence of a high correlation of these factors with the level of student interest in entrepreneurship. Personal development and the environment in which the personality, worldview and personal attitudes are formed play a significant role in the development of entrepreneurial competencies and intentions. Based on the results of our research, we have added parenting to personality factors, meant as outcome, not process.

Table 1. Factors in development of entrepreneurial competencies

Group of factors	Factors	Authors
Global	Digitization	Morris T. H., & König P. D., 2020 [22]
	Globalization	Benevolo C., Penco L. and Torre T., 2021 [23]
National	social values	Kim Y. T., & Heo C. M. 2021 [23]
	state regulation of entrepreneurship and regional policy	Li C. et al., 2020 [24]; Motoyama, Y. and K. Knowlton, 2016 [25].
	market conditions and taxation	Lovkova E., et al., 2018 [26].
	infrastructure development	Lovkova E., et al., 2018 [26].
Educational	higher education in entrepreneurship	Iglesias-Sánchez P. P., et al., 2021 [27]; Lv Y., Chen Y., 2021 [1].
	entrepreneurship courses, business education	Palos-Sanchez P.R., et al.,2019 [8].
	accelerators, hackathons and business incubators	Iglesias-Sánchez P. P., 2019 [27]; Li C. et al., 2020; Pesha A., Shavrovskaya M., & Caha Z.,2021 [7].
Intra-organizational	Motivation	Menzel et al. 2007 [10]; Park S. H., et al., 2014 [20].
	business climate and development strategy	Gerards R., van Wetten S., & van Sambeek C., 2020 [6]; Menzel et al. 2007 [10]; Park, 2014 [22]; Poduška Z. и др.[19], 2020; Zahra S. A., 1991 [11]
	corporate sustainability	Hyuk K. H., & Park W., 2019 [2].
	NWW	Gerards R., van Wetten S., & van Sambeek C., 2020 [6].
	organizational capabilities (information, finance, working hours)	Poduška Z. и др., 2020 [19].
Personal	motivation and intentions	Hyuk K. H., & Park W. 2019 [2]; Lv Y., ChenY., et al, 2021 [1]; Schmitt-Rodermund E., 2004 [21].
	Awareness	Schmitt-Rodermund E., 2004 [21].
	Openness	
	Extraversion	
	upbringing and origin	Schmitt-Rodermund E., 2004 [21]; Stewart W., Roth P., 2007 [28].
	Values	Stewart W., Roth P., 2007 [28].
	gender, age, education, experience	Parker S. C., 2011 [12]; Pawitan G. et al., 2018 [29]

Source: Author.

4 Conclusion

The current economic situation and technological progress, business globalization and global socio-political turbulence have an impact on the development of entrepreneurial competencies and the formation of entrepreneurial intentions both outside and inside companies. Business owners are looking for new opportunities to maintain the stable development, efficiency and competition of organizations. In view of the relevance of this topic, this paper identifies the main approaches to understanding and distinguishing entrepreneurship from intrapreneurship, which, although having many forms, have the following in common: the initiation of entrepreneurial activity within the company (both upward and downward), the employees with the appropriate set of competencies and motivation, and the use of the organization's resources for the implementation of such initiatives.

Based on the content analysis of publications dealing with the same topic, the main differences between the phenomena of entrepreneurship and intrapreneurship are identified in the risk management that the organization undertakes for the implementation of intrapreneurial initiatives and in the motivation of intrapreneurs, which differs from the motivation of entrepreneurs interested in self-realization.

Within the paper, several groups of factors for the development of entrepreneurial competencies are identified (see Table 1). The views of scientists to the issues of factors differ significantly; nevertheless, we believe that all the selected factors need to be equally taken into account, since only a holistic approach to the development of entrepreneurial competencies can lead to the expected growth in the indicators of entrepreneurial activity of the population of the countries.

Due to the influence of macro factors, a large number of risks arise in the manifestation of entrepreneurial initiatives and the start-up of new organizations. We agree with Iglesias-Sánchez P. P., et al., 2021; Lv Y., Chen Y., 2021. and other authors who highlight the unambiguous importance of educational factors for the development of entrepreneurial competencies and entrepreneurial intentions [1, 28]. The development of educational infrastructure and the introduction of courses and methods for teaching entrepreneurship in various areas of student training will become an additional driver for the development of entrepreneurship and the economy of countries as a whole. Our analysis shows that the key factors in the manifestation of entrepreneurial initiatives and encouraging of employee entrepreneurial thinking include the creation of an entrepreneurial climate and organizational policy aimed at strategic business development. A large number of studies show the influence of various demographic factors on the development of entrepreneurial competencies and the formation of entrepreneurial intentions, for example, Parker S. C., 2011 [12]; Pawitan G. et al., 2018 [29]. At the same time, we believe that the analysis of the influence of these factors on this phenomenon requires deeper and more extensive research and should involve the analysis of demographic factors in parallel with the socio-economic and cultural characteristics of different regions and countries. As for other personal factors, we tend to agree the authors whose work is presented in the final analysis, and who claim that the main thing is the motivation and intention to develop competencies and start up their own business. At the same time, it should be noted that it is possible to influence the formation of a person's motivation and intentions in the field of entrepreneurship through the implementation of targeted educational programmes, as mentioned e.g. in the study by Lv, Y., Chen, Y., 2021 [1].

In our opinion, the future is in the increased attention paid to the education system in order to achieve intentional development of entrepreneurial competencies, which concerns not only specialized areas of training, but also the inclusion of relevant courses and disciplines in the Bachelor's and Master's degree programmes. This initiative will allow us to recognize entrepreneurial talents and create conditions for the implementation of student and graduate entrepreneurial intentions, which will have a positive impact on the indicators of entrepreneurial activity of young people.

Our analysis of the factors for the development of entrepreneurial competencies can be applied in strategic programmes in order to ensure the development of intrapreneurship within companies, as well as in further scientific research on the influence of external and internal factors on the development of student entrepreneurial competencies and intentions.

Acknowledgment

The reported study was funded by RFBR, project number 19-29-07435

References

1. Y. Lv, Y. Chen, Y. Sha, J. Wang, L. An, T. Chen, L. Huang. How Entrepreneurship Education at Universities Influences Entrepreneurial Intention: Mediating Effect Based on Entrepreneurial Competence. *Frontiers in Psychology*. **12** (2021)
2. K. H. Hyuk, W. Park. Impact of Entrepreneurial Business Start-up Motivation, Entrepreneurial Spirit, and Entrepreneurial Competence Characteristics on Start-up Companies' Sustainability: Focusing on the Mediating Effect of the Start-up Companies' Business Performance. *Asia-Pacific Journal of Business Venturing and Entrepreneurship*. **14**(3), 59-71 (2019)
3. J. C. Rigtering, G. U. Weitzel, K. K. Muehlfeld. Increasing quantity without compromising quality: How managerial framing affects intrapreneurship. *Journal of Business Venturing*. **34**(2), 224-241 (2019)
4. T. Lans, M. Van Galen, J. Verstegen, H. Biemans, M. Mulder. Searching for Entrepreneurs among Small Business Owner-Managers'. *Proceedings of the ECER*. 1-22 (2010)
5. A. B. Elam, C. G. Brush, P. G. Greene, B. Baumer, M. Dean, R. Heavlow. *Global Entrepreneurship Monitor. 2020/2021 Global Report* [online]. Published by the Global Entrepreneurship Research Association, London Business School, Regents Park, London NW1 4SA, UK, 2021. Available at: https://www.gemconsortium.org/report/gem-20202021-global-report (2021)
6. R. Gerards, S. van Wetten, C. van Sambeek. New ways of working and intrapreneurial behaviour: the mediating role of transformational leadership and social interaction. *Review of Managerial Science*. 1-36 (2020)
7. A. Pesha, M. Shavrovskaya, Z. Caha. Comparative Analysis of the Level of Development of Entrepreneurial Competencies among Students in Russia and the Czech Republic. *SHS Web of Conferences*. **90**(1):02007 (2021)
8. P. R. Palos-Sanchez, P. Baena-Luna, A. C. Pena. Analysis of educational competences to evaluate entrepreneurial people. *Interciencia*. **44**(5), 291-297 (2019)
9. V. Bouchard, O. Basso. Exploring the links between entrepreneurial orientation and intrapreneurship in SMEs. *Journal of Small Business and Enterprise Development*. **18**(2), 219-231 (2011)
10. H. C. Menzel, I. Aaltio, Ulijn J. M. On the way to creativity: Engineers as intrapreneurs in organizations. *Technovation*. **27**(12), 732-743 (2007)
11. S. A. Zahra. Predictors and financial outcomes of corporate entrepreneurship: An exploratory study. *Journal of Business Venturing*, **6**(4), 259–285 (1991)
12. Parker S. C. Intrapreneurship or entrepreneurship? *Journal of business venturing*. **26**(1), 19-34 (2011)
13. B. Baruah, A. Ward. Metamorphosis of intrapreneurship as an effective organizational strategy. *International Entrepreneurship and Management Journal*. **11**(4), 811-822 (2015)
14. T. Vargas-Halabí, R. Mora-Esquivel, B. Siles. Intrapreneurial competencies: development and validation of a measurement scale. *European Journal of Management and Business Economics*. **26** (1), 86-111 (2017)

15. G. Pinchot. III. *Intrapreneuring: Why you don't have to leave the corporation to become an entrepreneur*. University of Illinois at Urbana-Champaign's Academy for Entrepreneurial Leadership Historical Research Reference in Entrepreneurship (1985)

16. H. Ma, T.Q. Liu, R. Karri. Internal corporate venturing: intrapreneurs, institutions, and initiatives. *Organizational Dynamics*. **45(**2), 114-123 (2016)

17. J. Augusto Felício, R. Rodrigues, V. R. Caldeirinha. The effect of intrapreneurship on corporate performance. *Management Decision*. **50**(10), 1717–1738 (2012)

18. Y. T. Kim, C. M. Heo. The Influence of Senior Entrepreneurship Competency and Start-up Support Policy on Entrepreneurship Intention: Focusing on the Moderating Effect of Mentoring. *Asia-Pacific Journal of Business Venturing and Entrepreneurship*. **16**(2), 109-121 (2021)

19. Z. Poduška, J. Nedeljković, D. Nonić, T. Ratknić, M. Ratknić, I. Živojinović. Intrapreneurial climate as momentum for fostering employee innovativeness in public forest enterprises. *Forest Policy and Economics*. **119**, 102281 (2020)

20. S. H. Park, J. N. Kim, A. Krishna. Bottom-up building of an innovative organization: Motivating employee intrapreneurship and scouting and their strategic value. *Management Communication Quarterly*. **28**(4), 531-560 (2014)

21. E. Schmitt-Rodermund. Pathways to successful entrepreneurship: Parenting, personality, early entrepreneurial competence, and interests. *Journal of vocational behavior*. **65**(3), 498-518 (2004)

22. T. H. Morris, P. D. König. Self-directed experiential learning to meet ever-changing entrepreneurship demands. *Education+ Training*. **63**(1), 23-49 (2021)

23. C. Benevolo, L. Penco, T. Torre. Entrepreneurial decision-making for global strategies: a "heart–head" approach. *Management Decision*. **59**(5), 1132-1157 (2021)

24. C. Li, N. Ahmed, S. A. Qalati, A. Khan, S. Naz. Role of business incubators as a tool for entrepreneurship development: the mediating and moderating role of business start-up and government regulations. *Sustainability*. **12**(5), 1822 (2020)

25. Y. Motoyama, K. Knowlton. From Resource Munificence to Ecosystem Integration: The Case of Government Sponsorship in St. Louis. *Entrepreneurship & Regional Development*. **28**(5–6), 448–470 (2016)

26. E. Lovkova, T. Kashicina, D. Savenkov, A. Kuzmina. Key vectors and concerns of small-sized entrepreneurship development in the Russian Federation. *Challenging the Status Quo in Management and Economics*. 19-25 (2018)

27. P. P. Iglesias-Sánchez, C. Jambrino-Maldonado, C. de las Heras-Pedrosa. Training entrepreneurial competences with open innovation paradigm in higher education. *Sustainability*. **11**(17), 4689 (2019)

28. W. Stewart, P. Roth. A meta-analysis of achievement motivation differences between entrepreneurs and managers. *Journal of Small Business Management*. **45**(4), 401–421 (2007)

29. G. Pawitan, M. Widyarini, C. B. Nawangpalupi. Moderating effect of demographic factors and entrepreneurial phase on the relationship between entrepreneurial competencies and innovation of ASEAN entrepreneurs. *Pertanika Journal of Social Science and Humanities*. **26**, 151-166 (2018)

Reputation Management in Times of the Pandemic - Empirical Analysis of Selected Health Care Providers Operating in the Slovak Market

Frantisek Pollak[1,*], *Peter* Markovic[1], Bozka Malinak[2], *Katarina* Belovodska[2]

[1]University of Economics in Bratislava, Faculty of Business Management, Department of Business Finance, Dolnozemská cesta 1/b, 852 35, Bratislava, Slovakia
[2] Acuity Engineering & Consulting Service Ltd, 10327 122 St NW, T5N1M1 Edmonton AB, Canada

Abstract. The issue of corporate reputation management is one of the key areas in the process of building trust between the supply and demand side of the market. The services sector, which, as a rule, works predominantly with intangible products, is particularly vulnerable to reputational damage. The presented study aims to examine the impact of the COVID-19 pandemic on the overall level of reputation of eleven Slovak teaching hospitals. The data sources used in the form of publicly available databases cover both the traditional factors of the brick-and-mortar market, as well as the factors related to the virtual Internet environment. The results of the analysis point to the fact that while the traditional level of reputation of the entities reflected changes in the market only to a relatively limited extent, the level of the online reputation of the analysed entities decreased significantly due to the pandemic. This finding underscores the fact about the fragility of reputation in the Internet environment. By comparing individual reputational factors in the pre-pandemic period with the situation during the peak of the third wave of the pandemic, the study offers a relatively compact knowledge base for formulating basic transformations for comprehensive research into the impact of a pandemic on corporate competitiveness.
Keywords: corporate reputation; reputation management; pandemic; COVID-19; teaching hospital

1 Introduction

Corporate reputation management is an important part of the overall management of corporate assets. Research [1-3] points to the fact that the issue of reputation is a crucial factor in the process of building trust between the organization and its customer. It does not have to be just a purely market environment. We also encounter the application of reputation management in corporate management in areas such as education or healthcare [4-5]. In the present study, we focused on the field of health care, or rather the area of health care providers. The study builds on the relatively extensive knowledge base of the authors [6-8], in which the issue of reputation management of entities operating (not only) in the health care sector during the period preceding the outbreak of the COVID-19 pandemic was thoroughly developed. However, the sharp onset of a pandemic in the first half of 2020 largely affected almost all areas of the market [9-11], and healthcare was no exception. On the contrary, the health care sector has come to the fore. The spring of 2020 was marked by a high degree of admiration and respect for healthcare professionals and the entire healthcare sector. With the aftermath of the pandemic, the initial enthusiasm of the demand side of the market gradually began to subside. The transition from the acute to the chronic part of the pandemic brought with it a relatively critical response from a large part of the market. This was based on both the overall congestion of the system and the relatively polarized market sentiment [12]. All these factors, whether it is the weak predictability of the situation at the beginning of the pandemic, or the significant rate of burnout resulting from the depletion of available resources (it should be noted here that it is both financial/material and human), have only exacerbated the overall fragility of the reputation. It is the fragility that is, according to the literature, the most characteristic feature of reputation [13-14]. The aim of the presented study is therefore to examine the effects of a pandemic on the reputation of healthcare providers. For the highest degree of authenticity and informative value of the knowledge produced, we chose eleven teaching hospitals that operate on the Slovak market of health care providers as a research sample. As this is a complex issue, we will examine both traditional and modern aspects of corporate reputation over time. The reference pre-pandemic basis for the analysis will be data from 2018. Traditional aspects of corporate reputation will be represented by assessments of entities by the state health insurance company based on the Patient Satisfaction Survey 2018 [15] and the Patient Satisfaction Survey 2020 [16]. Modern aspects will be represented by individual reputators synthesized within the parameter Total Online Reputation "TOR" measured in 2018 [17], which we will follow up with the current dataset prepared for the needs of the presented study.

* Corresponding author: frantisek.pollak@euba.sk

Given the research goal as well as the nature of the available data, it is therefore possible to formulate the following research question:

- Is it possible to identify significant changes in the reputation of entities in the pre-COVID-19 and COVID-19 periods?

The main research question can then be divided into two sub-questions based on the environment, namely:

- Have there been changes in the traditional reputation of subjects as expressed in patient satisfaction surveys?
- Have there been changes in the online reputation of entities?

From the point of view of the composition of the content of the presented study, we have chosen the traditional structure, where we will begin with a brief overview of the current state in the researched issues from the point of view of key concepts of the topic. Subsequently, we will move smoothly to the description of the objectives and methodological apparatus that is used in the study. Findings will be interpreted through summary tables; key facts will be visualized. The conclusion will summarize all the key findings and describe the main limitations and perspectives of the research. The present study is prepared with the aim of expanding the knowledge base for subsequent comprehensive research of the effects of a pandemic on corporate competitiveness.

1.1 Traditional reputation factors

We encounter reputation as a concept of the traditional world of bricks and mortar in the cross-sectional literature in various areas of science. The economic sciences view reputation as one of the components of a company's portfolio of intangible assets [18-19]. Management sciences, in turn, view the issue of reputation as a concept based on the tradition of protecting goodwill. In marketing, in turn, reputation is often confused with the image of the organization, even though reputation greatly exceeds the scope of image issues. Despite the apparent ambiguity in terminology, all concepts come together in a confidence-building process. Confidence integrates all the mentioned concepts in terms of content and scope, and from the point of view of science, it raises the issue to the extent of the philosophical concept [20]. When it comes to the issue of trust or even faith, we go thematically thousands of years into the history of the work of ancient philosophers. Given this fact, it can be stated that the factors influencing the issue in the traditional world are well known to all market participants. Unless the entity has direct knowledge of the active management of its reputation, it has the well-known know-how needed to ensure that it does not damage its reputation. Also, from the point of view of the media as a carrier of information, the reputation in its traditional understanding is relatively well controllable. Reputation management in its understanding as it was known from the end of the twentieth century is dominantly determined by so-called "hard" factors, such as the economic results of the organization, the rate, and extent of innovation, consistency in values, and attitudes, etc. We could include all these factors as "objective quality factors". In contrast, from the point of view of the digital environment, the issue of reputation is much more diverse.

1.2 Reputation factors in the online environment of the Internet

When it comes to the digital environment of the Internet, reputation dynamically multiplies the number of quantifiable as well as non-quantifiable variables needed for its qualified management. The Internet as a medium is usually unregulated in nature [21-22], the authenticity of messages distributed by the network is questionable. Whereas in the past it was necessary to control "only" the main information flows in the form of publicity in opinion-forming media for reputation management, today such an approach is insufficient. Social media has introduced the concept of open communities, which allow both sides of the market to interact in real-time [23]. In any case, a targeted approach creates considerable information asymmetry with enormous pressure on corporate resources. Control of these processes is also quite difficult. The number of variables is dynamic, new media are coming to the market, which is constantly creating demand for expanding the digital information portfolio of companies. If we consider that the unregulated nature of the Internet shortens communication channels and, on the other hand, constantly generates new communication platforms, we get into an unenviable situation as managers. That is why it is important to constantly generate new knowledge and continuously explore this dynamic issue. Organizations of all sizes must accept the same rules of the game [24-25] if they want to succeed in hectic and turbulent hypermedia times. However, as researchers, we must select a quantifiable number of variables, as it is not possible to perform an empirical analysis with non-quantifiable variables. For the purposes of this analysis, the number of quantifiable variables has been reduced to dominant media in the form of major industry players such as Google and Facebook. Authenticity and objectivity are, in our view, the key to online reputation management. From the point of view of factors that influence the reputation of subjects in the online environment, it can be stated that they take the form of a synthesis of factors of objective and subjective perceived quality.

2 Methodology

The presented study aims to examine the effects of a pandemic on the reputation of health care providers. Given the research goal as well as the nature of the available data, the following research question is formulated:

- Is it possible to identify significant changes in the reputation of entities in the pre-COVID-19 and COVID-19 periods?

The main research question can then be divided into two sub-questions based on the environment, namely:

- Have there been changes in the traditional reputation of subjects as expressed in patient satisfaction surveys?
- Have there been changes in the online reputation of entities?

For the highest degree of authenticity and informative value of the knowledge produced, we have chosen as a research set the entire segment of health care providers, these are entities that provide their services in the Slovak Republic. As a research sample, we selected eleven teaching hospitals, which represent key representatives of the industry in the research. As this is a complex issue, we will examine both traditional and modern aspects of corporate reputation in the run-up to and during the COVID-19 pandemic.

The traditional aspects of corporate reputation will be represented by a ranking of entities based on the Patient Satisfaction Survey 2018 [15] and 2020 [16] conducted annually by the state health insurance company, hereinafter referred to as "VšZP".

Based on the created knowledge base [6-8], modern aspects will be represented by a synthesis of individual partial online reputators within the parameter Total Online Reputation "TOR" measured in 2018 [26], which we will follow in the study by compiling an up-to-date overview. The indicator "TOR" represents a complex quantity calculated for each analyzed subject individually based on the following formula (1):

$$TOR = \frac{R_{ASA} + \sum_{i=1}^{n} R_i}{n+1} \quad (1)$$

Where:

TOR- Total online reputation v %

R_i- Reputation determinant (% score according to the i- reputation determinant)

R_{ASA}- ASA reputation determinant (% score according to the advanced sentiment analysis)

n- Number of determinants

From the point of view of the determinants that enter as R_i into the calculation, we take into account the following:

- Facebook rating;
- Google rating;
- Azet.sk rating.

From R_{ASA}'s point of view, we use for this case a two-stage sentiment analysis [7] of the first ten search results based on the official designation/name of the entity via Google search engine, where general Google search results are supplemented by search results in the "news" category.

The basic scientific methods used in the study are analysis and synthesis, we also work with elements of abstraction.

The data are processed via the Excel spreadsheet program of the MS Office 365 office suite.

For the purposes of data presentation, we use overview tables, which are graphically interpreted in specific cases.

3 Results and Discussion

Managing corporate reputation is a relatively complicated process, especially when unexpected variables enter the process. In the case of our research, such a variable was the COVID-19 pandemic, which created significantly specific conditions in the market of healthcare providers. In this chapter, we will gradually present the partial outputs of the analysis to answer the primary research question. Attention will be paid to both the traditional and digital markets.

3.1 The period before the outbreak of the COVID-19 pandemic

The year 2018 was chosen as the reference period for comparison, which represented a relatively stable reference point in the selected market. Based on data from the study [26] and interpretation in the study [17], it is possible to present in Table 1. complex state values of the monitored variables measured in the pre-pandemic period as follows:

Table 1. Offline and Online Reputation in 2018

Subject	VšZP ranking	TOR score (%)
University Hospital Martin	1.	83
Central Military Hospital SNP Ruzomberok	2.	80
Faculty Hospital F. D. Roosevelt Banska Bystrica	3.	72
Faculty Hospital Nove Zamky	4.	70
Faculty Hospital Nitra	5.	51
Faculty Hospital J.A.Reiman Presov	6.	62
University Hospital of L. Pasteur Kosice	7.	70
Faculty Hospital Zilina	8.	51
University Hospital Bratislava	9.	67
Faculty Hospital Trnava	10.	63
Faculty Hospital Trencin	11.	70

Source: Autor based on [15,26].

From the point of view of patient evaluation based on a survey conducted by the state health insurance company, it can be stated that nine out of eleven providers achieved a rating of "average", only two subjects, namely University Hospitals in Trnava and Trenčín in patient evaluation achieved a rating of "worst ". At the same time, none of the subjects in the sample achieved a "best" rating. Regarding the TOR, i.e., the overall level of the entity's online reputation, the situation is much more diverse. The first two subjects in the VšZP rank also occupy the first places in comparison based on online reputation, even in the same order. At the same time, the University Hospital in Martin achieved the highest coefficient based on the sum of the sentiment scores of the first ten Google search results, which is one of the partial parameters of the TOR score. The Central Military Hospital in Ružomberok has reached the level of 100% in terms of user ratings on the social network Facebook. Given the general findings, it should be noted that all subjects in the analysis of the pre-COVID-19 period reached at least 50% of their online potential.

3.2 Period during the COVID-19 pandemic

Based on the same methodology, we developed the following overview Table 2, through which it is possible to monitor the development of the monitored values of the analyzed subjects in the period during the COVID-19 pandemic.

Table 2. Offline and Online Reputation in 2021

Subject	VšZP ranking	TOR score (%)
Central Military Hospital SNP Ruzomberok	1.	71
University Hospital Martin	2.	80
Faculty Hospital F. D. Roosevelt Banska Bystrica	3.	66
Faculty Hospital Nove Zamky	4.	58
Faculty Hospital Zilina	5.	44
Faculty Hospital J. A. Reiman Presov	6.	32
Faculty Hospital Nitra	7.	41
Faculty Hospital Trnava	8.	59
Faculty Hospital Trencin	9.	61
University Hospital of L. Pasteur Kosice	10.	33
University Hospital Bratislava	11.	48

Source: Autor based on [16].

From the point of view of patient evaluation based on a survey carried out by the state health insurance company, it is possible to record an exchange between the first and second place for the first three providers. The best players on the market maintain their position in the long run. It is also possible to record the decline of two large market players, namely university hospitals in Prešov and Košice to the last ranks. A positive finding from the point of view of traditional reputation is the fact that none of the evaluated subjects achieves the rating "worst", while all subjects are evaluated by patients as "average" according to the methodology used. From the point of view of online reputation, the assumption of its fragility is confirmed. Subjects underwent a stress test during the pandemic, which was reflected in the sum of the results of the measured sentiments in the first ten Google search results. The difficult period of the fight against the pandemic also resulted in a decrease in customer ratings of official profiles of entities on the social network Facebook and Google ratings of profiles directly in the search results according to the official name of the entity. Only the University Hospital in Martin achieved the first quartile of evaluation according to the overall level of online reputation, while up to five out of a total of eleven subjects fell below the 50% level. In terms of possible causes of the decline in overall reputation, the analysis revealed both objective factors in the form of negative publicity related to the difficult pandemic situation, as well as possible managerial failures within individual levels of management of selected entities/health care sector in turbulent conditions of a graduating pandemic. Until the main research question can be answered, it is necessary to visualize and discuss in more detail the differences in the observed periods.

3.3 Comparison of pre-pandemic and pandemic period

By processing the empirical material produced in the analysis process, it was possible to obtain a more detailed overview of the impact of the COVID-19 pandemic on the reputation management of entities operating in the healthcare sector. In general, it is argued that online reputation is more prone to damage, as it is affected by factors that do not generally constitute the primary area of concern for the management of an entity. Also, the dominantly intangible nature of the factors influencing online reputations is largely more sophisticated. This creates additional requirements for resource allocation. The limited nature of resources and the unlimited nature of external factors were also reflected in the results of our research in terms of comparing pre-pandemic and pandemic indicators. The following Table 3 shows the shifts in the rankings of entities within the traditional and digital world as follows:

Table 3. Offline and Online Reputation- 2018 vs. 2021

Subject	**VšZP ranking** 2018/2021		**TOR score (%)** 2018/2021
University Hospital Martin	1./2.	(↓1)	83/80 (↓3%)
Central Military Hospital SNP Ruzomberok	2./1.	(↑1)	80/71 (↓9%)
Faculty Hospital F. D. Roosevelt Banska Bystrica	3./3.	(=)	72/66 (↓6%)
Faculty Hospital Nove Zamky	4./4.	(=)	70/58 (↓12%)
Faculty Hospital Nitra	5./7.	(↓2)	51/41 (↓10%)
Faculty Hospital J.A.Reiman Presov	6./6.	(=)	62/32 (↓30%)
University Hospital of L. Pasteur Kosice	7./10.	(↓3)	70/33 (↓27%)
Faculty Hospital Zilina	8./5.	(↑3)	51/44 (↓7%)
University Hospital Bratislava	9./11.	(↓2)	67/48 (↓19%)
Faculty Hospital Trnava	10./8.	(↑2)	63/59 (↓4%)
Faculty Hospital Trencin	11./9.	(↑2)	70/61 (↓9%)

Source: Author.

From the point of view of comparing the order of subjects based on the Patient Satisfaction Survey, it can be stated that the change in order occurred in nine out of a total of eleven cases. The leaders in the first half of the sample show a relatively stable position within the market. Based on the published data (since 2018, the survey no longer states the exact numerical value of the patient satisfaction index), it is possible to monitor the stabilization of the analyzed subjects at the value "average". While in 2018 only nine of the eleven subjects analyzed by us had the value "average", during 2020 all the analyzed subjects within the sample we selected had the evaluation "average". Thus, it can be stated that from the point of view of factors of the traditional world, the subjects of our selected sample managed the adaptation to the turbulent period of the pandemic relatively well. It is also possible, based on the analyzed data, to answer positively the first divided research sub-question. During the period under review, there were changes in the traditional reputation expressed by patient satisfaction surveys in nine out of a total of eleven cases. Regarding the online reputation of the subjects, the situation is considerably more complicated, the following Figure represents a graphical interpretation of the results across the observed period:

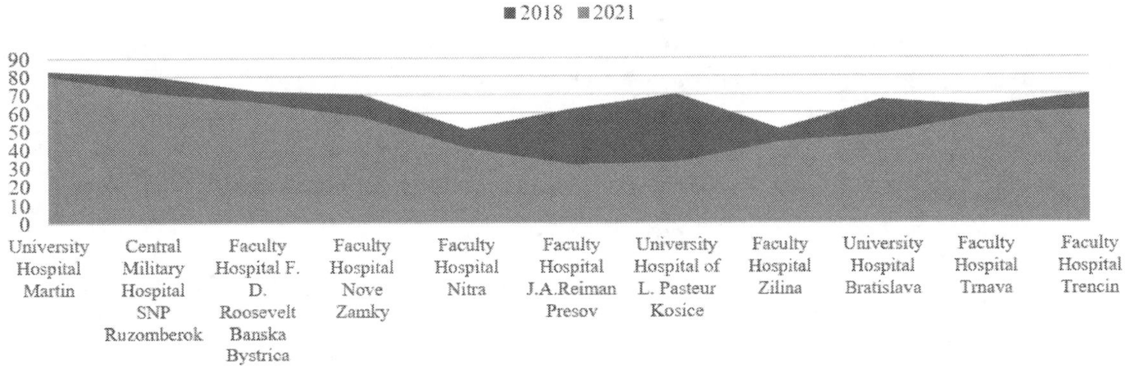

Figure 1. Total Online Reputation score (%)- 2018 vs. 2021

Source: Author.

In the monitored period, we record a relatively significant decline in online reputation for all entities within the sample we selected. The average value of the decline in the total online reputation is higher than 10%. In light of partial factors, this drop can be justified both by a decrease in the level of positive sentiments in search results (especially in search results in the Google "news" tab) and by a relative decrease in the overall user ratings of entities within the Facebook and Google platforms. Based on the results of the analysis, it is therefore possible to answer the second of the divided research sub-questions. During the pandemic, the subject's level of online reputation changed.

By answering both divided sub-questions, we get to the main research question, in which we asked whether it is possible to identify significant changes in reputation of the analyzed subjects in terms of facts in the period before the outbreak of the pandemic and during the COVID-19 pandemic. Level changes can be identified both in the case of a traditional reputation and in the case of an online reputation. The changes in the case of traditional reputation are less significant, overall, it is possible to state the stabilization of the reputation of the analyzed subjects towards the average. In the case of online reputation, the changes are significant, in all monitored cases it is a decrease in reputation below the reference level measured in the pre-pandemic period.

5 Conclusion

Corporate reputation is a relatively complex construct, the management of which requires a wide range of tools and procedures. Even in circumstances that can be described as "normal", it is relatively difficult to optimize all procedures so that the development of this intangible asset can be effectively assessed. The traditional world represented by brick-and-mortar tools makes it possible to optimize procedures for effective reputation management relatively predictably. In the presented study, we confirmed this assumption based on a comparison of two reputation rankings developed by a unified Patient Satisfaction Survey methodology, where over time subjects in our sample optimized procedures to stabilize reputation at its average level, despite adverse market factors. In contrast, we recorded a significant decline in the online reputation of entities, wherewith the onset of the pandemic, we observe within the sample a decline in almost all sub-indicators forming a complex parameter TOR. The unregulated nature of the virtual market, combined with the unpredictability of the pandemic era, has created a set of variables that are relatively difficult to control over time. Resource efficiency requires a thorough knowledge of the issue and current and relevant data. The present study has the ambition to contribute to the expansion of the knowledge base for subsequent comprehensive research into the effects of a pandemic on corporate competitiveness. From the point of view of limitations, it is necessary to point out the relatively high specificity of the market, with certain signs of catching up. Despite this fact, however, it is possible to approach the knowledge generated in the analysis as a general knowledge base, as the specifics of the digital customer largely eliminate market barriers and internationalize findings.

Acknowledgment

This article is one of the partial outputs of the currently solved research grant VEGA 1/0140/21 "Research on economically important factors of corporate reputation in the context of sustainable industry and low-carbon economy".

References

1. L. Hollebeek, K. Macky. Digital Content Marketing's Role in Fostering Consumer Engagement, Trust, and Value: Framework, Fundamental Propositions, and Implications. *Journal of Interactive Marketing*. **45**, 27–41 (2019)
2. O. Iglesias, S. Markovic, M. Bagherzadeh, J. Singh. Co-creation: A Key Link Between Corporate Social Responsibility, Customer Trust, and Customer Loyalty. *Journal of Business Ethics*. **163**(2), 151-166 (2020)
3. J. Soviar, M. Holubčík, J. Vodák, M. Rechtorík, F. Pollák. The Presentation of Automotive Brands in the On-Line Environ-ment—The Perspective of KIA, Peugeot, Toyota and VW in the Slovak Republic. *Sustainability*. **11**, 2132 (2019)
4. J. B. Ziemba, S. Arenberg, H. Reustle, M. E. Allaf, D. Haldeman. Consumers' Association of Hospital Reputation With Healthcare Quality. *Journal for Healthcare Quality*. **41**(4), 251-258 (2019)
5. G. Miotto, C. Del-Castillo-Feito, A. Blanco-González. Reputation and legitimacy: Key factors for Higher Education Institutions' sustained competitive advantage. *Journal of Business Research*. **122**, 342-353 (2020)
6. P. Dorcak, P. Markovic, F. Pollak. Multifactor analysis of online reputation as a tool for enhancing competitiveness of subjects from automotive industry. *Ekonomický časopis*. **65**, 173-186 (2017)
7. F. Pollák. *Online reputačný manažment v podmienkach stredoeurópskeho virtuálneho trhu*. Prešov: Bookman, (2015)
8. F. Pollák, P. Dorčák, P. Markovič. Corporate Reputation of Family-Owned Businesses: Parent Companies vs. Their Brands. *Information*. **12**(2), 89 (2021)
9. H. Hongwei, L. Harris. The impact of Covid-19 pandemic on corporate social responsibility and marketing philosophy. *Journal of Business Research*. **116**, 176-182 (2020)
10. Y. Y. Jiang, J, Wen. Effects of COVID-19 on hotel marketing and management: a perspective article. *International Journal of Contemporary Hospitality Management*. **32**(8), 2563-2573 (2020)
11. D. Y. Zhang, M. Hu, Q. Ji. Financial markets under the global pandemic of COVID-19. *Finance Research Letters*. **36**, 101528 (2020)
12. F. Pollák, R. Vavrek, J. Váchal, P. Markovič, M. Konečný. Analysis of Digital Customer Communities in terms of their interactions during the first wave of the COVID-19 pandemic. *Management & Marketing. Challenges for the Knowledge Society*. **16**(2), 134 – 151 (2021)
13. A. C. Schulz, S. Johann. Downsizing and the fragility of corporate reputation: An analysis of the impact of contextual factors. *Scandinavian Journal of Management*. **34**(1), 40-50 (2018)
14. G. Ordonez. Fragility of reputation and clustering of risk-taking. *Theoretical Economics*. **8**(3), 653-700 (2013)
15. VšZP. *Prieskum spokojnosti pacientov 2018* [online]. Available at: https://www.vszp.sk/poskytovatelia/indikatory-kvality/prieskum-spokojnosti-pacientov-2018/ (2018)
16. VšZP. *Prieskum spokojnosti pacientov 2020* [online]. Available at: https://www.vszp.sk/poskytovatelia/indikatory-kvality/prieskum-spokojnosti-pacientov-2020/ (2020)
17. R. Štefko, F. Pollák, B. Gavurová. Offline and online approaches to quality perception of Slovak university hospitals. *Marketing and Management of Innovations*. **4**, 107-115 (2018)
18. V. P. Rindova, I. O. Williamson, A. P. Petkova. Reputation as an Intangible Asset: Reflections on Theory and Methods in Two Empirical Studies of Business School Reputations. *Journal of Management*. **36**(3), 610-619 (2010)
19. S. Tadelis. What's in a name? Reputation as a tradeable asset. *American Economic Review*. **8**(3), 548-563 (1999)
20. C. Kobrak. The Concept of Reputation in Business History. *Business History Review*. **87**(4), 763-786 (2013)
21. M. J. Granger, D. L. Schroeder. Integrating the Internet into the business environment. *Industrial Management & Data Systems*. **94**(8), 37–40 (1994)
22. P. M. Stevens, K. P. Williams, M. C. Smith. Organizational communication and information processes in an Internet-enabled environment. *Psychology & Marketing*. **17**(7), 607-632 (2000)
23. R. T. Rust, W. Rand, M. H. Huang, A. T. Stephen, G. Brooks, T. Chabuk. Real-Time Brand Reputation Tracking Using Social Media. *Journal of Marketing*. **85**(4), 21-43 (2021)
24. F. Pollák, P. Markovič. Size of Business Unit as a Factor Influencing Adoption of Digital Marketing: Empirical Analysis of SMEs Operating in the Central European Market. Administrative Sciences. **11**(3), 71 (2021)

25. G. Appel, L. Grewal, R. Haid. A. T. Stephen. The future of social media in marketing. *Journal of the Academy of Marketing Science*. **48**, 79–95 (2020)

26. F. Pollák, P. Dorčák, P. Markovič, N. Svetozarovová, K. Belovodská. A Comprehensive Analysis of the Online Reputation of Healthcare Providers Operating on the Slovak Market. *37th International Conference on Organizational Science Development: Organization and uncertainty in the digital age*. **47** (2018)

The Fashion Industry and its Problematic Consequences in the Green Marketing Era - a Review

Nikola Sagapova[1,*], *Roman* Buchtele[2], *Radim* Dušek[1]

[1]Institute of Technology and Business, Faculty of Corporate Strategy, Okruzni 517/10, 37001 Ceske Budejovice, Czech Republic
[2]University of South Bohemia, Faculty of Economics, Studentska 13, 37005 Ceske Budejovice, Czech Republic

Abstract. This paper uses a narrative literature review to provide evidence of the problematic consequences of the fast fashion industry in the context of sustainability and to consider the option of the green marketing approach as a way of reducing negative impacts. Studies relevant to our research were identified through a search of Web of Science, Science Direct, Scopus and Google Scholar. In recent years, environmental sustainability has played an important role on the political agenda, but it is also driving innovation in business. The range of environmentally friendly products on the market is increasing, and consumers are showing interest in buying these types of products, which are usually more expensive. Unfortunately, fashion production is one of the least sustainable industries. Therefore, more and more fashion companies are trying to implement green approaches in their value chains. Green marketing in the context of the fashion industry essentially refers to products that are produced with respect to working conditions and the environment, and to the possibility of finding a compromise between business objectives and environmental issues.
Keywords: green marketing; fashion industry; environment; sustainability; CSR

1 Introduction

With the public's growing interest in the environment since the 1960s and companies' interest in improving their image through their environmental efforts, marketing communication in the form of green marketing is increasingly being used to appeal to environmentally conscious consumers [1]. Companies have realised that consumers are beginning to make environmentally conscious purchases. The increasing interest of consumers, politicians, investors and companies in the environment has led to a greening of business and the emergence of a new marketing concept known as 'green marketing'. It is marketing's response to the environmental impact of the “design, production, packaging, use and disposal of products” [2]. While in the initial phase of the development of 'ecological' green marketing, the focus was concentrated on specific environmental problems such as air and water pollution, excessive use of natural resources and dependence on products that are partially harmful to the environment. In the second phase of the development of 'environmental' green marketing, marketers focused more on reducing environmental damage through green consumer demand and competitive advantages for companies in the market. In the third phase of 'sustainable' green marketing, thanks to increased consumer awareness of global issues and the links between the environment, society and the economy, marketing targets sustainability and reflects the total environmental cost in product prices to create a sustainable economy and society [3]. There is a similar, but motivationally different classification and designation of the concept as environmental marketing, within which three different approaches emerged - ecological, green and sustainable marketing. The motivations for ecological marketing are moral impulses and a positive attitude towards the environment; green marketing is motivated by market pressure and sustainable marketing is motivated by the companies' long-term perspective on value preservation and profits, seeking a compromise between business objectives and environmental issues [4]. While environmental sustainability has taken an important role on the political agenda, it is also driving innovation in business. The number of environmentally friendly products that are available on the market is increasing and consumers are showing a growing interest in buying them and are willing to pay a higher price [5].

On the other hand, pseudo-green companies are entering the market to gain a competitive advantage through greenwashing practices without real concern and efforts for environmental protection [6]. Companies that engage in greenwashing attempt to convince consumers that the company or its products are environmentally friendly, when in fact they have a poor environmental performance, while at the same time communicating their good environmental performance. Such behaviour harm the consumer and investor trust in companies with products that actively care for the environment [7], leads to consumer confusion [8] and has a negative impact on green purchases, green brand image and loyalty even among environmentally conscious consumers [9,10]. Moreover, greenwashing is widespread in the fashion and textile industry [11]. These industries are also questionable in terms of sustainability and corporate social

* Corresponding author: sagapova@mail.vstecb.cz

responsibility (CSR) as such [12] and have been criticised for exploiting human and natural resources, environmental degradation, pollution and waste [13].

From an existential perspective, sustainability and sustainable production and consumption is no longer a luxury or a choice, but a necessity [14]. Therefore, this paper aims to obtain evidence of problematic consequences of the fast fashion industry in the sustainability context and to review the option of the green marketing approach as a way to reduce the negative impacts.

2 Methodology

This study uses a narrative literature review to gather available evidence of the consequences of the fashion industry in the sustainability context and its common practices and findings related to green marketing approaches. This method is preferred due to its possibility to find enough information, including the criticism, on a given topic in a relatively short time framework. Another reason for choosing this method is the perceived lack of information about the reality and the dark side of this industry within the scientific publications. To find relevant research studies, we defined the search topics. Based on previous expert opinions and background information, we start with the introduction of fast fashion and socio-economic, ethical and environmental contexts in the modern fashion industry. Then we continue with green marketing and sustainability in fast fashion to get a complete and perhaps critical overview of the issue and find possible gaps in current research and understanding of environmental awareness and moreover sustainability in this industry. Studies relevant to our research were identified through a search of Web of Science, Science Direct, Scopus and Google Scholar. The search period was not limited.

3 Results and Discussion

3.1 Fast fashion

The fashion industry has changed its dynamics in recent decades. The modern clothing market is dominated by 'fast fashion', which drives new trends, increases the number of fashion seasons and frequency of consumption, shortens the usability of items with no long-term value, and establishes offshore supply chains with cheap manufacturing to deliver large quantities of clothing quickly and at low cost to be priced competitively [15-18]. In the last two decades, fashion supply chains have shifted to developing countries, recently from Asia to Africa, promising socio-economic development to these countries, but ultimately fast fashion comes with a higher cost [19]. Due to production problems, and other economic, social, ethical and environmental impacts, companies are trying to promote and ensure sustainability in their value chains to improve performance, set sustainability criteria and conduct monitoring and auditing activities [20]. However, the audits are often ineffective, even fictitious, and do not reveal the actual unsatisfactory conditions [21,22].

3.1.1 Fast fashion and its socio-economic, ethical and environmental context

While the market pushes the industry to make clothes even cheaper, to cut corners, this has consequences in the form of declining product quality, and negative social and environmental effects [23]. The entire life cycle of inexpensive clothing involves risks to people and the environment, who bear the costs [24]. The key element for low-cost fashion production is cheap labour, which can be found in developing countries [20]. Cheap labour is associated with low wages and poor working conditions [24]. Extensive violations of workers' and human rights, long working hours, child labour and exposure to chemicals are commonplace in this industry [25]. Large companies are in a position to cut corners to make the greatest possible profit [26]. In this context, the global structure of industry almost forces companies to cut production costs by squeezing out suppliers, moving production abroad and contracting it out to workshops that also employ informal workers to gain a competitive advantage, but in this case, this international competition is more of a race to the bottom [27]. The textile and fashion industry is often criticised for involving practices of modern slavery [28]. Shifting production to developing countries lowers costs not only because of cheap labour and exploitation of human rights but also because of the lack of adequate social, environmental, safety and other legal standards regulating the fashion sector, making it the least regulated and least sustainable global industry [29,30]. The textile and fashion industry is one of the largest employers, but also the second-largest polluter after the oil industry [31]. The dyeing and processing of textiles produce a lot of heavily polluted wastewater, the discharge of which into the river is prohibited for environmental reasons [32]. However, in developing countries like Bangladesh, these polluted effluents are also recently discharged into rivers and streams to save the high costs of treatment chemicals and energy-intensive technologies [33]. This also leads to the pollution of surface water, groundwater and soils, causing significant negative externalities and affecting the lives of poor people and the livelihoods of farmers who own and cultivate polluted soils that lose value and fertility [34]. In contrast, cotton-based textiles require large amounts of water and pesticides, while polyester, for example, is made from petroleum [24]. Our clothing also releases chemicals and micro textile waste, and in the case of synthetic materials, microplastics can eventually enter the food chain. Due to the throwaway culture of fast fashion, over 2/3 of our textiles end up in landfills [35]. The fast fashion industry is obviously very wasteful and far from being sustainable or closed-loop, and clothing itself has gone from being a durable good to a daily purchase thanks to the fast fashion production system and business model [36]. One of the characteristics of the fast fashion industry is the limited protection of intellectual property, which often tempts designers to copy the designs and products of others [37].

Moreover, various accidents related to factories, such as the fire at the Tazreen factory in 2012 or the collapse of the Rana Plaza building in 2013. It was built without permits and serious examination of the building's statics and yet produced clothes for 29 mostly world-renowned brands. Even though a huge crack was discovered a day before the collapse, all people, mostly women and children, were forced to stay at their workplaces. Not just these tragedies have raised concerns about the functioning of the fast fashion system. Some other factories were later closed because of their poor construction. In addition to such accidents in industry, there have also been cases of unpaid wages for workers who work in factories producing goods for brands such as Zara, Next and Mango [38]. While fast fashion allows products to be delivered to retailers in a matter of weeks to meet consumer demand for new garments, overconsumption, industry practices and growing awareness of environmental and social consequences are providing the impetus for new ways of producing and consuming fashion [39]. The counterpart to fast fashion is so-called slow fashion, which promotes sustainable production and consumption of fashion [40]. However, fast fashion companies have also started to launch sustainable collections to increase the legitimacy of the company, the perception of CSR and trust in the brand in the eyes of customers [41]. Policymakers could promote sustainability in fast fashion through interventions, taxes or subsidies, as awareness of sustainability does not automatically lead to sustainable economic behaviour; customers should be more aware and update their knowledge so that they are not misled by false marketing communication from different fast fashion brands [42].

3.2 Green marketing and sustainability in fast fashion industry

Sustainability and ethics in the context of the fashion industry essentially refer to products that are manufactured with respect for working conditions and the environment, which can be a part of CSR [43]. The responsibility for sustainable fashion lies not only with the fashion industry, which must create conditions for sustainable production and remanufacturing, green marketing and green information exchange but also with fashion consumers, who can have a green attitude and improve their awareness, knowledge and education [42]. Marketing should provide a basis to encourage consumers to make sustainable purchasing decisions for longer-lasting quality clothing made from appropriate materials in appropriate production processes [44]. Yet, fast fashion is characterised by an increase in clothing consumption and a decrease in the trend cycle. It is also typical that consumers from the middle class buy these clothes, which are fashionable and cheap at the same time. However, the empirical results show the general trend that consumers have a greater awareness of environmental issues [45]. Unfortunately, consumers may not be aware of the costs of cheap fashion and the excesses of fast fashion need to be highlighted and communicated by policymakers, industry players and various stakeholders to guide consumers towards more sustainable fashion that fast fashion and its production, sourcing, design, marketing and consumption [37]. If sustainability is a goal, the fast fashion industry needs to change its practices and integrate marketing strategies that encompass the entire marketing mix (4 P's), resulting in reshaping the industry, reducing consumption, frequency of purchases, setting higher prices, less discounting, fewer collections per year, limiting the number of physical shops, extending the life of products through higher quality, active promoting of sustainability and value to reduce overconsumption and lower environmental impact [14]. Currently, consumer behaviour may also be influenced to some extent by the COVID 19 global pandemic. Fast fashion retailers use e-commerce solutions as their predominant business model at the time of the pandemic, so fast fashion clothing is easily available despite the situation [46]. The global impact of the fast fashion industry in terms of carbon emissions is enormous. Fast fashion production has doubled this century, resulting in a 30% increase in carbon emissions. This may suggest that the sector is becoming greener to some extent, but an absolute improvement in the environmental impact of fashion could be achieved by eliminating fast fashion as a business model [47].

The most common research is on the proper strategy of fast fashion companies. The aim is mainly to maximise profits or sales. An effective marketing strategy that could be explored by fast fashion marketers is linked to green marketing. As studied in Indonesia, companies that engage in green marketing and innovate their products can influence customers' willingness to make indirect purchases in the future through the company's image and CSR. From the consumers' perspective, the results show that CSR is the most important factor motivating consumers to make purchases as they are increasingly interested and concerned about the environment and sustainability [48]. Green marketing in fashion and claims such as organic or recycled fibres, sustainable manufacturing, donations to green organisations, fair trade and labour practices can increase companies' profits by reducing consumers' price sensitivity and increasing their willingness to pay [49]. However, based on research that focused on specifying the marketing strategy of fast fashion retailers in relation to consumer attitudes towards the sustainability of fast fashion products in the UK, one consumer segment was defined as not receptive to fast fashion products with sustainability attributes - the working-class female in the United Kingdom. Furthermore, this is a large segment in the UK. It was also found that among consumers of fast fashion products, their claimed knowledge of sustainability is likely to be higher than their actual knowledge because they are overconfident [42]. Fast fashion merchandisers also frequently enter into partnerships with non-governmental organisations (NGOs). They use this strategy mainly to mitigate negative reactions from environmentally-conscious consumer groups. The effects of such partnerships may be beneficial at first glance, but there is a good chance that they will fail and harm both parties. Fast fashion companies pursue a business model that aims to maximise profits, while NGOs cater to the higher social and environmental demands of the public. The partnership between fast fashion companies and NGOs has the advantage of increasing adaptability to change, flexibility, innovation and securing their own organisational identity. Obstacles may lie mainly in the area of possible opportunism and lack goal orientation [50].

Consumer behaviour in fast fashion from a sustainability perspective can also be looked at from the other side, i.e., the consumer side. The sustainable characteristics of products positively influence consumer behaviour in the fast fashion sector. On the other hand, consumer behaviour is more likely to be associated with positive implicit attitudes when sustainability is paired with luxury than with fast fashion. It was also confirmed that sustainability influences a higher level of perceived warmth. This also has a positive effect on consumer behaviour towards fast fashion products [51]. Some sustainable marketing activities play a key role in constructing a positive image of fast fashion and brand image. These might be economic, environmental, cultural and social activities, which can through improved image help the brand to survive on the market [52]. With luxury clothing, one can also observe the cultural environment in which the retailer operates. Sustainable communication for non-luxury apparel brands is more effective in a cultural environment characterised by higher awareness related to sustainability [53]. Similarly, environmentally conscious consumers pay more attention to content-based than associative environmental claims in advertising and respond more to communications from luxury brands than fast fashion brands, even though both luxury and fast fashion products are unsustainable and harmful to the environment [54].

4 Conclusion

The fashion industry is undoubtedly a key industry and one of the largest employers worldwide, but on the other hand, it is also one of the least sustainable industries in the world. The relocation of production to developing countries due to cost-cutting in many cases leads to the exploitation of human rights and the lack of adequate social, environmental, safety and other legal standards. Furthermore, the fashion industry is very wasteful and produces a lot of heavily polluted wastewater, so it is far from being an example of a sustainable industry. The reasons for this are obvious. Fast fashion dominates the clothing market of the 21st century. Fast fashion means increasing consumption, shortening the useful life of items supplied in large quantities and reducing manufacturing costs to increase price competitiveness.

In recent years, environmental sustainability has played an important role in the political agenda, but it is also driving innovation in business. The supply of environmentally friendly products on the market is increasing and consumers are showing interest in buying these types of products, which tend to be higher priced. Due to the aforementioned negative impacts, more and more fashion companies are trying to implement green marketing approaches in their value chains to improve performance and meet desired sustainability criteria. Green marketing in the context of the fashion industry essentially refers to products that are manufactured with respect to working conditions and the environment, which can be a part of CSR to find a compromise between business objectives and environmental issues. The sustainable approaches in the fashion industry are not only the responsibility of the main companies but also of the consumers, who should improve their awareness, knowledge and education on green attitudes. Effective green marketing should lay the foundation for changing clothing consumption and encourage consumers to make sustainable purchasing decisions for longer-lasting quality clothing made from appropriate materials in appropriate production processes.

References

1. S. Szabo, J. Webster. Perceived Greenwashing: The Effects of Green Marketing on Environmental and Product Perceptions. *Journal of Business Ethics*. **171**, 719-739 (2021)
2. M. Lampe, G. M. Gazda. Green marketing in Europe and the United States: An evolving business and society interface. *International Business Review*. **4**(3), 295-312 (1995)
3. K. Peattie. Towards sustainability: the third age of green marketing. *The Marketing Review*. **2**, 129-146 (2001)
4. Y. K. van Dam, P. A. X. Apeldoorn. Sustainable Marketing. *Journal of Macromarketing*. **16**, 56-56 (1996)
5. R. M. Dangelico, D. Vocalelli. "Green Marketing": An analysis of definitions, strategy steps, and tools through a systematic review of the literature. *Journal of Cleaner Production*, **164**, 1263-1279 (2017)
6. D. Jog, D. Singhal. Pseudo green payers and their greenwashing practices: a differentiating strategy for real green firms of personal care category. *Strategic Direction*. **35**(12), 4-7 (2019)
7. M. A. Delmas, V. C. Burbano. The Drivers of Greenwashing. *California Management Review*. **54**(1), 64-87 (2011)
8. Y.-S. Chen, C.-H. Chang. Greenwash and Green Trust: The Mediation Effects of Green Consumer Confusion and Green Perceived Risk. *Journal of Business Ethics*. **114**, 489-500 (2013)
9. I. Hameed, Z. Hyder, M. Imran. Greenwash and green purchase behavior: an environmentally sustainable perspective. *Environment, Development and Sustainability*. **23**, 13113-13134 (2021)
10. E.-J. Lee, H. Choi, J. Han, D. H. Kim, E. Ko, K. H. Kim. How to "Nudge" your consumers toward sustainable fashion consumption: An fMRI investigation. *Journal of Business Research*. **117**, 642-651 (2020)

11. G. Kaner. Greenwashing: How Difficult It Is to Be Transparent to the Consumer – H&M Case Study. In: C. Mukonza, R. E. Hinson, O. Adeola, I. Adisa, E. Mogaji, A. C. Kirgiz. *Green Marketing in Emerging Markets*. Cham: Palgrave Macmillan (2021)

12. D. Zimon, P. Madzik. The Influence of ISO 9001 & ISO 14001 on Sustainable Supply Chain Management in the Textile Industry. *Sustainability*. **12**(10), 4282 (2020)

13. C. D´Souza, A. J. Gilmore, P. Hartmann, V. A. Ibáñez, G. Sullivan-Mort. Male eco-fashion: a market reality. *International Journal of Consumer Studies*. **39**(1), 35-42 (2015)

14. R. Ciornea. Is the Transition to Bioeconomy a Sustainable Solution in Fast-fashion Industry, Considering the Overconsumption? – Premises for Future Research. *MID Journal*. **3**(1), 27-44 (2020)

15. V. Bhardwaj, A. Fairhust. Fast fashion: response to changes in the fashion industry. *The International Review of Retail, Distribution and Consumer Research*. **20**(1), 165-173 (2010)

16. S. A. Doyle, C. M. Moore, L. Morgan. Supplier management in fast fashion moving fashion retailing. *Journal of Fashion Marketing and Management*. **13**(3), 272-281 (2006)

17. K. Niinimäki, G. Peters, H. Dahlbo, P. Perry, T. Rissanen, A. Gwilt. The environmental price of fast fashion. *Nature Reviews Earth & Environment*. **1**, 189-200 (2020)

18. A. Joy, J. F. Sherry Jr., A. Venkatesh, J. Wang, R. Chan. Fast Fashion, Sustainability, and the Ethical Appeal of Luxury Brands. *Fashion Theory*. **16**(3), 273-295 (2012)

19. K. Khurana, S. S. Muthu. Are low- and middle-income countries profiting from fast fashion? *Journal of Fashion Marketing and Management*. doi:org/10.1108/JFMM-12-2020-0260 (2021)

20. D. Turker, C. Altuntas. Sustainable supply chain management in the fast fashion industry: An analysis of corporate reports. *European Management Journal*. **32**(5), 837-849 (2014)

21. A. F. Borthick, M. B. Curtis. Due Diligence on Fast-Fashion Inventory through Data Querying. *Journal of Information Systems*. **22**(1), 77-93 (2008)

22. A. G. Woodside, M. B. Fine. Sustainable fashion themes in luxury brand storytelling: The sustainability fashion research grid. *Journal of Global Fashion Marketing*. **10**(2), 111–128 (2019)

23. C. E. Henninger, M. Blazquez, R. Boardman, C. Jones, H. McCormick, S. Sahab. Cradle-To-Cradle Versus Consumer Preferences. In: The Fashion Industry. In S. Hashmi, A. I, Choudhury. *Encyclopedia of Renewable and Sustainable Materials*. Oxford: Elsevier (2020)

24. R. Bick, E. Halsey, C. C. Ekenga. The global environmental injustice of fast fashion. *Environmental Health*. **17**, 92 (2018)

25. J. Hobson. To die for? The health and safety of fast fashion. *Occupational Medicine*. **63**(5), 317-319 (2013)

26. M. N. Cruz. Is Fast Fashion a Concern of the Everyday Teen? *Student Research Symposium*. **8** (2017)

27. J. M. Bressán. From Neoliberal Fashion to New Ways of Clothing. *Socialist Register 2021*. **57**, 201-217 (2021)

28. A. V. Benstead, L. C. Hendry, M. Stevenson. Horizontal collaboration in response to modern slavery legislation: An action research project. *International Journal of Operations & Production Management*. **38**(12), 2286-2312 (2018)

29. X. Cherny-Scanlon, K. Agnes, Time for a Responsible Fashion Council? In S. Muthu, M. Gardetti (Eds.) *Green Fashion: Environmental Footprints and Eco-design of Products and Processes* (123-150), Springer, Singapore, (2016).

30. A. Clarke-Sather, K. Cobb. Onshoring fashion: Worker sustainability impacts of global and local apparel production. *Journal of Cleaner Production*. **208**, 1206-1218 (2019)

31. A. K. Pattanayak. Sustainability in fabric manufacturing. In: R. Nayak. *Sustainable Technologies for Fashion and Textiles*. Amsterdam: Elsevier (2020)

32. S. Sondhi. Sustainable approaches in effluent treatment: Recent developments in the fashion manufacturing. In: R. Nayak. *Sustainable Technologies for Fashion and Textiles*. Amsterdam: Elsevier (2020)

33. N. R. Khandaker, I. Afreen, D. S. Diba, F. B. Huq, T. Akter. Treatment of textile wastewater using calcium hypochlorite oxidation followed by waste iron rust aided rapid filtration for color and COD removal for application in resources challenged Bangladesh. *Groundwater for Sustainable Development*. **10**, 100342 (2020)

34. T. S. Devi, C. Ramasamy, S. Gurunathan. Farmers' Willingness to Pay to Avert the Negative Externalities of Pollution of Dyeing Industry in Tamil Nadu. *Agricultural Economics Research Review*. **23**(2), 245-252 (2010)

35. K. Shirvanimoghaddam, B. Motamed, S. Ramakrishna, M. Naebe. Death by waste: Fashion and textile circular economy case. *Science of The Total Environment*. **718**, 137317 (2020)

36. T. Brydges. Closing the loop on take, make, waste: Investigating circular economy practices in the Swedish fashion industry. *Journal of Cleaner Production*. **293**, 126245 (2021)

37. M. K. Brewer. Slow Fashion in a Fast Fashion World: Promoting Sustainability and Responsibility. *Laws*. **8**, 24 (2019)

38. T. Sádaba, P. SanMiguel, P. Gargoles. Communication Crisis in Fashion: From the Rana Plaza Tragedy to the Bravo Tekstil Factory Crisis. In: N. Kalbaska, T. Sádaba, T. Cominelli, F. Cantoni. *Fashion Communication in the Digital Age*. Switzerland: Springer (2019)

39. A. Buzzo, M. J. Abreu. Fast Fashion, Fashion Brands & Sustainable Consumption. In: S. S. Muthu. *Fast Fashion, Fashion Brands and Sustainable Consumption*. Singapore: Springer (2019)

40. T. Chi, J. Gerard, Y. Yu, Y. Wang. A study of U.S. consumers' intention to purchase slow fashion apparel: understanding the key determinants. *International Journal of Fashion Design, Technology and Education*. **14**(1), 101-112 (2021)

41. G. Miotto, S. Youn. The impact of fast fashion retailers' sustainable collections on corporate legitimacy: Examining the mediating role of altruistic attributions. *Journal of Consumer Behaviour*. **19**(6), 618-631 (2020)

42. B. Zhang, Y. Zhang, P. Zhou. Consumer Attitude towards Sustainability of Fast Fashion Products in the UK. *Sustainability*. **13**, 1646 (2021)

43. B. Shen, Y. Wang, C. Lo, M. Shum. The impact of ethical fashion on consumer purchase behaviour. *Journal of Fashion Marketing and Management*. **16**, 234-245 (2012)

44. H. Goworek, L. Oxborroe, S. Claxton, A. McLaren, T. Cooper, H. Hill. Managing sustainability in the fashion business: Challenges in product development for clothing longevity in the UK. *Journal of Business Research*. **117**, 629-641 (2020)

45. J. Jang, E. Ko, E. Chun, E. Lee. A Study of a Social Content Model for Sustainable Development in the Fast Fashion Industry. *Bridging Fashion and Marketing*. **3**(2), 61-70 (2012)

46. K. Bilińska-Reformat, A. Dewalska-Opitek. E-commerce as the predominant business model of fast fashion retailers in the era of global COVID 19 pandemics. *Procedia Computer Science*. **192**, 2479-2490 (2021)

47. G. Peters, M. Li, M. Lenzen. The need to decelerate fast fashion in a hot climate - A global sustainability perspective on the garment industry. *Journal of Cleaner Production*. **295**, 126390 (2021)

48. G. D. Tulangow, K. A. Kusumawardani. Green is the new black: the role of green marketing awareness and perceived innovation in the fast fashion industry. *The 4th International Conference on Family Business and Entrepreneurship* [online]. Available at: https://e-journal.president.ac.id/presunivojs/index.php/ICFBE/article/viewFile/1373/746 (2020)

49. C. Stall-Meadows, A. Davey. Green marketing of apparel: consumers' price sensitivity to environmental marketing claims. *Journal of Global Fashion Marketing*. **4**(1), 33-43 (2013)

50. S. Y. H. Liu, E. Napier, A. Runfola, S. T. Cavusgil. MNE-NGO partnerships for sustainability and social responsibility in the global fast-fashion industry: A loose-coupling perspective. *International Business Review*. **29**(5), 101736 (2020)

51. L. Grazzini, D. Acuti, G. Aiello. Solving the puzzle of sustainable fashion consumption: The role of consumers' implicit attitudes and perceived warmth. *Journal of Cleaner Production*. **287**, 125579 (2021)

52. J. Jung, S. J. Kim, K. H. Kim. Sustainable marketing activities of traditional fashion market and brand loyalty. *Journal of Business Research*. **120**, 294-301 (2020)

53. H. M. Kong, A. Witmaier, E. Ko. Sustainability and social media communication: How consumers respond to marketing efforts of luxury and non-luxury fashion brands. *Journal of Business Research*. **131**, 640-651 (2021)

54. T. Gogichaishvili, K. Enju, J. K. Sang. Environmental claims in online video advertising: effects for fast-fashion and luxury brands. *International Journal of Advertising*. **39**(6), 1–30 (2019)

Employee Engagement as a Factor of Internal Entrepreneurship

Marina Shavrovskaya[1,*], *Oksana* Borodina[2]

[1]Ural State University of Economics, Department of Labour Economics and HR Management, 8 Marta 62, 620144 Yekaterinburg, Russia
[2]Omsk State University of Economics, Department of Labour Economics, Mira 55-a, 644077 Omsk, Russia

Abstract. In recent years, there has been a growing interest in the topic of developing employee engagement both in the scientific community and in business. Managers are beginning to realize that not only the technologies used, but also the personnel who can be involved in work, submit ideas on how to improve it, or, conversely, silently carry out job descriptions, which can help to occupy leading positions. The purpose of this study is to analyze publications on employee engagement from 2017-2021, as well as the subjects of their study and research. The main hypothesis is that scientists are interested in various aspects of staff involvement, but of particular interest is how it is interconnected with the effectiveness of the organization. The results of this review form the basis for further discussion on approaches and tools for developing engagement as a factor of internal entrepreneurship.
Keywords: entrepreneurial competence; employee engagement; organizational effectiveness; intra-entrepreneurial behavior; internal entrepreneurship

1 Introduction

The topic of engagement is actively discussed in scientific circles and in business. So, an employee may like working in an organization, but it is not at all necessary that he will be ready to think about how to improve the organization and what to change in specific business processes. Therefore, engagement is not just job satisfaction, it is related to business performance and focuses on the types of behavior that contribute to achieving business results. So, it is proved in a study by Schneider, et al. that employee engagement significantly predicts the financial performance of an organization [1]. According to the approach of B. Shuck, J. L. Adelson, T. G. Reio Jr. the employee engagement scale consists of three sub-factors (cognitive, emotional and behavioral) and a higher-order factor (employee engagement) [2]. Various questionnaires are used to measure employee engagement, for example:

- Gallup (12 questions in the engagement questionnaire) [3]
- AON Hewitt (24 questions) [4]
- Hay Group (database of 200 questions, you can choose the suggested questions to create your survey) [3].

The authors analyze the relationship between performance measurement, performance management, employee engagement and productivity [5]. Min-Seong Kim, Dong-Woo Koo indicate that involvement in work significantly affects the innovative behavior of employees [6]. There are some studies showing a positive correlation between the intra-entrepreneurial behavior of employees, personal resources and involvement in work [7].

HR-managers play an important role in the development of both engagement and internal entrepreneurship [8], it is important that they increase employee engagement and productivity by initiating the inclusion of these indicators in performance studies [9]. Although, of course, it is worth noting that managers should help to develop employee engagement. As a result of the research, Min-Seong Kim, Dong-Woo Koo conclude that the direct supervisor plays a crucial role in encouraging engagement, innovative behavior and productivity [6].

A number of authors investigate the relationship between CSR and engagement, showing that the former contributes to an increase in the latter [10, 11].

Thus, the issues of engagement are analyzed from different sides, the authors associate it with the efficiency of the organization and productivity improvement, and one of the important roles in this is played by the development of internal entrepreneurship of employees.

* Corresponding author: marina_bel@list.ru

2 Methodology

This article provides a bibliographic analysis on the topic of employee engagement and internal entrepreneurship with an emphasis on publications from 2017 to 2021 in two databases: Web of Science and RSCI (eLibrary). During analyzing the publications, the definition of "employee engagement" was used to find its presence in keywords. The semantic analysis took into account the number of citations: RSCI - 5≤ citations≤13, WoS - 5≤ citations≤25. The authors also analyzed 30 of the most cited publications on the topic of engagement indexed in the databases WoS (25 publications) and RSCI (5 publications) on the subjects of research.

3 Results and Discussion

In total, 405 scientific publications have been published on the topic of employee engagement in the RSCI, 296 publications (59%) and 13 publications of them with a citation ≤5 have been published since 2017. There are 1256 publications on the topic of employee engagement in the WoS database, 874 (70%) during the study period.

Figure 1 shows a semantic analysis of publication keywords based on 13 publications of the RSCI (5≤ citations≤13) and 50 publications of the WoS database (5≤ citations≤25).

Figure 1. Word cloud of publication keywords on the topic of employee engagement

Source: Author.

It can be seen in Fig. 1 of the word cloud, there are not only synonyms of the concept of " engagement " such as loyalty, commitment, involvement, interest, efficiency, but also antonyms — detachment, indifference, burnout. This indicates the study of this concept from different sides, namely, researchers are looking for answers to such questions: how to analyze the level of engagement, how to involve staff in business processes, how to maintain the achieved level of involvement and thereby develop the company.

In addition, the frequent use of keywords (in addition to the search words "employee engagement") such as efficiency (productivity), communication, social responsibility, motivation and burnout, allow us to say that engagement has two aspects:

- basic competencies (personal characteristics) of employees which influence interest in their profession, passion for the process of activity, the desire to find meaning and pleasure from their work, involvement, initiative and personal interest (not only material) and responsibility for the results of their division and the company as a whole;

– what the company offers to the employee: a corporate culture aimed at a friendly atmosphere, top management that enjoys respect and trust, various tools that support initiative, and working conditions that affect job satisfaction.

Thus, two types of involvement can be distinguished:

1. Involvement in work – love for the work

2. Involvement in the company - association of oneself with the company

When a person loves what he does, but at the same time does not associate himself with the organization, sometimes it can bring good results. And if teamwork is important, this option can give a negative result.

Table 1 shows the distribution of the 30 most cited publications on the topic of engagement indexed in the WoS (25 publications) and RSCI (5 publications) databases by the subjects of research.

Table 1. The subject of research in 30 most cited publications on the topic of employee engagement

Subject of research	Total works	Average number of citations	Most cited publications
Research in the relationship of engagement with labor productivity and efficiency	7	60,6	B. Smith, 2017 [5] Min-Seong Koo, et al, 2017 [6] V. Wingerden, et al, 2017 [12] Fang Lee Cooke, et al, 2019 [13] H. Gordon, et al, 2018 [14] N. V. Gromova, 2018 [15] G. Reijseger, et al, 2017 [16]
Engagement research in particular companies	6	19,7	I. V. Rozdolskaya, et al, 2018 [17] Z. V. Yakimova, et al, 2018 [18] U. S. Borisova, 2017 [19] J. Milliman, et al, 2018 [20] Z. Xin, et al, 2019 [21] J. Meijerink, et al, 2020 [22]
Communication	6	44,5	K. Minjeong, et al, 2017 [23] K. Ruck, et al, 2017 [24] O. M. Karatepe, et al, 2018 [25] P. Holland, et al, 2016 [26] A. Ivanova, 2018 [27] J. A. Walden, et al, 2017 [28]
Others (burnout, well-being, CSR, engagement measurement, zero waste and engagement)	11	64,6	B. Shuck, 2018 [2] J. S. Gawke, et al, 2017 [7] V. Veleva, et al, 2017 [29] M. Willoughby, et al, 2017 [30] O. M. Karatepe, et al, 2020 [31] W. B. Schaufeli, et al, 2017 [32] C. Flammer et al, 2017 [10] T. Farid, et al, 2019 [33] G. Duthler, et al, 2018 [11] T. Lesener, et al, 2019 [34] R. Chaudhary, 2017 [35]

Source: Author.

7 articles out of 30 are devoted to the study of the relationship between engagement and work efficiency or organization, as well as labor productivity. The average number of citations of articles is 60.6, which indicates an interest in this topic in the scientific community. These issues, one way or another, are related to internal entrepreneurship, the level of which determines the ability to offer and generate ideas for improving existing business processes, as well as creating new products and services, which, in turn, contribute to efficiency growth. In the RSCI database, articles are mainly devoted to the study of employee engagement in a particular organization and the number of citations is quite small. A number of scientists study the relationship between engagement and internal communication (6 articles out of 30 with an average number of citations of 44.5). In addition, the authors are interested in the relationship between engagement and the well-being of staff, CSR, as well as the measurement of engagement and how to involve employees in "zero waste".

During the analysis of the research, the authors identified a number of interesting discussion points that are important for understanding engagement. For example, there is an opinion that employee engagement is a tool to improve the efficiency of the company [36], which is actively used in many organizations. And there is a position, and it is close to the authors, that employee engagement is a management philosophy that forms a culture of participation and helps to involve employees in continuous improvement and increases their personal interest in the success of the organization [17]. It is noteworthy that despite the fact that the formation of a corporate culture that promotes employee engagement is a complex and lengthy process, the approaches in the tools differ significantly.

Employee engagement should not be viewed through the prism of ideology, as passion. Enthusiasm is easier to create than engagement, but working with emotions gives a short-term effect. True engagement is formed at the expense of values, not agitation, and provides the company with a long-term competitive advantage. Among the recommendations for increasing staff engagement, the following can be highlighted:

- hire already involved candidates with "burning eyes"
- set specific goals for employees that correlate with the success of the organization
- be generous not only with bonuses, but also with praise
- form, first of all, the involvement of managers who, by their example, will involve others in the company's business processes
- form positive emotions and communication in the team
- develop employees, give them the opportunity to feel their importance to the company.

It is worth noting that when measuring employee engagement, it is important to remember that despite the fact that companies strive to increase engagement rates, too high results may be based on negative factors, reflecting "social desirability". In other words, employees may answer questions under pressure or out of fear that they will let down the manager or the service. Taken into account this fact, it is necessary to carefully interpret the data from the engagement surveys.

5 Conclusion

The involvement of personnel in the work is the most important condition for the effective functioning of organizations of various types. The statement that a highly qualified specialist is the most valuable asset of the organization is beyond doubt. However, the effectiveness of an employee's work depends on many factors: his interest in working effectively, awareness of his value in the organization, the degree of satisfaction with the content of work, the corporate culture, prospects for personal and professional growth, the availability of working conditions and many other factors. Therefore, increasing the involvement of personnel in the work of the organization becomes the most important task of managers and the company as a whole. The high speed of technology change, the increase in the volume of information, the need for people (especially, the youth) to realize in existing conditions, as well as the need for opportunities to implement their ideas, requires the use of new approaches to human resource management.

The authors analyzed publications on employee engagement for the period of 2017-2021, as well as the topics of their study and research. We have confirmed the hypothesis that scientists are interested in various aspects of staff involvement, e.g. communication, social responsibility, motivation and burnout. The authors research how employee engagement is interrelated to the effectiveness of the organization. The probability that an involved employee will submit more ideas to improve business processes and become an internal entrepreneur is higher than an uninvolved one. But there are a number of issues that still need to be analyzed further, for example, whether it is possible to immediately accept involved employees, and how to form a culture that will support this involvement, or how to direct managers to work in this direction.

References

1. B. Schneider, A. B. Yost, A. Kropp, C. Kind, H. Lam. Workforce engagement: What it is, what drives it, and why it matters for organizational performance. *Journal of Organizational Behavior.* **39**(4), 462-480 (2018)
2. B. Shuck, J. L. Adelson, T. G. Reio Jr. The employee engagement scale: Initial evidence for construct validity and implications for theory and practice. *Human Resource Management.* **56**(6), 953-977 (2017)
3. A. A. Tokareva, S. G. Baronene. Methodology of research on the involvement of university employees. *University Management: Practice and analysis.* **23**, 1-2 (2019)
4. A. Hewitt. *Trends in Global Employee Engagement*[online]. Available at: https://www.business.com/images/content/58a/d9f712f87b1207f720dee/0-0-/ (2012)
5. M. Smith, U. S. Bititci. Interplay between performance measurement and management, employee engagement and performance. *International Journal of Operations & Production Management.* **37**(9), 1207-1228 (2017)
6. M.-S. Kim, D.-W. Koo. Linking LMX, engagement, innovative behavior, and job performance in hotel employees. *International Journal of Contemporary Hospitality Management.* **29**(12), 3044-3062 (2017)
7. J. S. Gawke, M. J. Gorgievski, A. B. Bakker. Employee entrepreneurship and work engagement: A latent change score approach. *Journal of Vocational Behavior.* **100**, 88-100 (2017)

8. A. Pesha, M. Shavrovskaya, O. Borodina. Development of Entrepreneurial Competencies of Employees within the Framework of internal Entrepreneurship Programs in the Region. *SHS Web of Conferences.* **90** (2021)

9. W. R. Carter, P. L. Nesbit, R. J. Badham, S. K. Parker, L.-K. Sung. The effects of employee engagement and self-efficacy on job performance: a longitudinal field study. *The International Journal of Human Resource Management.* **29**(17), 2483-2502 (2018)

10. C. Flammer, L. Jiao. Corporate social responsibility as an employee governance tool: Evidence from a quasi-experiment. *Strategic Management Journal.* **38**(2), 163-183 (2017)

11. G. Duthler, G. S. Dhanesh. The role of corporate social responsibility (CSR) and internal CSR communication in predicting employee engagement: Perspectives from the United Arab Emirates (UAE). *Public relations review.* **44**(4), 453-462 (2018)

12. J. V. Wingerden, D. Derks, A. B. Bakker. The impact of personal resources and job crafting interventions on work engagement and performance. *Human Resource Management.* **56**(1), 51-67 (2017)

13. F. L. Cooke, B. Cooper, T. Bartram, J. Wang, X. Mei. Mapping the relationships between high-performance work systems, employee resilience and engagement: A study of the banking industry in China. *The International Journal of Human Resource Management.* **30**(8), 1239-1260 (2019)

14. H. J. Gordon, E. Demerouti, P. M. Le Blanc, A. B. Bakker, T. Bipp, M. A. M. T. Verhagen. Individual job redesign: Job crafting interventions in healthcare. *Journal of Vocational Behavior.* **104** (2018)

15. N. V. Gromova. Employee engagement is the main reserve for improving the efficiency of modern companies. *Bulletin of the Russian University of Economics. G. V. Plekhanov.* **6** (2018)

16. G. Reijseger, M. C. W. Peeters, T. W. Taris, W. B Schaufeli. From motivation to activation: why engaged workers are better performers. *Journal of Business and Psychology.* **32**(2), 117-130 (2017)

17. I. V. Rozdolskaya, M. E. Ledovskaya, Y. A. Mozgovaya. Creating a culture of engagement as a method of motivating staff in the process of implementing the strategic goals of the organization. *Bulletin of the Belgorod University of Cooperation, Economics and Law.* **5** (2018)

18. Z. V. Yakimova, A. S. Pushkin. Dynamics of the level of staff involvement depending on the length of service in the organization. *Azimuth of scientific research: Economics and Management.* **22** (2018)

19. U. S. Borisova, L. N. Vasilyeva. Involvement of the organization's personnel: a sociological analysis. *Society: sociology, psychology, pedagogy.* **5** (2017)

20. J. Milliman, A. Gatling, J. Kim. The effect of workplace spirituality on hospitality employee engagement, intention to stay, and service delivery. *Journal of Hospitality and Tourism Management.* **35**, 56-65 (2018)

21. Z. Xin, M. Liang, X. Feng, X. Bo. How social media usage affects employees' job satisfaction and turnover intention: An empirical study in China. *Information & Management.* **56**(6) (2019)

22. J. Meijerink, A. Bos-Nehles, J. de Leede. How employees' pro-activity translates high-commitment HRM systems into work engagement: The mediating role of job crafting. *The International Journal of Human Resource Management,* **31**(22), 2893-2918 (2020)

23. K. Minjeong, S. Minjung. How symmetrical employee communication leads to employee engagement and positive employee communication behaviors: The mediation of employee-organization relationships. *Journal of Communication Management.* **21**(1), 82-102 (2017)

24. K. Ruck, M. Welch, B. Menara. Employee voice: an antecedent to organizational engagement? *Public Relations Review.* **43**(5), 904-914 (2017)

25. O. M. Karatepe, A. Ozturk, T. T. Kim. The effects of organizational and personal resources on stress, engagement, and job outcomes. *International Journal of Hospitality Management.* **37**(4), 858-879 (2018)

26. P. Holland, B. Cooper, C. Sheehan. Employee voice, supervisor support, and engagement: The mediating role of trust. *Human Resource Management.* **56**(6) (2016)

27. A. Ivanova. Research of Corporate portal resources in Personnel Engagement Management. *Human Resources and Intellectual Resources management in Russia.* (2018)

28. J. A. Walden, E. H. Jung, C. Y. K. Westerman. Employee communication, job engagement, and organizational commitment: A study of members of the Millennial Generation. *Journal of Public Relations Research.* **29**(5), 1-17 (2017)

29. V. Veleva, G. Bodkin, S. Todorova. The need for better measurement and employee engagement to advance a circular economy: Lessons from Biogen's "zero waste" journey. *Journal of Cleaner Production.* **154**, 517-529 (2017)

 https://doi.org/10.1051/shsconf/202213501012

30. M. Willoughby, M. Parsons, P. Boxall, G. Cheung. Factors predicting Registered Nurses' intentions to leave their organization and profession: A job demands-resources framework. *Journal of advanced nursing.* **74**(5) (2017)

31. O. M. Karatepe, H. Rezapouraghdam, R. Hassannia. Job insecurity, work engagement and their effects on hotel employees' non-green and nonattendance behaviors. *International Journal of Hospitality Management.* **87** (2020)

32. W. B. Schaufeli, et al. An Ultra-Short Measure for Work Engagement The UWES-3 Validation Across Five Countries. *European Journal of Psychological Assessment* (2017)

33. T. Farid, S. Iqbal, S. Castro-González, J. Ma. Employees' perceptions of CSR, work engagement, and organizational citizenship behavior: The mediating effects of organizational justice. *International Journal of Environmental Research and Public Health.* **16**(10), 1731 (2019)

34. T. Lesener, G. Burkhard, C. Wolter. The job demands-resources model: A meta-analytic review of longitudinal studies. *Work & Stress.* **33**(1), 76-103 (2019)

35. R. Chaudhary. Corporate social responsibility and employee engagement: can CSR help in redressing the engagement gap? *Social Responsibility Journal.* (2017)

36. S. Marisa, U. S. Bititci. Interplay between performance measurement and management, employee engagement and performance. *International Journal of Operations & Production Management.* **37**(9), 1207-1228 (2017)

Proactivity as a Factor of Satisfaction with Distance Learning during Lockdown from the Perspective of Students' Perception and Preferences

Nadezhda Sivrikova[1,*], *Tatyana* Ptashko[1], *Elena* Chernikova[1], *Artem* Perebeynos[1], *Elena* Kharlanova[1] and *Fedor* Ivanov[1]

[1]South Ural State Humanitarian Pedagogical University, Department of Social Work, Pedagogy and Psychology, Chelyabinsk, Russian Federation, Lenin prospect 69, Russia

Abstract. During the first wave of COVID-19, universities in many states switched to distance learning. Students faced the challenge of moving quickly from face-to-face to distance learning. This created the prerequisites for analysing this new experience for Russian universities. There are many studies analysing satisfaction with online learning. Nevertheless, they were conducted in the period of voluntary transition to this form of learning and outside the context of proactivity. This paper attempts to address this kind of research gap and answer the question of the correlation between satisfaction with distance learning during the lockdown and students' attitudes towards proactivity. After switching to distance learning in the spring of 2020, students answered questions in an online questionnaire. 386 people (212 women and 174 men) took part in the survey. The age of participants was 17 - 21 years. The results of the study showed that 1) students do not have negative feelings about distance learning; 2) students still prefer face-to-face learning, and distance (online) learning is perceived as additional or temporary measure; 3) during the lockdown, students showed a mild tendency to react reactively to the situation; 4) proactive attitudes of students affect the degree of their satisfaction with distance learning.
Keywords: learning satisfaction; proactive behaviour; reactive behaviour; distance learning; pandemic; proactive attitude; coronavirus

1 Introduction

The experience with the COVID-19 pandemic has accelerated the digital transformation of education. In many states, lockdown was introduced in connection with the outbreak of the disease, and educational institutions switched to distance learning. This led to the fact that at the initial stage of the fight against the spread of the virus, educational institutions faced many difficulties - from "falling" due to server overloads to "falling" due to teacher fatigue [1]. From the point of view of the administration, the teachers' work schedule should have become less demanding; however, in practice, universities faced the opposite situation. Students also faced many difficulties. This led to anxiety among parents who were afraid that children would not be able to master the educational program on their own independently. As a result, teachers "bombarded" with the students'work, papers, and homework of students, which had to be quickly checked and commented on in detail, providing feedback to the students.

Currently, the benefits of information and communication technologies in learning are not disputed, unlike distance learning [2]. The obvious advantages of this form of learning include flexible schedule, relatively low costs, the acquisition of additional competencies in the use of information and communication technologies. Typical barriers to the introduction of distance learning are e.g. bias, negative experience, technical illiteracy and insufficient technical equipment, gender specifics of Internet use, the gap between the theory and practice of e-learning [3]. Researchers believe that many educational institutions, especially in third world countries, are not prepared to solve problems related to the supply of high-quality distance courses [4]. However, preliminary studies in developing countries have shown that distance learning can meet the needs of students, including in educational institutions where this technology was not previously used [5].

During the COVID-19 outbreak, distance learning was seen as the best option for ensuring the safety of students and their environment [6]. In Russia, the coronavirus pandemic has led to the mobilization of existing Internet technologies in education: communication infrastructure, digital learning resources, convenient teaching tools, effective teaching methods, educational organizations, support services for teachers, parents and students, and a system of interaction between the government and schools. All schools and institutions of higher education were pushed online. Life itself

* Corresponding author: bobnv79@mail.ru

started a large-scale experiment, the short and long-term consequences of which have yet to be evaluated by scientists. However, the undeniable consequence of the experience gained was the readiness and easy switching of universities to distance learning during the second and third wave of COVID-19. Subsequently, quarantine measures were not as large as in 2020. They were applied locally in individual educational institutions or for individual educational groups.

As a result, distance learning has now become a familiar tool to prevent pandemic situations without suspending the educational process. But only few studies are devoted to assessing the effectiveness of this form of education in a situation where the transition to it is not voluntary for students. The forced nature of the transition to distance learning can have a significant impact on their attitude towards this form of education, causing protest among students and their parents.

2 Literary research

Satisfaction with distance learning is defined as the student acceptance of this format and the level of comfort associated with its use [7]. Researchers consider global satisfaction (general attitude to distance learning) and differentiated satisfaction (i.e., satisfaction with the sides of individual actors in distance learning). An example of the latter is the blended e-learning satisfaction (BELS) model. This model includes the following factors: self-effectiveness, performance productivity, system functionality, content features, interaction, and psychological climate [8].

Satisfaction with the educational process consists of three components: cognitive, emotional, and behavioural. The cognitive component consists in assessing the positive and negative aspects of distance learning. The emotional component is a positive or negative attitude towards the experience of distance learning and its individual elements. The behavioural component is expressed by the readiness to continue distance learning or to recommend similar training to other people.

Researchers from different countries have repeatedly studied the degree of student satisfaction with distance learning. Until recently, however, these studies have been conducted predominantly in mixed education [2,9,10] or continuing education in the workplace [11-13]. A large number of studies deal with the analysis of the effectiveness of individual e-learning tools [3,14-16]. Previous studies show that students have a positive attitude towards e-learning. For example, the study by R. R. Codilan indicates that students appreciate the availability of e-learning [15]. According to another study, university students in Malaysia have a positive attitude towards e-learning and intend to practice it for educational purposes. [17] The experience of introducing distance learning elements at the Siberian Federal University shows that 15% of its students and 11% of its teachers are satisfied with its academic effectiveness [17]. Most previous studies were conducted in the context of voluntary consent to e-learning. Only recently, studies appeared that analyse the level of satisfaction with distance learning during lockdown. Their results are contradictory. For example, the results of studies show that distance learning of physiotherapy aroused negative feelings among students in Australia [19], while in Italy, students were as satisfied with distance learning of physiotherapy as in the case of face-to-face learning [20].

According to research in Russia, the negative attitude towards distance learning during the lockdown is due to the lack of students′ self-organization skills, the inability to take responsibility for their education [1]. This suggests that the level of student activity is an important internal factor of satisfaction with distance learning during the lockdown period. This term describes the ability of a person to distinguish himself/herself from other human beings. It consists of the ability to take the initiative to set and achieve their own goals in a difficult situation, even if nothing in the inner circle provokes them to do so [21].

Proactiveness correlates with motivation and positive results of the activity. For example, the level of proactivity in everyday life is associated with the satisfaction with activities [21]. This suggests that the level of student proactivity is related to the level of satisfaction with distance learning.

Thus, distance learning seems to be a feasible option to face the COVID-19 pandemic. But this requires a positive attitude of students. Therefore, proactivity can be a resource that contributes to the success of adapting to distance learning. The aim of this study is to examine the correlation between the level of proactivity and satisfaction with distance learning during the lockdown and to empirically demonstrate this fact.

3 Methodology

3.1 Study design

The research question is whether proactivity has an impact on the satisfaction with distance learning during lockdown. The questionnaire survey was conducted using the online resource Google Forms after the first lockdown in Russia in June 2020. Calls for student involvement in the study were made through joint student council groups. Participation in the survey was voluntary.

The participants were students (aged 17 to 21 years) from state universities of Russia. A total of 386 students participated in the survey. The research sample consisted of 174 male students (45%) and 212 female students (55%).

To answer the research question, the following hypothesis was formulated: There is a statistical correlation between proactivity and satisfaction with distance learning during the lockdown. The hypothesis is based on proactivity as a prerequisite for self-determination in an unfamiliar situation. Based on this, students who set educational goals are expected to be more satisfied with distance learning during the lockdown.

3.2 Instruments

The questionnaire consists of 4 sets of questions. The first set of questions is aimed at identifying satisfaction with distance learning. The set consists of 5 items measurable on 5-point Likert scale. The second set of questions is aimed at determining the cognitive component of satisfaction with distance learning. It contained 8 items where students were asked to choose one or more of the proposed options. Students can give their own answer too. The third set includes questions aimed at assessing the behavioural component of satisfaction with distance learning. The set consists of 2 items measurable on 5-point Likert scale. The fourth set of the questionnaire contains 2 questions describing the level of student proactivity in learning. The first question concerns the changes in student activity during the lockdown. The second issue concerned is the level of achievement of learning goals during the lockdown.

In order to verify the reliability of the questionnaire, we calculated Cronbach's alpha. The result obtained confirms the adequate internal consistency of the tool for this study (Table 1).

Table 1. Reliability indicators of scales satisfaction with distance learning

Scale	Cronbach's Alpha	Number of items
The emotional component of satisfaction with distance learning	0.744	5
The cognitive component of satisfaction with distance learning	0.68	8
The behavioural component of satisfaction with distance learning	0.94	2

Source: Authors' calculation – SPSS.

In order to achieve the objectives of the study, correlation analyses were performed using the statistical software SPSS v.23.

4 Results and Discussion

The results of the study show that the activity of students during the period of forced isolation did not change (X = 2.87). 9.6% of respondents noted that their activity increased during the pandemic. The number of students who noted a decrease in their activity was 15.5% (Table 2).

In general, students believe that distance learning did not affect the level of achievement of their educational goals (X = 3.01). Some participants (10.4%) noted that distance learning had a positive effect on this (Table 2).

Distance learning created obstacles in achieving educational goals for 11.7% of the respondents. Educational activity "decreased" in the case of 15.5% of the respondents, "rather decreased" for 20.7% of the respondents, "did not change" for 34.2% of the respondents, "rather increased" in 19.9% of the respondents, and "definitely increased" in 9.6% of cases. Distance learning "has clearly contributed" to the achievement of their educational goals in 10.4% of the cases, "rather contributed" to their achievement for 23.6% of the respondents, "did not affect the achievement of the goals" for 32.6% of the respondents, "rather created an obstacle to their achievement" for 21.8% of the respondents, "thwarted their achievement" for 11.7% of the respondents.

Table 2. Level of student proactivity during lockdown

Likert scale	Activity change	Achieving the Goal
1	15.5%	10.4%
2	20.7%	23.6%
3	34.2%	32.6%
4	19.9%	21.8%
5	9.6%	11.7%
Number of respondents	386	386

Mean	2.87	3.01
Minimum	1	1
Maximum	5	5

Source: Authors' calculation – SPSS.

It can be concluded that during the lockdown, students showed proactive behaviour to varying degrees. In the research sample, about 30% of the students showed a proactive attitude in the process of distance learning, and the same number showed a tendency to reactive behaviour. The remaining 30% did not notice any changes in their own activity during the pandemic. A similar situation was in assessing the achievement of educational goals.

Students´ satisfaction with distance learning during the lockdown period is shown in Table 3.

Table 3. Level of satisfaction with distance learning during lockdown

Likert scale	Distance Learning Satisfaction Components		
	emotional	cognitive	behavioural
1	13.0%	9.6%	16.3%
2	31.6%	24.9%	18.1%
3	35.0%	28.0%	20.2%
4	20.5%	30.1%	19.9%
5	0%	7.5%	25.4%
Number of respondents	386	386	386
Mean	2.99	3.24	3.34
Std. deviation	0.91	1.11	1.39
Minimum	1.0	1.00	1.0
Maximum	4.8	5.0	5.0

Source: Authors' calculation – SPSS.

At the same time, the emotional component of the analysed phenomenon was rated the lowest.

About 30% of the respondents tend to have rather a positive attitude towards distance learning, but the same percentage tend to have rather a negative attitude (see Table 3). About 37.6% of the respondents believe that distance learning was organized at a sufficient level. However, 28% of the respondents found it difficult to say how well distance learning was organized during the pandemic. 34.5% of the respondents participating in the study believe that the organization of distance learning was unsatisfactory. About 45% of the respondents demonstrated a negative behavioural attitude towards distance learning.

The results of the study show that the students who took part in the research found the idea of switching to distance learning during the pandemic positive (40%) or rather positive (26.5%). This idea was perceived negatively by 10% of the students, and rather negatively by other 12.7% of the students. However, 60.6% of the respondents consider distance learning a forced measure caused by the need to prevent the spread of the pandemic. 37.7% believe that they have fully mastered and can continue their studies in the form of distance learning. At the same time, the idea of combining distance and classical forms of education was divided: 25.4% of the respondents perceive this idea positively, 20.9% rather positively, 13% negatively and 20.9% rather negatively.

It can be concluded that the students show a neutral attitude towards distance learning. The forced transition to online learning during the pandemic did not arouse positive emotions in them. They were able to adapt and are ready to switch again to distance learning if necessary. According to the students, distance learning represents a complement or an additional element to the traditional form of study. Students do not see distance learning as an equivalent alternative to traditional forms of learning. The experience the students gained during the pandemic did not arouse their interest in distance learning, because it was associated with many problems and difficulties.

The Spearman correlation coefficient was used to determine the relationship between the variables under review. Table 4 presents a matrix of correlations between the parameters of satisfaction with distance learning and student activity during the pandemic.

Correlations exist between proactivity and satisfaction elements of distance learning during the lockdown. According to the data obtained, the level of activity and achievement of the goal correlates with three components of satisfaction with distance learning (emotional, cognitive, and behavioural). In this case, correlation coefficients are higher in the case of the relationship between the degree of achievement of educational goals and satisfaction than between the degree of activity and satisfaction.

In fact, students who noted an increase in their level of activity during the pandemic were more satisfied with distance learning than students who noted a decrease in their level of activity. And students who claimed that distance learning contributed to achieving their educational goals were more satisfied with distance learning during the pandemic than students for whom distance learning represented an obstacle to achieving educational goals.

Thus, the study confirms the existence of a direct relationship between the level of proactivity and satisfaction with distance learning during the pandemic.

Table 4. Correlations between proactivity scores and life satisfaction components

Test parameters		Distance Learning Satisfaction Components		
		emotional	cognitive	behavioral
Activity change	r_s	0.41	0.37	-0.36
	p	0.0001	0.0001	0.0001
	n	386	386	386
Achieving the Goal	r_s	0.69	0.58	-0.68
	p	0.0001	0.0001	0.0001
	n	386	386	386

Source: Authors' calculation – SPSS.

The study enabled the identification of the students' attitude towards distance learning in the case of lockdown. During the pandemic, satisfaction with distance learning was moderate rather than high. Students consider this to be a forced temporary measure. They prefer the traditional (face-to-face) form of learning. A similar pattern can be seen in Italy [20] and Australia [19].

There are data that indicate a more negative attitude of students. In Ukraine, the level of satisfaction with the content of the course, its form, and teaching methods within distance learning is significantly lower than those shown in this study (1% - 7%) [3].

Such discrepancies in data can be explained by several reasons: differences in the organization of distance learning, the level of readiness of students and teachers to switch to distance learning format, the possibility of choice. In the universities participating in this study, the transition to distance learning was rapid and forced. Not all teachers and students were ready for this form of interaction. Teachers created e-learning materials under pressure of time. At the same time, new distance learning technologies (for example, Zoom conferences) were being developed. As a result, more than 50% of the students faced technical problems, and about 30% of the students (48.8% of them in their last year of study) complained about insufficient computer literacy of teachers. About 20% of the students mentioned their own computer illiteracy. The generation gap concerning the use of technology by digital natives (students) and digital immigrants (teachers) may have affected this case [22]. This has affected the learning process and results. Reading literature replaced lectures; seminars were replaced by individual work and performance of practical tasks. This form of distance learning corresponded rather to self-study and caused resistance from the side of students. However, a positive fact is that more than half of the respondents saw their laziness as an obstacle they had to fight with. This obstacle is a major problem in the organization of self-study. The lack of technical opportunities for distance learning was perceived negatively by 22% of the students.

The experience of the general forced immediate transition to distance learning shows that at this stage universities were able to ensure maximum security for students by switching to distance learning. However, the quality of such education was at a satisfactory level only and did not meet the optimistic forecasts made earlier. This is partly due to the internal attitude of students.

The study shows the reactive position of most students in the situation of forced transition to distance learning. This is evidenced by the fact that only 16.8% of the students see this experience as a new opportunity, 12.3% strive to achieve their educational goals, and 7.1% of the students noticed their increased activity. This corresponds to the data on the prevalence of proactive attitudes among students, specifically 8% of young people aged 17 - 24 years, more often expressed by men (10%) than by women (8%) [23].

On the other hand, according to the results of the study, the level of proactivity is associated with the level of satisfaction with distance learning. This confirms the hypothesis that the ability to work independently, the ability to set their own educational goals contribute to a more positive perception of the forced transition to another form of learning. This is reflected in the fact that proactive students more often show a positive attitude towards e-learning, are able to highlight the constructive aspects of distance learning, and are ready to study from home.

5 Conclusion

The Covid-19 pandemic showed that the transition of education to distance learning during the lockdown is a reality of Russian higher education. This situation indicates the need to find conditions and resources to make the transition to distance learning as comfortable and effective as possible. Indirect data from previous studies indicate that satisfaction with distance learning depends on student proactivity. This paper focused on confirming this assumption through an analysis of students' subjective perceptions and preferences. In this context, we tried to answer the question of whether there is a correlation between proactive attitude and the level of satisfaction with distance learning during the lockdown. The results showed that the change in activity level and the presence of own educational goals during the lockdown correlates with the level of satisfaction with distance learning. This is true for all three components of satisfaction: emotional, cognitive, and behavioural.

In the future, we plan to study the impact of proactivity on satisfaction with distance learning in a situation of voluntary choice. It is important to study the impact of experience on this process too. The transition to distance learning was carried out during other waves of Covid-19 too. Therefore, the perceived satisfaction might change.

References

1. M. A. Lukashenko. Self-management skills as a basis for the effectiveness of e-learning. *Psychology*. **30**, 61–70 (2021)

2. T. C. Herrador-Alcaide, M. Hernández-Solís, R. Sanguino Galván. Feelings of satisfaction in mature students of financial accounting in a virtual learning environment: an experience of measurement in higher education. *International Journal of Educational Technology in Higher Education*, **16**(20) (2019)

3. S. Dovbenko, R. G. Naida, V. M. Beschastnyy, H. V. Bezverkhnia, V. V. Tsybulska. Problem of Resistance to the Introduction of Distance Learning Models of Training in the Vocational Training of Educators. *International Journal of Learning, Teacher, Educational Research*. **19**, 1–12 (2020)

4. H. Abuhassna, M. A. Z. M. Zakaria, N. Yahya, A. B. M. Kosnin, W. M. Al-Rahmi. Examining students' satisfaction and learning autonomy through web-based courses. *International Journal Advanced Trends in Computer Science and Engineering*. **9**, 356–70 (2020)

5. D. Sulisworo, D. A. Kusumaningtyas, T. Handayani, E. Nursulistiyo. The learning design and student's response to physics online learning in rural school of Indonesia. *International Journal of Innovative Technology and Exploring Engineering*. **9**, 4999–5006 (2019)

6. G. Basilaia, M. Dgebuadze, M. Kantaria, G. Chokhonelidze. Replacing the Classic Learning Form at Universities as an Immediate Response to the COVID-19 Virus Infection in Georgia. *International Journal of Research Applied Science & Engineering Technology*. **8**, 101–8 (2020)

7. R. Rachmawati, E. Octavia, S. Dewi Herawati, O. Sinaga. Culture, Environment and E-learning as Factor in Student Performance (Case Studies in Management Accounting Study Programs). *Universal Journal of Educational Research*. **7**, 72–8 (2019)

8. J. H. Wu, R. D. Tennyson, T. L. Hsia. A study of student satisfaction in a blended e-learning system environment. *Computers & Education*. **55**, 155–64 (2010)

9. L. Zhou, S. Wu, M. Zhou, F. Li. 'School's Out, But Class' On', The Largest Online Education in the World Today: Taking China's Practical Exploration During The COVID-19 Epidemic Prevention and Control As an Example. *SSRN*. **4**, 501–19 (2020)

10. A. Al-Awamleh. Students' satisfaction with blended learning programmes in the Faculty of Physical Education. *Science of Education Today*. **9**, 37–47 (2019)

11. J. Rodríguez-Santero, J. J. Torres-Gordillo, J. Gil-Flores. Confirmatory Factor Analysis of a Questionnaire for Evaluating Online Training in the Workplace. *Sustainability*. **12**, 4629 (2020)

12. M. H. Song, T. I. Han. A study on the learning satisfaction and work utilization of the teacher safety e-learning. *International Journal of Information and Education Technology*. **9**, 909–17 (2019)

13. Gopinathan, S. Praveen Kumar, Pothumani. Customer satisfaction in career enhancement with the usage of E-Learning software. *International Journal of Recent Technology and Engineering (IJRTE).* **8**, 466–8 (2019)
14. C. L. Bian, D. L. Wang, S. Y. Liu, W. G. Lu, J. Y. Dong. Adaptive learning path recommendation based on graph theory and an improved immune algorithm. *Koreanstudies Information Service System.* **13**, 2277–98 (2019)
15. R. R. Codilan. Efficiency and Acceptability of a Customized e-learning Platform in a Philippine State University and College. *Indian Journal of Computer Science Engineering.* **10**, 83–8 (2019)
16. D. Padmo, L. Sri Ardiasih. The utilization of mobile devices for improving access to online learning for distance education's students. *Turkish Online Journal of Distance Education.* **20**, 147–61 (2019)
17. M. M. Alamri, V. M. Al-Rahmi, N. Yahaya, A. M. Al-Rahmi, H. Abualrejal, A. M. Zeki, Q. Al-Maatouk. Towards adaptive e-learning among university students: By applying technology acceptance model (TAM). *International Journal of Engineering Advanced Technology.* **8**, 270–6 (2019)
18. O. Smolyaninova, E. Bezyzvestnykh. Implementing Teachers' Training Technologies at a Federal University: E-portfolio, Digital Laboratory, PROLog Module System. *International Journal of Online and Biomedical Engineering (iJOE).* **15**(4), 69 (2019)
19. L. Ng, K. C. Seow, L. MacDonald, C. Correia, A. Reubenson, P. Gardner, A. L. Spence, S. Bunzli, B. I. R. De Oliveira. eLearning in Physical Therapy: Lessons Learned From Transitioning a Professional Education Program to Full eLearning During the COVID-19 Pandemic. *Physical Therapy.* **101**, 1–9 (2021)
20. G. Rossettini, T. Geri, A. Turolla, A. Viceconti, C. Scumà, M. Mirandola, A. Dell'Isola, S. Gianola, F. Maselli, A. Palese. Online teaching in physiotherapy education during COVID-19 pandemic in Italy: a retrospective case-control study on students' satisfaction and performance. *BMC Medical Educaction.* **21**, 1–7 (2021)
21. F. Lieder, G. Iwama. Toward a formal theory of proactivity. *Cognitive, Affective, & Behaviroal Neuroscience.* **21**, 490-508 (2021)
22. S. Dickinson. Post-traumatic growth in the twenty-first century: how current trends may threaten our ability to grow after trauma. *The Journal of Positive Psychology,* **16**(4), 1–9 (2020)

Adaption of Marketing Mix for Selected Country

Barbora Szolgayová[1,*]

[1]Institute of Technology and Business, Faculty of Corporate Strategy, Okružní 517/10, České Budějovice, Czech Republic

Abstract. Using the SWOT analysis, it was found that Kofola Plc. is in a sufficiently strong position to enter the foreign market, specifically the Austrian market, and the competition in the Austrian market was further analysed. It showed that there are three major competitors in the market. Using the PEST analysis, information concerning the exchange of goods and exports, or the financial situation of Austrian citizens was found. The company's marketing mix on the Czech market and its subsequent adaptation to the Austrian market were described. The greatest need for adaptation was identified in the marketing communication and also in the product itself, specifically packaging, which will require a language mutation in the composition. Also, some change in appearance to match the marketing communication will be needed. The marketing communication will focus mainly on the unique taste of Kofola, which could be an attraction for potential customers, mainly because Austrians already like the herbal drink made by Almdudler, which is also one of the company's competitors, together with Coca - Cola Austria and PepciCo.
Keywords: adaption of marketing mix; marketing mix; international marketing; marketing strategy; analysis; marketing

1 Introduction

This article on the topic of adaptation of the marketing mix for a selected country deals with the issue of the marketing mix of the traditional Czech company Kofola Česko Slovensko Plc., given that the portfolio of this brand is very extensive, this article will focus on one of the subsidiaries, Kofola Plc. (hereinafter referred to as Kofola). The marketing mix will be adapted to one of the neighbouring countries to the Czech Republic, namely the Republic of Austria. This company already operates in several foreign markets. Austria was chosen for several reasons. The first is a certain similarity between the two countries - the adaptation of the marketing mix will not be as difficult as, for example, in third world countries. Furthermore, due to the considered distribution routes, neighbouring countries are considered the most suitable. Kofola has already been operating on the Czech, Slovak and Polish markets for many years, so the German and Austrian markets are possible. After considering both options, the Austrian market was evaluated as a more suitable option.

2 Literary research

The selected topic of this article is extensive and contains many definitions and concepts. A large amount of quality literature is written on the topic of the marketing mix, and overall, a wide range of information is available from different sources.

From foreign sources, the author Philip Kotler and his book Principles of Marketing [1] definitely should be mentioned. The basics of marketing or marketing mix are clearly explained in this book. Another very interesting author is Seth Godin [2]. In 2018, he published the book This is Marketing, where the main idea of his approach is that big marketers do not use consumers to solve society's problems but use marketing to solve other people's problems.

A group of authors calling themselves 50 Minutes has published The Marketing Mix [3], which will provide the reader with a practical introduction to each component of the so-called 4P (product, price, promotion and distribution). The main attraction is the guarantee of understanding the basics of the marketing mix in just 50 minutes. The work Marketing obchodní firmy [4] by M. Zamazalová is one of the few books that is directly intended for Czech readers and describes the conditions on the Czech market and contains many practical examples from Czech companies. In the article The 4P Classification of Making Mixing Revisited from 2015, the authors W. van Waterschoot and CH. van den Bulte [5] deal with the basic concept of the so-called 4P and its revision. The author Roxana Codita [6] in the book Contingency factors of marketing-mix standardization focuses on the issue of standardization of the marketing mix in connection with Central and Eastern Europe. The Concept of 'Marketing Mix' and its Element is an article describing the mix in its extended concept of "8P" (4P + people, processes, physical records, productivity and quality) by M. T. Khan [7]. The second edition of Marketing in Tourism [8], written by the author Dagmar Jakubíková, deals with marketing tools and the customer's

* Corresponding author: 16477@mail.vstecb.cz

view is underlined here. In the article by Efthymios Constantinides [9], titled The Marketing Mix Revisited, it is possible to find a summary of literature reviews dealing with marketing, mix or relationship marketing.

The work Strategic Marketing [10] also by Dagmar Jakubíková, in which she generally deals with the issue of strategic marketing, from the management process to situation analyses. Already the fifth edition of the publication Marketing Strategy and Management [11] by Michael Baker includes the classic concept of the so-called "4P", but also new research, current examples or case studies, and also offers several perspectives on the creation and implementation of marketing strategy.

A number of authors also deal with international marketing, such as Tobias Richter [12], who wrote Management International Marketing Mix. It focuses on international marketing strategy and the related marketing mix of world markets. Other authors dealing with the same issues are, for example, Bohumír Štědroň, who, together with other authors, wrote the book International Marketing [13], in which he deals mainly with simple models, concepts and deciding how and which foreign market to enter. Přikrylová and Machková also deal with modern marketing communication [14,15].

2.1 Research questions

The above-mentioned Machková [16] is also the author of the publication International Marketing, in which she deals with strategic trends. The publication Basics of Marketing by the author Miroslav Karlíček [17] describes the basic issues of marketing, what strategy to choose, what price to set, how to create the right product or how to adapt the marketing mix. Destination Marketing Management is a publication by Monika Palatková [18], who focuses on marketing strategy, especially the marketing mix and its adaptation, and the organization of marketing management, and last but not least, Gerald A. [19], who wrote the book International Marketing and Export Management in 2005, focused on international marketing and export-related management processes.

This article will focus on the basic concept of the marketing mix and its adaptation to the selected country, Austria.

1. How many competing companies operating in Austria provide a similar product?

2. Which area of the 4P marketing mix will be needed to be most adapted for the selected market?

3 Methodology

For the adaptation of the marketing mix, as already mentioned in the introduction, the author chose one of the neighbouring countries of the Czech Republic, namely the Republic of Austria. Austria has been a very important trading partner of the Czech Republic for a long time and vice versa. Austria is ranked among the ten richest countries in the world due to its high GDP per capita. Despite its relatively small area and population, this is a very interesting market with strong domestic purchasing power. The Austrian market is very advanced in terms of marketing and global companies are there. It can be really challenging to enter the Austrian market. Companies offering products with high added value, but only while maintaining a price advantage over the competition, can succeed.

For a successful marketing campaign, it will be very important to choose the right promotion and also the marketing preparation, which is a key point for success in this market, as well as in similarly developed markets. The mentality of Austrian citizens is similar to the Czech mentality, so it should not be difficult to empathize with the needs and wishes of these customers, as for example with the German neighbours.

As for the costs of promotion, they are of course higher than in the Czech Republic, especially in terms of the media. This is also the main reason why they have heard so little about Czech products in this country so far and also because some companies may underestimate the importance of advertising and promotion on the Austrian market.

3.1 Methodology of the research

The topic of this paper is the adaptation of the marketing mix to the selected country. The marketing mix of Kofola Plc. will be adapted. This company is already very well known to me, thanks to the author's bachelor's thesis, which was focused on the marketing communication of this company.

The author determined the main goal of this work, namely, to answer two research questions (see the research questions chapter). First of all, she would like to create a SWOT analysis and find out what situation the company is in, whether it will be in a strong enough position to enter a new market. She also evaluated the company's current marketing mix in the Czech Republic. The author would also like to create a competition research survey and PEST analysis. It predicts and analyses the future influences that will affect the company in the Austrian market, so that she has an idea of what the company will face in the market. Subsequently, an adapted marketing mix for the Austrian market will be created. After addressing this whole issue, clear answers to the research questions asked should emerge.

3.2 Introduction of Kofola Plc.

Kofola Plc., (hereinafter referred to as Kofola), as already mentioned, falls under the parent company Kofola ČeskoSlovensko Plc, which is one of the largest producers of soft drinks in several European markets. This company was founded in the 1950s. In 1957, a pharmaceutical company was given the task of developing and subsequently producing a soft drink that can replace foreign cola drinks. Two years later, KOFO syrup was developed from which Kofola is still made.

3.3 Research methods

SWOT analysis

This chapter will identify the strengths and weaknesses of the company, as well as its opportunities and threats, which will provide an overview of the current situation of the company. The SWOT analysis will point out those areas on which the company will be able to rely in the future, and vice versa, areas that need to be improved or eliminated.

Strengths and weaknesses show the internal situation of the company and show factors that can affect the company positively or negatively and affect the future prosperity of the company at the same time.

Strengths

- strong position on the Czech and Slovak market
- financial stability
- traditional Czech brand
- brand knowledge
- marketing activities
- Kofola drink composition
- favourable price compared to the competition
- strong customer support
- Kofola on tap
- strong local brand
- wide portfolio

Kofola has a very strong position on the Czech and Slovak markets (see Fig. 1). According to internal materials, the company has been developing in a positive direction in recent years, 2018 was a growth year for the company. It is a traditional Czech brand that has strong ties with customers. The drink became known also thanks to sophisticated marketing activities, especially in terms of advertising. The composition of this beverage has certain advantages over competing cola drinks, it does not contain phosphoric acid and contains less sugar and caffeine. The price is also a bit more favourable than the competition's, and it is also a certain plus that it is possible to buy Kofola on tap.

Weaknesses

- long-term number two on the Czech market in the area of cola drinks (see Fig. 1)
- the association of kofola and communism
- brand knowledge only in the Czech Republic and Slovakia
- pressure from retail chains on margins
- very little chance that Kofola will get to restaurants where its competitor already operates
- climate dependence
- taste

Kofola has long been the second largest player on the Czech market in the field of cola drinks, and its strongest competitor is Coca-Cola (see Figure 1). Another disadvantage is the association of the Kofola brand with communism, which may discourage some customers although the company is wholly owned by different owners than before. As far as foreign countries are concerned, the Kofola brand is known only in the Czech Republic and Slovakia, in other countries where the company operates under the name Hoop. Unification under the same name could strengthen the company's position in other markets.

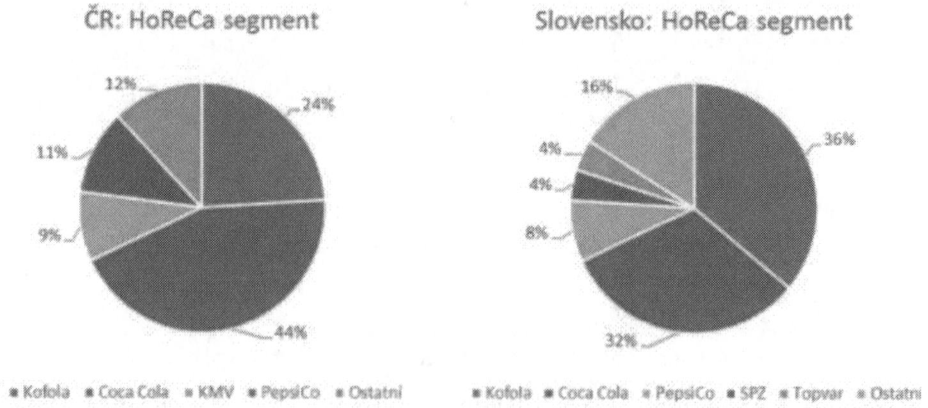

Figure 1. Kofola market shares

Legend: ČR – Czech Republic, Slovensko - Slovakia

Source: Author based on Kofola [20]

From the above picture, it is clear that Kofola has a strong market position, but Coca - Cola is still a big competitor for it. Currently (2020), Kofola 's shares increased to a record 27% in the Czech Republic and 40% in Slovakia, mainly at the expense of Coca - Cola.

Probably the problem of most of the beverage industry is the dependence on the weather in the summer months, Kofola states in its press releases that the years with long winters or floods have affected the company's profit. Also, its taste, which is very specific, may not be appreciated by everyone.

The other two areas of SWOT analysis affect the company from the external environment, which will be described in more detail in the PEST analysis.

Opportunities

- entry into new foreign markets
- advertising campaigns
- new technologies
- expansion of the product portfolio
- moving production closer to the target markets
- increase in market share
- social events

One of the opportunities is to enter a new foreign market, given that the companies already operate in several. Kofola mainly focuses on television advertising, but the strength of the Instagram social network and the addressing of so-called influencers will also appeal to a mass of potential customers. The expansion of the product portfolio can be considered another opportunity, for example, new flavours, or participation in social events such as festivals.

Threats

- financial crisis
- cheaper substitutes
- strong competition
- the ever-increasing trend of a healthy lifestyle
- fines for non-compliance with deliveries to customers (quality, time or quantity)
- rising energy prices
- deposit on PET bottles
- change of legislation

Coca-Cola, a competitor, can be considered the biggest threat. Rising energy prices can also have a big impact on the company's future in the form of rising input prices, which would then make the final product more expensive. Another negative that could affect the company is the introduction of deposit PET bottles or fines for non-compliance with deliveries to customers in the required time, quality or quantity.

Marketing mix of Kofola Plc. on the Czech market

Product

Kofola drink has a very specific taste, and it is very different from other cola drinks on the Czech market. It contains less sugar and caffeine than competing products. Kofola comes with a new flavour every year, for 2020 it is Kofola Med, which falls into the Christmas limited edition.

The Kofola drink can be purchased in several types of packages and sizes, which vary according to flavours. Of course, the largest selection is at Kofola Original, which is sold in a can with a volume of 0.25, in a glass bottle also with a volume of 0.25 l and 0.33 l, then a 0.5 l can, which can also be purchased in the variants of Kofola Citrus and Kofola Sugar free. As for PET bottles, the range of flavours is more extensive. Kofola Original, Apricot, Melon, Raspberry, Blackberry and Gooseberry can be purchased in a PET bottle with a volume of 0.5 l. It is possible to buy Kofola Original and with Guarana flavour in a 1 l PET bottle. The flavours Original, Apricot and Melon can be found also in a PET bottle 1.5 l and the last variant as far as PET bottles are concerned, the one with a volume of 2 l can be purchased with Kofola Original, Cherry, Citrus and Sugar Free. Kofola is also sold for tap in 20 l and 50 l barrels.

According to the author's opinion, the packaging is very important for the product, but also its size. The Kofola drink is produced and distributed in a suitable range of packaging.

Price

As already mentioned, Kofola's price is slightly lower than that of the strongest competitors on the Czech market. The price of a 2-l bottle is around CZK 20–30, depending on the store and whether there is currently a special offer for the given product. The same is true in restaurants where Kofola on tap is served for approximately CZK 5-6 per 1 dcl, for example, Kofola with a volume of 0.4 l is cheaper than the competing Coca - Cola drink 0.33 l in a glass bottle. This price is determined by each restaurant itself, the purchase price is around 1.8 CZK per 1 dcl, so the margin is relatively high.

Keeping the price lower than the competition is the company's long-term pricing strategy.

Promotion

Kofola's marketing communication is very original and targeted directly at Czech consumers. Jannis Samaras was behind the marketing communication for a long time. He managed to put together an excellent team. In 2018, he was replaced by Daniel Buryš, who continues his work.

Communication campaigns are aimed mainly at two groups of people: young people - students between the ages of 15 and 25, who want to have fun and enjoy life. The second group are older people around the age of 50 and older, when Kofola reminds them of their youth and remembers its beginnings and its fame in the 1970's.

In its communication campaigns, Kofola places particular emphasis on love, wit and originality. This is evidenced by the advertising slogans or the very motto of the brand - " When you love her/it, no need for reason." *(Note: hardly translatable wordplay*). Or also the slogan for the fiftieth anniversary - "Love is no science. Kofola has known this for 50 years. "At the beginning of April 2019, Kofola released a completely new slogan for a new advertisement -" There is enough love for everyone. "

The company mainly uses TV commercials, POS materials in restaurants, social networks and its website www.kofola.cz to communicate with its customers.

Distribution

Since 2015, Kofola has launched its own direct distribution of its products, especially the hotel, restaurant, and catering segment. Thanks to direct distribution, Kofola is gaining a closer and better relationship with its customers. Direct distribution has been operating in Slovakia since 2009, and thanks to the success it had there, the Czech Republic was also inspired.

As already mentioned, Kofola is distributed mainly to shops, restaurants and in the summer months to a large number of refreshment kiosks, such as near cycling routes, swimming pools, campsites and popular tourist destinations in general. It also focuses more on middle- and lower-priced restaurants as part of the competition.

In 2011, Kofola tested a complete novelty, a self-service vending machine, which should reinforce the phenomenon of lemonade on tap. The machines were applied to places where, for some reason, it is not possible to use the classic bar, which is located, for example, at some branches of the city library in Prague, where it was called Kofolík.

PEST analysis of the Republic of Austria

Political and legal factors

The Czech Republic joined the European Union in 2004, so the rules for exchanging goods are governed by the rules set out in Community legislation. Trade between EU countries is not subject to any tariff or non-tariff barriers, which used to make cross-border transport more difficult.

Goods traded are only subject to excise duty and VAT, these taxes are adjusted according to individual EU members. In Austria, VAT at the basic rate is 20% and at the reduced rate it is 10% and 13%, but in the current coronavirus situation, taxes have been reduced in certain sectors till the end of 2020.

Overall, the two countries are very similar in terms of political and legal law, given that they are both members of the EU and neighbouring each other. Which in itself is a guarantee of similarity.

Economic factors

Austria is one of the richer countries in terms of GDP per capita. Austria ranks 17th in the World Bank with 52,557 mez. $ and the Czech Republic in 39th place with 37,530 mez. $. The amounts are in a hypothetical currency that would have the same value in that location as the U.S. dollar at the time.

The average wage in the Czech Republic was CZK 34,125 in 2019, calculated at € 1302 at the current CNB exchange rate; this year it rose by more than CZK 1,000 to CZK 35,402 for the three quarters, which is € 1,350. On the other hand, the average salary in Austria was approximately € 2,400, converted to Czech crowns it was CZK 62,919, with an average of 14 salaries being paid in Austria.

Due to the current crisis, the unemployment rate has been rising steadily for the last six months, currently in November to 9.5% (7.3% in 2019) from 8.7% in October (7.0% in 2019). The annual inflation rate was 1.3% in November 2020, the same as in October. Prices slowed down mainly for housing and services, alcoholic beverages, and tobacco, etc. However, inflation in food and non-alcoholic beverages remained stable.

Social cultural factors

Entrepreneurship is always influenced by the socio-cultural factors of the country. We can include all population demographics in these factors.

Demographic trends

The population of Austria by 2020 is just over 9 million and the population density is about 109 people per square kilometre. The number of citizens living in the countryside is gradually growing, currently it is 43%, in 1955 it was 7% less. In 2019, the population of men was about 4.4 million and 4.5 million women.

Ethnic composition and languages

The majority of the population is Austrian 94%, then there are Germans and Turks. In Austria, German is the official and mother tongue, but there are some differences from other German-speaking countries. There are two basic dialects in the country, Alamannisch (German) and Bairisch (Bavarian). Among other things, the original minorities also speak other languages, such as Hungarian or Slovenian.

Religion

More than half, specifically 64.1% of the population, belong to the Roman Catholic Church and 3.8% to one of the protestant churches, and about 180,000 Christians are members of the Orthodox Churches. 8,140 citizens profess Judaism.

Technological factors

Austria has made a lot of progress in research and development in the last approximately twenty years. In 1994, the research quota was 1.53%, rising to 3.19% in 2020, making Austria the strongest R&D in Europe. In the ranking of EU research spending in 2018, Austria is in second place behind Sweden (3.31%) with 3.17%. In a global comparison, Austria ranks 7th and is therefore one of the best destinations in the field of research and development.

Austria is one of the few European countries that has exceeded the stated goal of EU research policy. However, it sets more ambitious goals: research strategy of the Federal Government considers Austria to be a leader in innovation in Europe in 2020.

Competition research on the Austrian market

A very strong competitor for the companies is Coca-Cola HBC Austria (see Fig. 1), which is one of the largest companies in the non-alcoholic beverage industry in the country and belongs to the Cola-Cola Company and has been operating in Austria since 1929.

Another strong competitor is Almdudler, which produces a herbal drink similar to Kofola. It is the most popular drink right behind Coca-Cola.

Pepsi is another cola drink that could come to mind for everyone in the competition of cola drinks as far as the Czech Republic is concerned, but in Austria it is not among the leading brands of carbonated soft drinks. Pepsi does not have its own branch in Austria, it is present only through Vöslauer Mineralwasser GmbH, which owns the brand's license.

4 Results

4.1 Evaluation of SWOT analysis using IFE and EFE matrices

The IFE matrix, which evaluates the company's internal factors, and the EFE matrix, which evaluates external factors, will be used to evaluate the SWOT analysis.

For each point of the IFE matrix, points in the range of 1-2 for the weak points and 3-4 for the strong ones will be assigned, and the weight in the range of 0.00 - 1.00, however, the sum must be equal to 1. It is necessary that the strengths and weaknesses are the same number, so the main ones were chosen according to the author's judgment.

Subsequently, both values will be assessed according to the IE matrix.

Table 1. IFE matrix

Strengths	**Points**	**Weight**	**Total**
Strong position on the Czech and Slovak market	4	0,15	0,6
Financial stability	4	0,1	0,4
Traditional Czech brand	4	0,08	0,32
Marketing activities	4	0,12	0,48
Drink composition	3	0,03	0,09
Favourable price	3	0,05	0,15
Customer support	4	0,06	0,24
Weaknesses	-	-	-
Number two on Czech market	1	0,03	0,03
Association with communism	1	0,02	0,02
Brand knowledge only in the Czech Republic and Slovakia	2	0,1	0,2
Pressure on margins	1	0,05	0,05
Restaurants - competitor	1	0,09	0,09
Climate dependence	1	0,08	0,08
Specific taste	2	0,04	0,08
Total	-	**1**	**2,83**

Source: Author.

The EFE matrix is composed of opportunities and threats that affect the company within the external environment. As with the IFE matrix, points and weights will be allocated and the number of opportunities must be equal to the threats and vice versa.

Table 2. EFE matrix

Opportunities	**Points**	**Weight**	**Total**
Entry into new foreign markets	4	0,15	0,6
New technologies	3	0,05	0,15
Expansion of the product portfolio	3	0,05	0,15
Moving production closer to the target markets	3	0,04	0,12
Increase in market share	3	0,05	0,15
Advertising campaigns	4	0,15	0,6
Social events	3	0,05	0,15
Threats	-	-	-
Financial crisis	2	0,09	0,18
Strong competition	2	0,07	0,14
Trend of a healthy lifestyle	1	0,06	0,06
Fines for non-compliance with deliveries	1	0,07	0,06
Deposit on PET bottles	2	0,04	0,08
Change of legislation	2	0,06	0,12
Cheaper substitutes	1	0,07	0,07
Total	-	**1**	**2,63**

Source: Author.

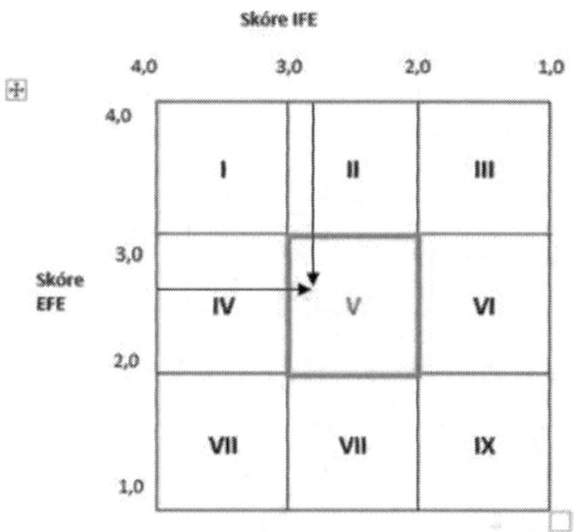

Figure 2. Matrix IE

Legend: Skóre - score

Source: Author.

According to the results of the IFE matrix 2.83 and the EFE matrix 2.63, the company is located in quadrant number V. in the IE matrix, which means maintenance and strengthening. Entering into new markets and development of new products is recommended. It follows that the company is strong enough to enter the Austrian market.

4.2 Design of an adapted marketing mix for the Austrian market

The marketing mix is the basis for the strategy of a company that wants to expand into foreign markets. Thanks to the adaptation of the marketing mix, we will get important answers to the questions that the company must address when entering new foreign markets.

Product

As far as the product is concerned, four product modifications need to be considered, namely changing the name, flavour, size of the packaging and the packaging itself.

A change of name is not necessary in this case, "Kofola" in German does not mean anything specific, so there is no need to change the name and if the drink will be sold under the same name as in Slovakia and the Czech Republic, it can only bring good feedback, because Austrians traveling to these countries may already know this drink.

Kofola is just as original, thanks to its unique taste - a mixture of herbs and licorice. As already mentioned, it contains less sugar than competing drinks, which could provoke positive reactions, but also negative ones, that the product is of lower quality due to the content of artificial sugar substitutes. Kofola is Kofola due to its unique taste, so the author believes that the taste should remain unchanged. Other flavouring variants that Kofola is enriched with are possible, such as Kofola Gooseberries, etc. In terms of positive feedback, the ever-increasing trend of a healthy lifestyle could be a plus for the drink due to the content of herbs and low sugar. To begin with, Kofola should launch both Original and Sugar-free variants on the Austrian market.

As in the Czech Republic and Austria, the standard bottle sizes are 0.5 l, 1 l, 1.5 l and 2 l. The appearance of the packaging, not only due to the communication policy, but also a certain presentation of the product, will need to be changed. The appearance, colour structure and style should be preserved in order to unify this drink with the Czech Republic and Slovakia. The author believes that it should be clear from the packaging that it is a herbal drink, so some should be depicted on the packaging (the specific composition of these 14 herbs is known to only a handful of people, so it is not possible to state anything specific, but from the picture it should be clear that these are herbs / natural substances) which would make Kofola different from other cola drinks and could be more interesting, given that Austrians already like herbal drinks but are not associated with cola. Of course, there is also a translation of the composition, which should no longer contain anything in the Czech language, but only in the German language.

Price

The Austrian market is dominated by the large company Coca-Cola Austria, which could be said to determine the pricing policy. Other smaller companies follow it, or according to it, at least orient themselves when setting prices. The two-litre PET bottle of Coca-Cola sells for about 2.25 EUR, a bottle for 1.5 l for 1.85 EUR, 1 l for 1.49 EUR and 0.5 l for about 1 EUR. Prices are of course indicative depending on the retail chain and any special offers.

Given that Coca-Cola Austria is the market leader, choosing a similar pricing policy seems the most appropriate. If the price were lower than the competition, the Kofola drink could give a cheap and lower quality impression. Also, the price cannot be the same as in the Czech Republic, due to different incomes. For these reasons, a price of EUR 1.25 - 1.49 per litre of Kofola is proposed. This price should be sufficient to make sales in Austria profitable.

Communication

Marketing communication is very key to success on the foreign market, given that Kofola is still completely unknown to Austrian customers, the right choice is the way to success. Many companies that tried to enter the Austrian market were unsuccessful due to underestimated marketing communication.

The Kofola promotion campaign should be based on its unique taste, which makes Kofola an original product. As already mentioned, Austrian customers like the herbal drink Almdudler. In the case of Kofola, this is a kind of combination of cola and herbs, so consumers should see it as a kind of compromise between competing Coca-Cola and Almdudler drinks.

Although marketing communication in the Czech Republic is based on the long tradition of this brand and love, in this case the most appropriate way of communication is not tradition and nostalgia for Austrian consumers, this image would not have to communicate anything. Promotion should focus on the natural origin of the product, unique taste, quality and, last but not least, love. It is in the Czech Republic that romantic advertising spots are very popular, which could ensure a certain popularity on the Austrian market as well.

The main motto of the advertising campaign could be the phrase "Kofola - mit Kräutern angereicherte Räder", which can be translated as "Kofola - wheels enriched with herbs" or "Kofola - der mit Liebe und Kräutern", translation "Kofola - with love and herbs "

The communication mix can be divided into several basic forms, namely advertising, sales promotion, public relations, and direct sales. Due to the fact that Kofola is still an unknown product, investing relatively large amounts in, for example, television advertising is risky. The company is in its infancy in this market, so the author finds the following as a suitable form of communication:

Advertising - press advertising

- Press advertising should focus on magazines and periodicals rather than traditional dailies. Examples of magazines in which the advertisement could be found are, for example, Woman, Gusto, Maxima Magazin, Moments or Fizzz.
- Advertising on the Internet would complement the printed advertising, of course there would be the creation of an Austrian website, with a domain name such as kofola.at, where the customer could find any information about the product, its advantages, history and origin. Then there are the various banners on the website, which will be in the same spirit.

Sales promotion - tastings

- The product tasting could kick-start retail sales, which are controlled in Austria by three Rewe Group Retail companies, which include stores such as Billa, Penny Market or Merkur, Internationale Spar Centrale with Spar, Eurospar and Interspar, and third is Aldi. Kofola would also be sold in these stores, as Kofola is unknown, it would be risky to go on sale in restaurants and other snack establishments.

Public relations - sponsorship

- Sponsorship is very popular in Austria and all types of PR in general. The Austrian nation as well as the Czech nation loves cultural and sports events, so sponsoring a music festival such as the Donauinselfest in Vienna or the Filmfestival Rathausplatz could be the right choice. As far as sports activities are concerned, the Vienna City Marathon is held every year in Vienna, which is attended by thousands of athletes and spectators every year.

Direct marketing - online marketing

- Advertising on social networks is one of the best places to gain new customers and is also a huge player in interpersonal communication. Creating a Facebook profile, but mainly an instagram profile is the basis. Subsequent addressing of influencers who would promote the Kofola drink and organize competitions, such as foodblogger Stefanie Anich.

Distribution

There are several forms for entering the foreign market. Commercial representation, capital-intensive inputs, which can be, for example, licenses or mergers and acquisitions, which fall under capital foreign investment. As finding a suitable partner could be very difficult, none of the non-capital inputs is suitable for Kofola. These inputs could be considered if Kofola succeeds in the Austrian market, given the complexity of the inputs. The author finds the choice of entry through a sales representative or distributor the most suitable for the beginning. Kolarik & Leeb GmbH, which has been cooperating with the Czech company Budějovický Budvar since 192, could be a potential distributor, and other possible distributors can be found on the website of the Austrian Chamber of Commerce www.wko.at. The author finds this choice to enter the foreign market the most suitable due to low risk and low entry costs. After a possible success, Kofola ČeskoSlovensko can build another subsidiary in Austria.

5 Discussion

The following answers to the research questions emerged from the above analyses and surveys.

1. *How many competing companies operating in Austria provide a similar product?*

As for the competing companies that have their own branch in Austria, the biggest competitor for Kofola Plc is the company Coca-Cola Austria, which is the market leader. Furthermore, Almdudler, which produces and sells a beverage which, with certain factors, resembles Kofola, as it contains a mixture of herbs.

If companies operating on the Austrian market through another company were also taken into account, this would be another competitor, PepciCo. However, all carbonated soft drinks that may be in competition with society may be taken into account.

It could be said that there are three major competitors in the Austrian market that provide similar.

2. Which area of the 4P marketing mix will need to be most adapted for the selected market?

Due to the fact that Austria is in several ways very similar to the Czech Republic, such a large adaptation of the marketing mix was not necessary. However, some changes were necessary, the biggest one could be the change in marketing communication, as well as the product, specifically the packaging, which required modifications at first glance, namely the language version.

6 Conclusion

This work is focused on the adaptation of the marketing mix to the selected country, in order to analyse both the Austrian market - competition, market entry and PEST analysis of this country, and the company's position in the Czech market and assess whether it is in a strong enough position to expand into foreign markets and subsequently design an adapted mix for the Austrian market.

The first point was the processing of the SWOT analysis of Kofola and the subsequent evaluation using the IFE, EFE and IE matrices, thanks to which it was evaluated that the company is in a sufficiently strong position to enter new markets. Furthermore, the basic marketing mix of the 4P company was described. It was subsequently adapted to the Austrian market, which resulted in an answer to one of the research questions asked, "Which area of the 4P marketing mix will need the most adaptation to the selected market?". In some respects, Austria is a very similar country to the Czech Republic, which also emerged from the PEST analysis. It was not necessary to significantly adapt all components of the mix, but the greatest need is marketing communication, which will focus more on unique taste than tradition and love like in the Czech Republic. Of course, the packaging will need to be adapted to the language version and adapted to the marketing campaign.

The next step was an analysis of competition in the Austrian market, which revealed that there are three major competitors for Kofola as, Coca-Cola Austria and Almdudler, which produces a herbal drink similar to Kofola and possibly PepsiCo, but which operates so far only represented on the Austrian market.

Given that the size of the Czech market is insufficient for such a large company, it could be said that it is necessary to expand into foreign markets. Kofola has already managed and completed it in several foreign markets, whether in the east or in the neighbouring countries of the Czech Republic, so Austria is another good opportunity for expansion.

References

1. P. Kotler, G.Armstrong. *Principles of Marketing*. United Kingdom: Pearson Education Limited (2017)
2. S. Godin. *This is marketing: You Can't Be Seen Until You Learn to See*. New York: Portfolio (2018)
3. 50 Minutes.com. *The marketing mix: Master the 4 Ps of marketing*. Brussels: Plurilingua Publishing (2015)
4. M. Zamazalová. *Marketing obchodní firmy*. Praha: Grada Publishing a. s. (2009)
5. W. Van Waterschoot, CH. van den Bulte. The 4P Classification of Maketing Mix Revisited. *Journal of marketing*. **56**(4), 83-93 (1992)
6. R. Codita. *Contingency factors of marketing-mix standardization*. Berlin: Gabler Verlag (2011)
7. M. T. Khan. The Concept of 'Marketing Mix' and its Elements. *International Journal of Information, Business and Management*. **6**(2), 95–107 (2014)
8. D. Jakubíková. *Strategický marketing: strategie a trendy*. Praha: Grada Publishing a. s. (2013)
9. E. Constantinides. The marketing Mix Revisited: Towards the 21st Century Marketing. *Journal of Marketing Management*. **22**, 407-438 (2006)
10. D. Jakubíková. *Marketing v cestovním ruchu: jak uspět v domácí i světové konkurenci*. Praha: Grada Publishing a. s. (2012)
11. M. Baker. *Marketing strategy and management*. United Kingdom: Palgrave and Macmillan (2014)
12. T. Richter. *International marketing mix management: Theoretical Framework, Contingency Factors and Emprical Findings from World-Markets*. Berlin: Logos Verlag (2012)
13. B. Štědroň a kol. *Mezinárodní marketing*. Praha: C. H. Beck (2018)
14. J. Přikrylová, H. Jahodová. *Moderní marketingová komunikace*. Praha: Grada Publishing a. s. (2019)
15. H. Machková, E. Černohlávková, A. Sato, J. Malý, J. Sedláček. *Mezinárodní obchodní operace*. Praha: Grada Publishing a. s. (2014)
16. H. Machková. *Mezinárodní marketing*. Praha: Grada Publishing a. s. (2015)
17. M. Karlíček, M. Machek, M. Novinský, Z. Chytková, J. Dvořák, P. Bartoš, J. Koudelka, L. Tyll, R. Tahal, J. Mařík, D. Říha, P. Pištělák. *Základy marketingu*. Praha: Grada Publishing a. s. (2018)
18. M. Palatková. *Marketingový management destinací*. Praha: Grada Publishing a. s. (2011)

19. A. Gerald, J. Strandskov, E. Duerr. *International Marketing and Export Management*. United Kigdom: Pearson plc. (2005)
20. Kofola. Konsolidovaná výroční zpráva za rok 2018. [online]. Available at : https://investor.kofola.cz/wp-content/uploads/2019/05/08-annual-report-2018-kofola-ceskoslovensko-cz-_70eab829fa24c9478021e58e117881f7.pdf

Evaluation of Level and Quality of HR Business Partnering in HR Department of Production Company

Kristýna Šindelková[1,*]

[1]Institute of Technology and Business, Faculty of Corporate Strategy, Okružní 517/10, České Budějovice, Czech Republic

Abstract. The objective of this study was to understand the quality and level of HR Business partnering within HR department of a production company. In relation to HR Business partnering, several activities are defined, which are closely interconnected with the HR department of the company. In connection with the determined activities, there are defined skills, abilities, and qualities a modern HR officer should have in the context of the individual processes and HR Business partnering. The main benefit of this study consists in determining innovative techniques in the process of human resources management, which may help increase efficiency and improve the transformation of individual corporate processes.

Keywords: HR Business partnering; Human Resource Management; personnel management; HR department; social responsibility; sustainable development

1 Introduction

The process of human resources management is not limited to managers that lead their subordinates; it includes other employees working in the production company on the basis of employment contract. An important part of the process of human resources management in an HR department is to specify the direction and goals in organizing human capital. This would enable the maximum use of both corporate and personal goals of employees of a selected company.

The key problem in HR Business partnering consists in the importance of the change which leads to new position in the company. HR Business partner sees his/her role as a shift from routine and administrative activities to far more diverse, consultancy activities, with a close interconnection of strategic processes and activities. HR Business partner is expected to achieve self-realization, commitment and increased engagement, which is often a symbol of prestige and it is a priority required from HR Business partners. A big problem for HR Business partner is the transformation of his/her position concerning the preparedness and ability of adaptation to changes in order to achieve long-term efficiency and coherence in the HR department as well as in other departments of the company. The ability of transformation should be based on the intention, change, precise definition of HR Business partner´s role and its content with regard to the technical prerequisites and organizational changes of employees in the HR department. What is important for each HR Business partner is his/her qualitative competence, preparedness for changes in the tasks performed and in measures with personnel character. Long-term effective persuasion of managers represents a great challenge in HR Business partnering, where communication between these entities as well as among all stakeholders plays an important role.

2 Literary research

Key benefit for processing theoretical knowledge is the use of scientific articles and scholarly literature dealing with the issue of innovative techniques in modern Human Resources (HR). In the problematic area of HR, it is necessary to focus on new, innovative knowledge that is changing and adapting within the global environment in HR department. However, it is necessary to build on the theoretical knowledge which pave the way for new theories and knowledge in HR Business partnering. This knowledge enables the improvement of partial steps and the level of HR management. According to the author of the paper, the key professional literature is the publication by Vojtovič [1], which deals in detail with personnel management in the conditions of scientific progress with the concept in production management, with subsequent practical application in personnel management.

Morris et al. [2] argue that the major changes in Human Resource Manager (HRM hereinafter) are caused by managerial experience, which are constantly improved and are closely interrelated, depending to a certain extent on new organizational forms in the company. The author believes that findings published by Battaglio [22]are very innovative, as the author focuses on the issue of human resources and the use of employee hard skills with subsequent promotion of the access along with the soft skills in HR department.

[*] Corresponding author: 21275@mail.vstecb.cz

The specification of the development of human relations, which have an impact on new roles in the HR department, can be considered very innovative. The models of sustainable human resources management (SHRM) are constantly developing; in the literature, several useful models are proposed, e.g. the model by Zaugg et al. [3], which identifies the success factors in the perspective of the organization and employees. The author also defines the fact that the empirical research depends on the sustainability of HRM in practice. The study also systematically expands the concept of HRM, which is based on conceptual and qualitative case study. Järlström et al. [4] state that any organization needs a basic concept capable of satisfying all stakeholders in the process of HR Business partnering. The study by De Prins et al. [5] focuses on the perspective of HR in the long term; Kumar et al. [6] deal with specific models of sustainable HRM. The issue of HR Business partnering as a model is further addressed by Hunter [7], who points to the changes and challenges concerning the role of HR Business partner. The author provides an overview of newly-created models in providing services for the functioning of HR, such as changes in the development of shared services or outsourcing.

There are many authors who deal with the issue of roles in the HR sector, e.g. Dhanpat et al. [8], who identified seven major topics indicating that HR officers must take over the roles of strategic business partners and talent managers, which requires personal competencies, such as being a creative innovator, change agent, strategic partner, and enablement of people that need to be hired in the area of human resources. So far, there have not been many studies focused on Generation Y with a subsequent connection of HR department. Pechová [9] concludes that a vision is very important in HR department, as it can serve as a basis for evaluating various characteristics of a creative person and other criteria for evaluating innovative technique of VAC (The virtual assessment centre).

The precise specification of the role of experts in the area of human resources is addressed by Ulrich and Brocbank [10], who define various roles of employees working in the area of HR. According to Bakuwa [11], HR officers are professionals who have a number of roles within HR. The most common role is human capital developer, who supervises and works individually or in a team to ensure employee satisfaction so that they are successful in the near future. Industry 4.0, referred to as the current digitization trend, is related to the automatization of production in production companies and the changes in labour requirements brought by new technologies of practices and work opportunities, as added by De Ruyter et al. [12]. Ammari et al. [13], employees need to update their competencies according to the requirements of new industries.

Many employees need to be retrained in order to be able to adapt to this transformation. The functional expert, formerly known as administrative expert, takes a new role of an expert related to the human capital practice, which are a key to the HR value [11]. According to Sharma [14], HR officer shall participate in the development of HR systems in order to achieve their proper functioning in the organization. Experts in the area of human resources as functional experts contribute to the competitiveness of the organization in terms of its human capital by selecting and recruiting the best candidates.

Methods and procedures for filling the vacancies are important for the adaptation of personnel information system, which serves as a basis for the application system and the support for personnel work. Network-oriented personnel management of human resources directed by strategies for various technologies plays an important role [15]. Decision-making on human resources is essential, as stated by Stewart a Brown [16] or Mathis [17], who focus on the current issues in the area of Human Resource Management. The author considers human resources planning to be successful if strategically targeted at the specific needs of the organization. The author determines the strategic planning of human resources and emphasizes the role of HR officers as business partners who are of crucial importance for strategic planning and procedures within HR department. The importance of the HR Business Partners as the strategic role for organization is addressed by Kenton and Yarnall [18].

International human resources management provides a critical evaluation of the current international human resources management. This is confirmed by Armstrong and Taylor [19] and the theories and models in human resources management. There are also new trends that change constantly with human resources management and that need to be considered. As Horváthová, Bláha and Čopíková [20] state, these trends most often concern the application of ethical principles within human resources management, personnel controlling, or diversity management. The research of the authors is related to generation Y and its efforts for engagement and clarification of the role of HR department in employee development. The trends are also connected with the innovations in modern recruitment. Tegze [21] focuses on the practices of modern recruitment, which serve as a basic pillar for finding employees.

The author of the paper submitted applies the above issue in the evaluation of the level and quality of HR Business partnering in an HR department. The main objective is to present and analyse new measures in HR Business partnering in a selected company and thus streamline the transformation of processes and their partial information in HR Business partnering. The main output of the study is to define the prerequisites that enable a long-term and effective innovation of trends in modern human resources management. For the purposes of the paper, the following two research questions and one hypothesis were formulated:

RQ1: Which skills and competencies does an HR manager need in business partnering and in HR department of the company Mondi Bupak?

RQ2: What business processes or activities are interconnected with HR Business partnering in the company Mondi Bupak?

H1: Social responsibility and sustainable development of the company Mundi Bupak helps develop and participate in other activities within corporate processes and in HR department.

A prerequisite for answering the research questions is targeted development of competencies for current and potential future processes in the company Mondi Bupak. The main goal is to improve and implement partial changes in a HR department.

3 Methodology

The company selected as a test sample for analysis within an HR department is Mondi Bupak, s. r. o. based in České Budějovice. Mondi Bupak is one of the leading manufacturers of high-quality corrugated cardboard. The company has a long tradition in manufacturing and specializes in manufacturing innovative packaging solutions for a wide range of industries, including food and consumer industries. The portfolio of the products of the company includes standard slotted cardboard and shape-cut packaging, as well as packaging with high-quality multi-colour flexo-printing, packaging with applied tear tape, reinforcement tape, and sealing tape, moisture and oil-resistance barrier packaging, water-resistant corrugated cardboard gluing.

Company policy of Mondi Bupak consists mainly in the product quality, environmental care, and occupational safety. To achieve the above objectives, which are included in the company policy, the company uses an integrated quality system. A part of the system is an experienced sales and service team, which provides customers with full support and logistics services that ensure both current and above-standard requirements. Above-standard requirements concern direct deliveries as well as deliveries ensured through external warehouse.

The environment of the company is based on the concept of sustainability, which is also the strategy of the company. The corporate environment encourages the growth of values, which may ensure the success of the company within the global challenge of sustainability. Human resources department, as well as other departments in the company promote open communication and cooperation, which facilitate to address risks and opportunities with the aim to generate values for all stakeholders in the long run and tackle complex global challenges, which other competitors are only able to tackle separately. Performance is a priority for the company in terms of its focus on lean processes and quality management programmes which increase the productivity and efficiency. The company applies long-term processes of continuous improvement, which are focused on increasing productivity, efficiency, reduction of waste and ensuring lean processes. Mondi Bupak also focuses on supporting organic growth and strengthening cost competitiveness. The key activities in the area of human resources management establishes the safety of employees at the administrative level as well as in potentially hazardous environment, such as in production. The long-term goal is to encourage critical thinking of employees and thus improve their skills, which are part of the development of agile and motivated employees. This enables further implementation of the company´s strategies and thus manage the success of the company in a sustainable way.

The method of observation was used as a primary method for evaluating HR Business partnering in the Human Resource department of the manufacturing company Mondi Bupak, which enables monitoring the processes in the company and identify the intensity of individual activities of the Human Resource department. The observation method enabled the introduction of the interconnection of individual personnel processes in the HR department and their systematic summary and application to the dependence of phenomena in the HR Business partnering. A significant starting point was the analysis of key activities, which facilitate the transformation of the processes and sub-activities within the company. The application of the observation method was followed by the application of the relational method, which enabled the analysis of the dependence between HR Business partnership, HR department, and corporate processes. The method of observation and the relational method are focused on the interconnection of the processes in the following areas:

- Challenges, strategies, management culture,
- Market, supply and customers,
- Operation, partners and suppliers,
- Management structure and activities,
- Sustainability approaches and practices,
- Education and development of employees,
- Corporate social responsibility,
- Partnership,
- Communication in organization.

Considering the above areas, a structured interview with the head of the HR department is conducted. The questions in the interview are formulated with regard to the aforementioned processes in the areas determined based on the results of the observation and relational method. The structured interview serves as a source for collecting corporate data and information about corporate processes. Individual questions in the interview are focused on the areas such as employee satisfaction (e.g. motivation, financial benefits, etc.), interconnectedness of individual processes (e.g. selection of suppliers), employee involvement in decision-making processes in order to improve the operation of the organization and processes concerning the HR department.

The structured interview is processed using the content analysis, which enables a systematic, objective and quantitative analysis of the interview outputs. The content analysis also enables sorting out the data obtained through the interview and systematic identification of the key data, which are subsequently used for relational analysis.

4 Results

In the globalized world, sustainability within HRM and manufacturing techniques in a manufacturing company is assumed to be related to the proper materials and practical use of assimilation only. The sustainability of traditional roles and tasks of a HR department often shift to HRM competencies in order to maintain the human resource base in the organization. HRM significantly influences organizational viability of human resources in the company. Recently, there have been a number of studies relying on researchers in order to maintain human resources management and strategic potential for HRM.

Within its global strategy, Mondi Bupak focuses on human resources, specifically, recruiting, retaining and development of human resources in order to gain competitive advantage over other manufacturing companies through human resources management.

The method of observation applied in the structure of the management identified management activities based on excellent production, successful customers, sustainable development, quality solutions and inspiring people. In terms of processes interconnectedness in the organization and in the HR department, the observation method and documentary analysis identified sustainability approaches and practices, which were subdivided into 6 groups and further specified by the parameters below.

Table 1. Sustainable approaches and practices in Mondi Bupak

Parameters	Specification of parameters
Safety of employees and suppliers	Avoid completely fatal accidents related to the work of employees and suppliers
	Prevention of injuries with long-term consequences
	Reduction in the number of work-related accidents
	Greater employee security in professional and private life
Qualified and engaged employees	Cooperation with employees in order to create a better workplace
Fairness at workplace	Promotion of fair working conditions at workplace
Supplier retention, responsible procurement	Promoting the transparency of supplier change, promotion of fair working conditions with key suppliers
Relationships with communities	Strengthening social values in communities by means of stakeholders and meaningful investment
Propose solutions that would create value for customers	Promote sustainable and responsibly manufactured products

Source: Author.

Based on the observation method, an interview with the head of the HR department, HR manager Arnošt Vydra, was conducted. The first part of questions was aimed at employee satisfaction. The methods used by the company to determine the satisfaction of its employees include a regular survey of employee satisfaction or Employee Survey. The survey includes several sub-categories, and employees comment on various topics. The survey is followed by the evaluation and presentation of results. Subsequently, action plan "You said, we did" is prepared. The results of the survey are compared with the results from the previous surveys conducted earlier. Employee satisfaction and implementation of the action plan are thus greatly emphasized. Based on the results, there were built new employee showers, the remuneration system was changed, the system of passing information and communication between individual departments was improved, etc. Another method is a regularly conducted development interview with a supervisor - PDR (Performance Development

Review), which, in addition to the evaluation of the previous year, results in creating an educational business plan for the next year and setting the objectives for further development.

The second area concerned employee motivation and benefits. In terms of determining the interconnectedness of individual processes, the author of the paper considers very important to identify the attitude of the company towards its employees in the form of their motivation and stimulation. An assumption is that individual motivational factors enable the company to ensure the smooth running of the corporate activities in individual processes. The outputs of the structured interview in the area of employee motivation and benefits include non-financial benefits regulated by the company in the collective agreement, which is based on the social development and care for employees. The main benefits provided by the company include: reduced working hours, extra week of holiday, healing programmes (massages, pedicure, vaccination), social programme, corporate catering, team sports support, benefits aimed at employees´ children (children´s camps, ski courses), corporate events, "Making a difference day" – a one-day event to raise environmental awareness, waste sorting, chemical substances disposal, occupational health and safety, fire protection.

The third part of questions in the interview focused on the stimuli which enable to ensure employee work-life balance. Employees are offered flexible working hours when possible. All employees can make use of services offered by contract masseurs or pedicurist (stressed parts of body, necessity to wear safety toe shoes in production, etc.). Within extended health care offered by Mondi Bupak, full vaccination against tick-borne encephalitis and subsequent revaccination, and flu vaccine are covered. As for child-care benefits, the company contributes financially twice a year to activities organized by schools, interest organizations, sports clubs, etc. This refers mainly to children´s camps, sports courses, ski courses. Mondi Bupak contributes 65% of the total voucher price for multi-day stays in the Czech Republic, and CZK 2,500 in the case of stays in foreign countries. This benefit can be used by employees for all their children. As for winter season activities, the company rents an ice rink for employees and their children. The work-life balance is supported by providing an extra week off for their employees during Christmas holidays, when the production is interrupted so that employees could enjoy this holiday with their families and friends. This benefit helps motivate employees and reduces staff turnover as well as employee absenteeism.

The fourth part of questions dealt with the activities beyond the applicable legislation in terms of occupational safety and health protection. For Mondi Bupak, employee safety is a priority. As part of occupational safety, Mondi Bupak is proud of the fact that no person who enters or work on company premises has ever been injured or damaged. This thus applies to the safety of employees, suppliers, or visitors. Every year, the company organizes a "Making a difference day", an all-day event aimed at raising environmental awareness, sorting waste, disposal of chemical substances, and occupational safety and fire protection. During the year, the company also carries out interactive training beyond the applicable legislation. The company tries to prevent work-related injuries to eliminate them. Interactive training has a form of exchanging information with other companies.

The remaining part of questions concerns the involvement of employees in the decision-making processes related to the improvement of the operation of the company. The company has a database of employee improvement proposals. The employees can express their opinion and make specific improvement proposals. The company also conducts regular surveys of employee satisfaction, holds regular meetings of employees and management of the company, during which interesting suggestions and improvement proposals are made. Another way to involve employees in the decision-making processes is Trust box, through which the employees can express their views and opinions.

The research methods applied have brough results concerning the interconnectedness of the processes with the HR department activities. Based on the results, the author of the paper was able to compile a scheme of interconnectedness of activities in connection with the HR department (see Figure 1). A key element in HR business partnering with the HR department is sustainable development and corporate social responsibility, on which other corporate processes and activities are dependent.

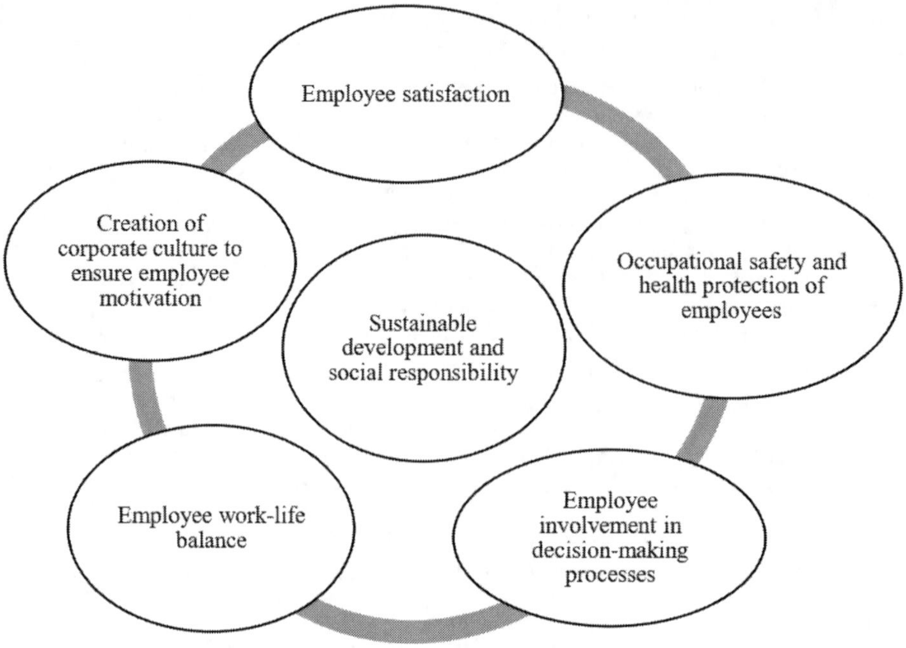

Figure 1. Scheme of interconnectedness of activities in relation with HR department

Source: Author.

Individual elements in the scheme were determined with regard to the strategic values of the company Mondi Bupak, which are based on the values supporting sustainable and responsible cooperation with customers, partners, and employees in the organization. The aim is to increase the social value of communities by means of stakeholders using global framework, which enables addressing local priorities. There are also supported steps to promote greater transparency and functioning of the conditions that serve as a basis for creating reliable, inclusive and sustainable supply chain. The corporate policy aims at qualified and engaged employees; therefore, it strives to build a culture that would inspire, involve and develop all employees, thus achieving the development of their potential. The workforce of the company is based on the diversity of job positions.

The documentary analysis shows that Mondi Bupak carries out a number of activities focused on responsible growth. Specifically, this refers to the reduction of the impact of its activities on the environment, emission and waste reduction, promotion of responsible forestry and development of relationships based on cooperation, which is critical to future success and strengthening a zero-accident safety culture.

Based on the theoretical skills and observation method in the company, the author determined the competencies and skills each HR manager in business partnering and HR department would need. HR manager competencies and skills need to be applied to the specific interconnectedness of the processes in the company, as already determined by the author. The scheme of the competencies and skills created by the author is shown in Figure 2.

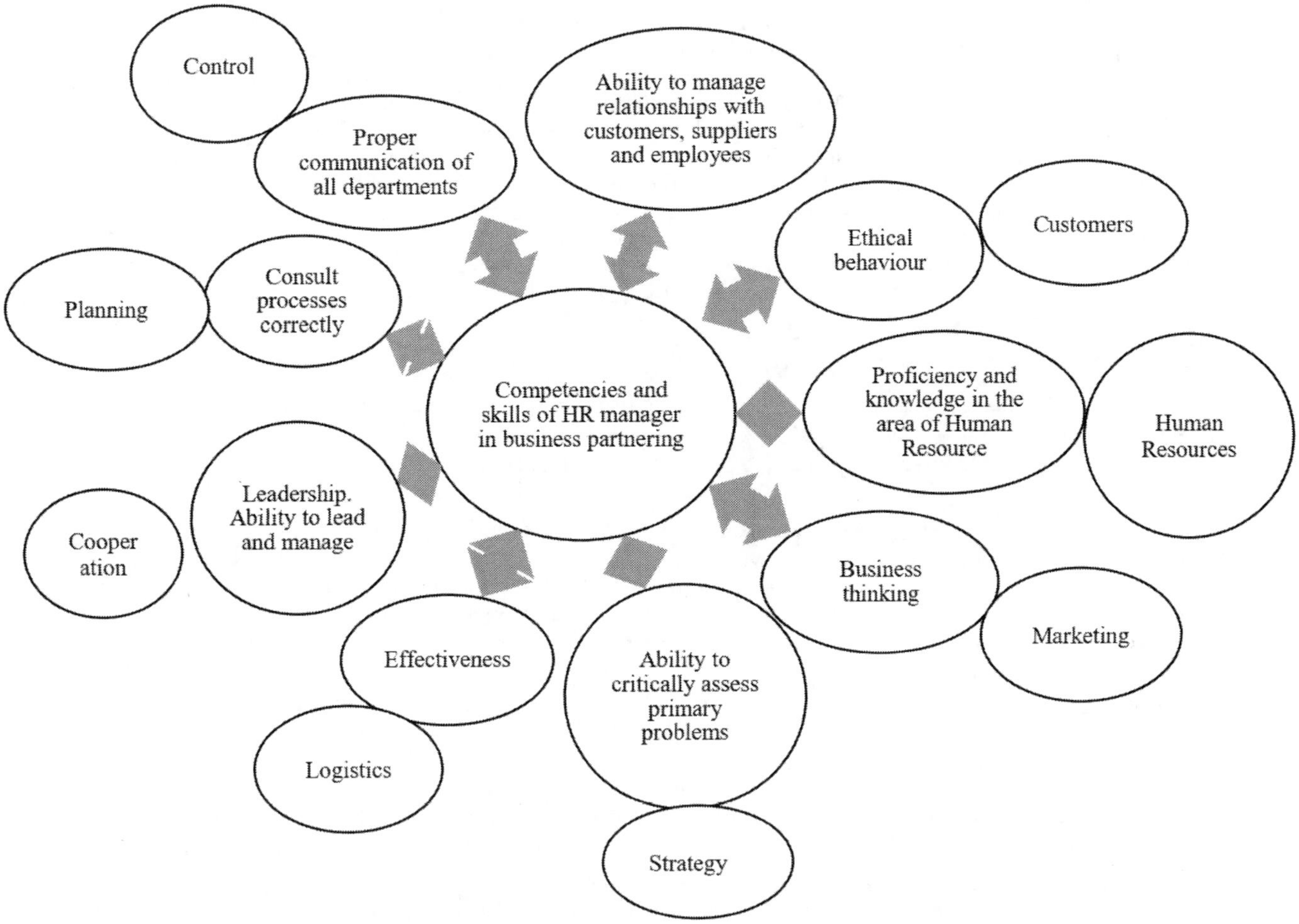

Figure 2. Necessary competencies and skills of HR managers in business partnering in Mondi Bupak with regard to the interconnectedness of individual processes

Source: Author.

Corporate processes were determined by the author on the basis of the observation method. It applies to the processes which are key in the company according to the author. The key element is determined competencies and skills of a HR manager in business partnering. The company emphasizes good communication in the management of relationships with customers, suppliers and employees, with a focus on the business process. The company strives for processing the orders as fast as possible and for meeting the requirements of a given customer. This is closely related to ethical behaviour, which is embedded in the Code of Ethics of the company Mondi Bupak, emphasizing solidity, goodwill and ethical behaviour along with sustainable development and social responsibility. Human Resources, specifically HR management, is a key process in the company, which deals with recruitment and adaptation of employees, employee satisfaction, and regular training in order to prevent staff turnover. By means of a structured interview with the head of the HR department, the author found that even during the COVID-19 pandemic, staff turnover in the company remains minimal, as employee satisfaction is a good prevention of high staff turnover. Proficiency in the process of HR management is very important for an HR business partner; without it, it is not possible to understand and adapt to other processes in the company. HR business partner should have business thinking, which is important from the perspective of the international operations of the company together with timely order processing. HR business partner should also be able to identify, evaluate and solve primary problems. Both primary and secondary problems are linked to other departments, logistics most often, where effectiveness plays an important role. Effectiveness might ensure the correct production, proper storage of products, removal and transport of the products to customers.

An important part of corporate processes is cooperation, especially in terms of proper management and leadership, which helps motivate, stimulate and manage employees in all corporate departments. HR business partner must be able to plan and consult processes, starting from processing orders through coordinating production in terms of logistics to removal of finished products. Precision, transparency and the ability to process individual requirements related to contracts play an important role. Last but not least, the competencies of an HR business partner must be based on controlling, which includes proper communication of all departments.

To improve effectiveness and interconnected of processes, HR managers in business partnering should use chatbot, which is able to digitize them. Specifically, it is HR Mobile, which allows HR managers to approve various matters using smart phone. This trend helps develop corporate culture and share knowledge and skills both from the side of managers and employees. Another possible trend in HR is the app Leadership Journey, which is intended for the development of managerial competencies and E-skills. Digitization and robotization of HR paperwork plays an important role in current e-learning in cooperation with applicants. The current trend applied in human resources management is the method of 360-feedback, which is gaining in popularity. The method is based on the principle of collecting data on an employee or a team from ore evaluators or assessors. Due to using various resources, it becomes very objective. The model of 360-feedback has a wide range of application. Most often, it is in the development of employee training and allocating bonuses. Data collection has a form of anonymous questionnaires, where the questions enable the identification of key areas for the development of individuals and the whole departments. According to the author, this method is the most effective one for improving the transfer of information between individual corporate processes ensured mostly by HR business partners.

5 Discussion

The processed results presented in the schemes enabled the specification and transparency of the interconnectedness of individual activities. The schemes include the competencies and skills of HR managers in Business partnering with a focus on key corporate processes. At the beginning of her research, the author formulated two research questions and one hypothesis.

The hypothesis was focused on social responsibility and sustainable development of Mondi Bupak, which enables the company to develop and participate in other activities within corporate processes and HR department. As seen in Table 1 and the scheme in Figure 1, the hypothesis was confirmed. In Table 1, the author defines sustainable approaches and practices with a focus on employee safety, qualification, engagement, responsibility, and generating values. The hypothesis can also be confirmed based on the scheme in Figure 1, which shows sustainable development and social responsibility participating in corporate processes and HR department.

The first research question deals with the skills and competencies HR managers need in business partnering and HR department of Mondi Bupak. These competencies are presented in the scheme in Figure 2. Based on the research methods and theoretical knowledge, it can be stated that HR managers must be able to manage relationships with customers, suppliers and employees; must act in line with the principles of ethical behaviour; must have professional knowledge and skills in the area of Human Resources; must be capable of business thinking; must be able to critically assess primary problems, think effectively, manage and lead people, consult processes and communicate correctly with all departments.

The second research question dealt with the interconnectedness of corporate processes and HR business partnering. This is also presented in Figure 2 along with the individual competencies and skills. Based on this method, key processes were determined, where the competencies and skills of HR business partners are applied. Specifically, this refers to processes such as business, customers, Human Resources, marketing, strategies, logistics, cooperation, planning, control.

The results also define innovative trends or practices in the process of human resources management, which may help improve effectiveness and transformation of individual corporate processes. Specifically, these include HR Mobile, application Leadership Journey, and 360-feedback. According to the author, the most effective one is the 360-feedback method, which is able to streamline and improve the transfer of information between corporate processes, which are a part of HR business partnering.

6 Conclusion

In general, the role of HR business partner assumes the knowledge of business, strategic thinking, managerial skills and language skills. However, HR business partner must have other competencies and skills, which are related to key corporate processes, as in the case of the company Mondi Bupak. One of the methods which turned out to be suitable for the identification of the HR business partner competencies and key processes was the method of observation. For the purposes of the paper, a structured interview with the head of the HR department was conducted, which enabled to outline the possible problems in human resources management. The results of the interview, together with the content analysis and documentary analysis enabled to determine sustainable approaches and practices and social responsibility of the company Mondi Bupak. Social responsibility is very important in the current globalized world, playing an important role in both internal and external corporate culture. Good corporate culture and employee satisfaction improves the quality and efficiency in performing the assigned tasks. This fact is intertwined in all corporate departments and processes within business partnering.

HR manager must be able to ensure quality communication between the management and employees, promote team work, involve all employees in business and have an overview on the interconnectedness of all key processes. HR business partner´s competencies and skills intervene in business and operation of the company. The dedicated role of HR Business Partner is currently a matter of course for many international companies. This is also due to innovative thinking an HR business partner acquires thanks to the overview of corporate processes. Nowadays, HR Business partnering is primarily about mutual trust, support, and service among all employees in the organization. The process of human resources management is intended for all HR Business partners which lead their employees as well as workers who have a great influence in the key processes of a given company. HR business partnering sets the orientation and goals of a given organization in the area of corporate culture. Based on this fact, HR Business partner implements a suitable practice in the area of motivation and human capital management so that the corporate and personal goals of employees converge to the maximum extent possible.

References

1. S. Vojtovič, S. *Koncepce personálního řízení a řízení lidských zdrojů.* Praha: Grada Publishing a. s. ISBN 978-80-247-3948-9 (2011)
2. J. Morris, J. Hassard, R. Delbridge, et al. Understanding managerial in the modern Japanese firm: The influence of new organizational forms and changing human resource management practices. *Economic and Industrial Democracy.* 1-51 (2019)
3. R. J. Zaugg, A. Blum, N. Thom. Sustainability in Human Resource Management. *Evaluation Report. Survey in European Companies and Institutions.* 1-25 (2001)
4. M. Järlström, E. Saru, S. Vanhala. Sustainable human resource management with salience of stakeholders: a top management perspective. *Journal of Business Ethics*. **152**(3), 703-724 (2018)
5. P. D. Prins, L. V. Beirendonck, A. D. Vos, J. Segers. Sustainable HRM: Bridging theory and practice through the 'Respect Openness Continuity (ROC)'-model. *Management Revue. Socio-economic Studies.* **25**(4), 263-284 (2014)
6. A. Kumar, P. Bhaskar, S. P. Nadeem, M. Tyagi, J. A. Garza-Reyes. Sustainability Adoption through Sustainable Human Resource Management: A Systematic Literature Review and Conceptual Framework. *International Journal of Mathematical, Engineering and Management Sciences*. **5**(6), 1014-1031 (2020)
7. I. Hunter. *HR Business Partners.* New York: Routledge. ISBN 978-05- 660-8625-0 (2016)
8. N. Dhanpat, Z. P. Buthelezi, M. Joe, T. V. Maphela. The role of human resource professionals. *SA Journal of Human Resource Management* [online]. South Africa: Aosis. **18**, 1-11 (2020)
9. J. Pechová, J. Personnel trends in a globalised world. *SHS Web of Conferences*. **74**, 1-7 (2020)
10. D. Ulrich, W. Brockbank. *The HR Value Proposition.* Boston: Harvard Business Review Press. ISBN 97814221148051 (2005)
11. R. C. Bakuwa. Exploring the HR professionals' employee advocate role in a developing country: The case of Malawi. *Australian Journal of Business and Management Research*. **2**(12), 39–48 (2013)
12. A. D. Ryuter, M. Brown, J. Burgess. Gig work and the fourth industrial revolution. *Journal of International Affairs*. **72**(1), 37–50 (2019)
13. D. Ammari, A. G. Özüdoğru, E. Ergün, Students. How Industry 4.0 Changes Business: A Commercial Perspective. *International Journal of Commerce and Finance*. **4**(1), 84–95 (2018)
14. D. Sharma. Roles for HR professionals. *Integral Review: A Journal of Management*. **5**(2), 1–11 (2012)
15. M. Šikýř. *Personalista pro manažery a personalisty.* Praha: Grada Publishing, a. s. ISBN 978-80-271-9528-2 (2016)
16. G. L. Stewart, K. G. Brown. *Human Resource Management: Linking Strategy to Practice.* Praha: Grada Publishing a. s. ISBN 978-11-949-2177 (2020)
17. R. L. Mathis, J. H. Jackson. S. R. Valentine, P. Meglich. *Human Resource Management.* Nelson Education, LTD. ISBN 978-1-305-50070-9 (2016)
18. B. Kenton, J. Yarnall. *HR: The Business Partner, Futhering the journey.* Abington: Routledge. ISBN 978-1-856-17847-1 (2010)
19. M. Armstrong, S. Taylor. *Řízení lidských zdrojů.* Praha: Grada Publishing, a. s. ISBN 978-80-247-9883-7 (2015)
20. P. Horváthová, J. Bláha, A. Čopíková. *Řízení lidských zdrojů nové trendy*. Praha: Management Press s. r. o. ISBN 978-80-7261-430-1 (2016)
21. J. Tegze. *Jak hledat zaměstnance v 21. Století: Techniky moderního recruitmentu.* Praha: Grada Publishing, a. s. ISBN 978-80-271-2362-9 (2018)
22. P. Battaglio. The future of Public Human Resource Management. *Public Personnel Management.* **49**(4), 499-502 (2020)

Technical Analysis of Selected Stocks Time Series Based on Value Screening

Michal Tlustý[1,*]

[1] Institute of Technology and Business, Faculty of Corporate Strategy, Okružní 517/10, 37001 České Budějovice, Czech Republic

Abstract. The paper deals with the issue of value screening and shares. For the selection of a company, value screening with specific criteria was used. From a total of 6 companies, the one with the most traded stocks was selected, namely the company Apple Inc. Subsequently, the analyses of trend and moving averages were performed for the 50-day, 100-day, and 200-day period, as well as the analysis using RSI indicators. Specific changes detected were graphically illustrated and described in detail in graphs. In the following chapter, Discussion of Results, specific indicators were explained, which emerged within the analysis of the shares, and proposals of measures for a broker were made, which recommended the trader to buy or give up the position. The objective of the paper, which consisted in the performance of value screening using specific criteria, select a suitable company, analyse its time series, and propose a specific measure, was achieved.
Keywords: technical analysis; shares; technical indicators; value screening; stock markets

1 Introduction

Stock market and stocks in general are currently a very topical issue, mainly due to the coronavirus pandemic. A lot of companies have suspended their activities or their operations have been 50% reduced, which decreases their value on stock markets. Such measures may be necessary given the current situation but it can represent a real problem in the long run. Small companies which do not have sufficient capital for surviving months without any income in the situation when fixed costs still need to be paid, will start to go bankrupt. This is followed by failure of other smaller and micro enterprises, which could produce and sell their product but since the large enterprise which had been buying 90 % of their production had to suspend the operation, they go bankrupt as well. However, every pandemic will end one day, and this situation appears to be suitable for investors in stock markets, who might take advantage of these declines.

The time series shows the changes in the prices of a specific company stocks over time and helps see the context that might not be obvious at first glance. Such information is often presented in the form of a graph, which enables predicting the future development of the data under review and if necessary, to prepare for it in advance. Nevertheless, it should be born in mind that a 100% reliable prediction is not possible and might not correspond to the real development. In general, it can be said that the more and the higher quality data available, the more accurate the prediction will be.

The paper aims to use appropriately defined value screening to identify a specific company and analyse its time series using suitable tools. In conclusion, its development is predicted and investment recommendations are made.

2 Literary research

The key to each investment success as well as to the success of any project in general are input data. Without information, it is not possible to be successful in business in the long run, and this particularly applicable in the case of investing in stock markets. There are many ways to analyse and evaluate the input data [1], one of them being the so-called value screening. This consists of two basic components, screening as such and value investing [2]. Screening is a tool that enables traders to process a large volume of data they have to deal with and makes it easier for them to sort. In simple terms, screening helps shareholders to find the information in the vast amount of data they are able to process manually [3]. Screening is basically a new tool. Although there was a certain definition and theory of screening about twenty years ago, screening was not feasible in practice due to the fact that the Internet was not that widespread. This changed about ten years ago when the Internet underwent a major development and value screening has become a basic tool for the evaluation of stock market data. However, its price was very high; therefore, it was available only for very rich investors, who were able to convert the investments in the tool into benefits and screening was custom-made on their data. Currently, the Internet is at a high level practically all over the world and screening as such is available for free to all [4].

* Corresponding author: michaltlusty@email.cz

No novice trader can execute the trade of the century or buy for extremely high prices immediately after entering the stock market. Any novice investor should have a reasonable capital and buy or sell extremely carefully [5]. Therefore, investors in the market can be classified into three individual groups. The first group includes the so-called "small players". These are beginner investors who do not have large capital and are not able to significantly influence the whole market. Their motto should be "Trend is your friend". The second group consists of the so-called "big players", mainly multinational companies that trade billions of dollars every day. An example of big players are e.g. the companies Goldman Sachs or JP Morgan [6]. The third group are the so-called "Brokers", whose task is to move the market and who represent intermediaries between the trader and the stock market [7].

It is important for any trader to be able to analyse trends correctly and deduce clues and indications from the results which would help them to decide whether to buy or not [8]. As mentioned above, one of the rules is "trend is your friend", which means that trend should be monitored and followed. A trader who follows market trends usually makes a profit, while a trade who bucks the trend at all costs usually loses [9]. The length and stability of trend followed by a trader also plays a role. If the market is stable over a longer period, the investor´s chances of making a profit increase. In practice, it is thus important to find long-term and steadily growing trends that a trader might make profit from for many months or years and follow them to the very end [10].

Trends need to be analysed from several different perspectives. If the trend curve grows for a long time and suddenly turns downwards, the highest value the curve has achieved is referred to as "high". Conversely, if the trend decreases gradually and then it suddenly turns upwards, the lowest value is referred to as "low" [11]. The development of the trend in the graph can be divided into three different directions. The first one is the so-called growing trend which indicates that the increasing values of stocks outweigh declining stock prices in the long run; for a trade, this trend behaviour is a signal for investing [12]. The second situation is when decreasing values outweigh the growing values, making the trend decrease in the long run, which is a signal for a trader not to invest or to perform further analysis [13]. The third possible behaviour is the so-called sideway trend, which provides the information that both decreasing and increasing values are approximately at the same level and the trend thus does not show extreme differences. In such a case, it is recommended waiting for other signals, not investing or selling [14].

Stock trend can be divided from the time perspective into a long-term, medium-term, and short-term trend. Long-term trends are the most important ones for investors and play the most important role in market analysis. If long-term trend is growing, both short-term and medium-term trends tend to be growing as well, but it is not a [15]. Short-term trend represents a shorter period than a long-term trend, which means that although a short-term trend is decreasing, the overall long-term may still be growing. A short-term trend describes the behaviour of trend in the order of days or hours and is very time-sensitive. Short-term trends must also include the so-called fresh information, which may completely change its direction, although long-term and medium-term trends are still growing [16].

In general, there are two types of financial analyses available to traders. The first one is fundamental analysis, which deals with the data that can be obtained from the macro-economic environment of the company, state, etc., such as gross domestic product, unemployment rate in percentage or the current political situation in the state [17]. The second type is technical analysis, which deals mainly with trends, RSI indicators, moving averages, and other analyses [18]. The basic difference between fundamental and technical analysis is that technical analysis assumes repeats itself. In practice, this means that if long-term trend is growing, technical analysis assumes that the trend will still be growing based on the historical data [19]. This is only a simplified example and technical analysis can predict the repetition of a large number of successive events from the past. For example, if situation 1 occurred in the past and situation 2 followed, if situation 1 is occurring right now, based on the historical data, technical analysis predicts that situation 2 will follow [20].

In order to achieve the objective of the paper, three hypotheses are formulated:

1) The application of specific value screening criteria will find a maximum of 4 companies.
2) The technical analysis indicators will recommend buying or selling the position immediately after the beginning of the monitored period.
3) RSI indicators will recommend sell or buy, which will prove to be potentially correct in the next period

3 Methodology

Each analysis and prediction of stock time series require data in a specific form. Data suitable for the analysis of time series are available in many forms for free on internet websites. The most common and the most suitable type of data are data arranged in tables divided into six columns. In each row of a specific value of the stock, it is necessary to enter the date on which the value was measured, as well as open and close (the prices at which a stock began and ended trading), and high and low (the maximum and the minimum price). Some statistical datasets also contain the average closing price (adj close) and the volume of traded stocks (volume). These data are mostly used in a graph which shows the movement of stock prices. Graphs can be prepared for a short period (e.g. one hour) or for longer periods (e.g. ten years). Graphs can have a form of a stock chart or a line graph. The difference between a stock chart and a line graph is that stock chart is intended to be used for showing the course of a time series. Most commonly, it shows volume, open, high, low, and close.

Data for the purpose of the paper are obtained from the website Yahoo! Finance. The data show the time series of the last year (from 1 November 2019 to 1 November 2020). The values are presented in US dollars. First, it is necessary to choose a specific company, whose data are to be analysed. For this purpose, a basic screening of joint-stock companies on the US stock exchange is performed, which enables to narrow down the options using the pre-selected criteria to several companies, out of which the one which trades the highest volume per day is selected. A total of 6 criteria for screening are set according to the following arguments: The first criterion is market cap (market capitalization), which provides information about the overall value of a company. The specific selected value is a large company with a market capital of USD 10 billion or more (+Large [over $10bln]). The second criterion is and its specific value vas selected so that the given companies were among the 500 most important publicly listed companies. Another criterion indicates the percentage of earnings the company pays on dividends, which means that the more it pays, the less money it has for investments. Due to the fact that a stable and growing company is sought, the selected value is up to 50 %. Another parameter is the sector the company operates in. For the purposes of this paper, only the technology sector is selected, since financial companies, companies operating in health care and other less interesting sector need to be excluded from the dataset. The next criterion is the average volume of trades per one day. In this case, the value of more than one million is selected in order to prevent the situation that all stocks are bought by a small group of traders and this is reflected in the price of stocks. The last parameter concerns the return on investment, where the value of 25 % and higher is selected in order to ensure the company is profitable and growing.

Table 1. Screening criteria

Criterion	**Criterion value**	**Meaning**
Market cap	>USD 10 billion $	Stable company
Index	S&P 500	Important, publicly listed company
Payout ratio	<50 %	Investment in innovations
Sector	Technology	Information on the company is publicly available
Average daily volume	>1 mil	Stability of traded stocks
Return on investment	>25 %	Profitable and growing company

Source: Author based on finviz.com.

As already mentioned, there are many methods to analyse stock time series and investors should choose a method that would be suitable, meets their expectations, and that would be applicable in the market the investors are active in. For the purposes of this paper, two main indicators are selected, specifically moving averages and RSI indicators.

One of the basic methods for time series analysis is the method of simple moving average (SMA). It is an arithmetic average which changes over time. The most common time periods used in moving average are 30, 50 and 100 days. The basic principle of moving average (and difference between MA and arithmetic average) is that with each day, the analysed data are added a new value, i.e. the information from the current day and at the same time, the oldest value is removed. That is why the average is referred to as "moving". The fact that two data values change every day makes moving average a tool that includes irregular data and whose calculation is different and updated every day. The number of values always represents the number of days for which the moving average is calculated. This method can be used for any dataset which changes over time; in the case of stock markets, price values are most commonly used. Moving averages are recommended especially in the case of analysing a longer time series and trend functions are not sufficient for its analysis. Simple moving averages are calculated as follows:

$$SMA = \frac{(P_1 + P_2 + ... + P_n)}{n} \quad (1)$$

Where:

SMA = Value of simple moving average

P = stock price at a specific time (most often the closing price)

n = overall number of values (mostly days)

Given that the calculation is based on the historical data, it is a lagging indicator. For this reason, there is another type of moving average, an exponential moving average (EMA), which works on the same principle as simple moving average, but the new and updated data are assigned higher weight that the older data. This does not mean that exponential moving averages represent a better method than simple moving average. This always depends on the preferences of investors and the type of market they operate in.

In practice, there are other strategies which combine two or three moving averages. If two moving averages, one of them is "slow" and the second one is "fast". Slow moving average includes more values that fast but for the same time period. In practice, this means that both slow MA and fast MA are calculated for the period of 30 days but slow MA describes one value per day, i.e. 30 values, while fast moving averages describes 24 values a day (one per hour), i.e. 720 values. In this case, fast MA describes the trend curve more in detail, and various crossovers of these two averages provide specific indications that help investors in making investment decisions. If there are three moving averages (fast, slow and medium), the principle is the same, but there are more trend crossovers and the indications are even more specific.

Another method used for analysing stock markets are RSI indicators. It is a universal tool that can be used for any market and any time period. The market analysis using RSI indicators are mostly based on closing prices and for the calculation, it is necessary to choose a relevant period. Periods indicate how many time periods of specific length the calculation includes. For example, if P = 12 and the calculation of RSI indicators is required on the hourly charts, the calculation shall include the values for the last twelve hours. The formula for calculating RSI indicators is as follows:

$$RSI = 100 - \frac{100}{(1+RS)} \tag{2}$$

Where:

RSI = value of RSI indicator

RS = average profit from closing price / average loss from closing price

The formula shows that it is necessary to know the value of RS prior to calculating. If the calculation was carried out on the above example of twelve hours, it is necessary to determine how many times, in the overall period of twelve hours, the value was higher than its previous value and subsequently, the number is divided by the total number of values (twelve in this case). Next, it is necessary to determine how many times the value was lower than its previous value and this number shall be divided by twelve; afterwards, the first determined value shall be divided by the second determined value. If the number of values indicating that the closing price was higher than its previous price exceeded the number of values indicating that the closing price was lower than its previous price, the value of RS is greater than one.

The resulting value of RSI indicators ranges between 0 and 1, where 1 means 100 %. The main purpose of using RSI indicators is that they may help in deciding whether to buy or sell the position in the market. In practice, it is generally stated that the value of RSI indicators exceeds 30 % upwards, the price has got into oversold and buying is recommended. Conversely, if the value of RSI indicators exceeds 70 % downwards, the prices has moved from overbought and its further decrease is expected; it is thus recommended to sell the position. If the resulting value of RSI indicators ranges around 30-70 % not exceeding any of the limits, it is recommended waiting and monitoring the development. However, experts also recommend not selling buying immediately after the values exceed one of the above limits. If any of the limits is exceeded, the investor should wait until the stock price gets from oversold or overbought and sell or buy afterwards.

After the analysis of moving averages and RSI indicators, the results are processed in the form of a graph with a detailed description of the results and the behaviour of all indications is explained for the investor.

4 The development process of strategic management of chinese start-up enterprises

As the first step, value screening of all joint-stock companies was performed. Using the above criteria, a total of six joint-stock companies were found which corresponded to the selected parameters. The first company was a company with the ticker AAPL. This is a ticker of the company Apple Inc., which is based in California and specializes in the production of hardware and software. Their best-known products include e.g. Mac computers, smart phones (IPhone), or tablets (IPad). The second company is a company with a ticker ACN - Accenture plc. The company is based in Ireland and specializes in management consulting, technology services, digital services, and cybernetics. Another company has a ticker CDNS (Cadence Design Systems, Inc.) and is based in San Jose. Its main products include computer software, e.g. chip systems, printed circuit boards or integrated circuits. The ticker of another company is CTXS, it is the company Citrix Systems, Inc. based in Florida, which specializes in software, specifically applications, cloud or virtualization of computers. The fifth company has a ticker TER, Teradyne, Inc. The company is based in Massachusetts and specializes mainly in producing special testing devices. The company is a supplier of many companies with a long tradition, e.g. Samsung or Intel. The last company in the sample is HP Inc. with a ticker HPQ. The company produces computer technology from software through hardware to complete computer systems.

Given that out of the six companies only one of them can be selected for the analysis, the choice is narrowed down using the value of average volume of trades per day. This means that the company with the highest value of the volumes of trade is subject to analysis. In this case, it is the company AAPL - Apple Inc.

The above internet resource provided a dataset of the company Apple Inc. (AAPL) containing data for the period of one year (1 November 2019 – 30 October 2020). The dataset was processed using mathematical calculations described in the methodology part of the paper. All additional calculation and graphs were processed using the software Microsoft Excel. As the first step, RSI indicators were calculated using the period of 14 days. The limit of the oversold market is 30 %, while the value of 70 % was selected as a limit for overbought market. In the next step, moving averages were calculated, specifically 50-day, 100-day and 200-day moving averages. All of the values were based on the closing price. As the last step, relevant graphs were created.

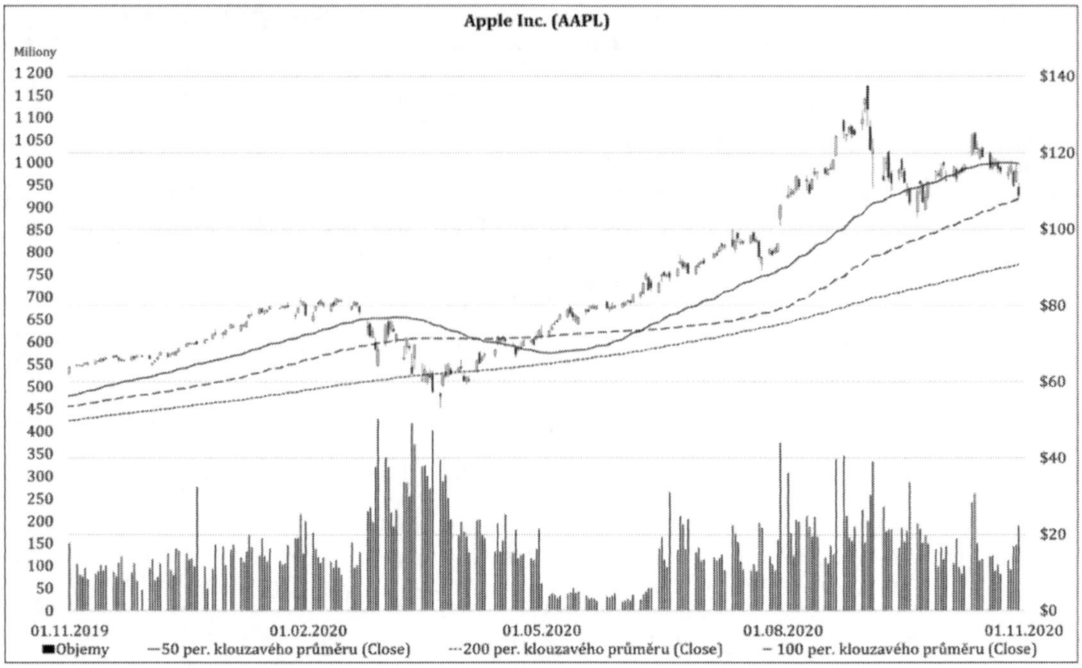

Figure 1. Development of trend, moving average, values of trade volumes

Source: Author.

Legend: objemy – volumes, xper. Klouzavého průměru – x-day moving average.

The above graph describes a trend, moving averages and the volume of trades of Apple Inc. It follows from the graph that from the beginning of the monitored period (1 November 2019), the values showed a stable growth, which is a frequent phenomenon when analysing time series of similar companies. This stable growth continued until the mid-February 2020, when the coronavirus started, which was naturally reflected in the prices of positions. In the period 2/200 - 3/2020, the prices fell sharply below the initial value from November 2019. However, the crisis has had a relatively manageable impact on technology-oriented companies, which does not apply to companies in other sectors. Apple Inc. as a stable technology company has sufficient financial reserves and such a sharp drop cannot cause its failure. On the contrary, it has happened many times that such a drop helped companies and their stock prices grew much faster in the next period than they would have if there had been no drop. This is also evident from the course of the graph. Approximately from 5/2020, the company´s stock prices started to rise steadily at a much faster pace than at the beginning of the year until 9/2020, when the second wave of coronavirus started. This was followed by a sharp drop in price but as can be seen from the next development, the company was at least partially prepared for this and the prices started to rise immediately in the next period.

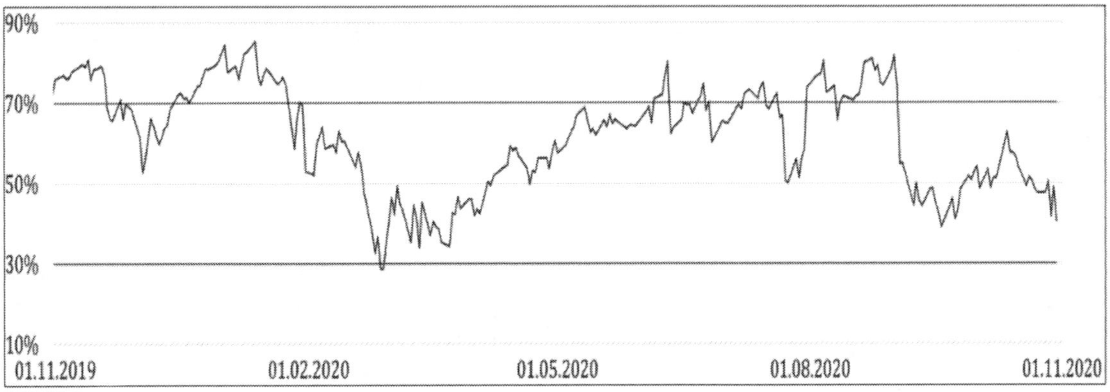

Figure 2. RSI indicators

Source: Author.

The course of the course in the graph describing the RSI indicators is logically very similar as in the previous trend graph. At the beginning of the monitored period, the value of RSI indicators was slightly above 70 % until the end of the year 2019. At the beginning of the year 2020, the price fell below 70 % for a few days but then it sharply increased again up to 80 %. This situation lasted until about 2/2020 and then, as shown in the price analysis, with the coronavirus crisis, the prices fell sharply, which was naturally reflected in the development of RSI indicator values, which fell to 30 %. In the next period, the company tried to repaired the damages caused by the drop and the values of RSI indicators grew steadily. In 9/2020, with the second wave of the coronavirus crisis, the values of trends and RSI indicators fell suddenly; however, the drop was not as sharp as during the first wave, since the company had had time to prepare at least partially for the situation.

5 Discussion

In general, it can be argued that the values of stock prices of almost all companies have fallen since the beginning of the year 2020 due to the coronavirus pandemic. Some experts claim that these have been the lowest values for the last seven years. This drop has caused huge financial problems or even failure of many small enterprises. The company under review, Apple Inc, is not an exception; the coronavirus crisis has affected this company as well. However, a stable company, which Apple Inc undoubtedly is, will be able to recover from the crisis without any major difficulties.

The first part of the results shows the development of Apple Inc stock price and 50-day, 100-day and 200-day moving averages were calculated, which were logically marked as fast, medium, and slow MA. The above graphs indicate that all three types of moving averages did not cross each other until 3/2020 and nearly copied the development of stock prices. This has changed after the aforementioned drop in mid-February 2020, when the stock prices fell sharply. In 4/2020, the fast 50-day MA, which responds most quickly and sensitively to price changes, crossed down the medium 100-day MA. As specified in the methodology chapter of the paper, if fast moving average crosses down the slower MA, it is recommended to sell the position, since in such a case, a drop in prices usually follows. However, this happened only partially in this case, since these technical parameters are focused mainly on a long-term development, in which the fundamental events, such as the coronavirus crisis, are not reflected much. In simple terms, technical parameters work best when there are no events that would have a worldwide effect. Government interventions can also significantly influence technical indicators. In the next period, between June 2020 and July 2020, the situation stabilized and fast MA crossed up the medium MA, which is a signal for a trader to buy positions. Further development shows that in the summer of 2020, when the first wave of the coronavirus crisis was over and government measures were cancelled worldwide and the world was slowly returning back to normal, technical indicators were not influenced by any fundamental events, the recommendation of buying the position would be correct because in the next period, there was a sharp increase in stock prices until September 2020. From the point of view of the moving average, there was no other indicator in the monitored period.

The second part of the results dealt with the analysis of RSI indicators. At the beginning of the monitored period, in November 2019, their values were slightly above 70 %, which did not produce any sign for buying or selling; according to the experts, the ideal reaction is to wait for further development. The first clear sign appeared at the beginning of 2020, when the values of RSI indicators were below 70 % and suddenly sharply crossed the limit of 70 %. In such a case, the ideal reaction is to sell the position in the market; according to the next development, this is the right solution, as if the trader had sold the position at the beginning of 2020, they would avoid the sharp drop in stock prices between 2/2020 and 3/2020. In mid-March 2020, the values of RSI indicators were slightly below 30 % and rose again. This means that they crossed up the limit of 30 %; in such a case, the ideal reaction is to buy in bulk. If a trader reacted to this indicator and made a purchase in March 2020, this would bring them a high profit, since in the next period, stock prices increased sharply. In 6/2020, the values of RSI indictors crossed up the limit of 70 % and fluctuated around this value for some time. In 8/2020, they fell down sharply, crossing down the limit of 70 %. This indicator recommends giving up the position. If a trader did this, it would again bring them profit, since a month later, the second wave of the coronavirus started and stock prices fell sharply again.

In November 2020, neither moving averages nor RSI indicators produced any signals that would recommend selling or buying positions. The ideal reaction thus appears to be not to change the share portfolio and hold the position, or do not buy and wait for further development and other indicators.

5.1 Evaluation of hypotheses

According to the first hypothesis, the determined criteria of value screening will find a maximum of 4 companies. This hypothesis can be rejected because the value screening found a total of 6 companies, out of which the one with the most traded stocks was selected. In this case, it was the company Apple Inc.

The second hypothesis was that the technical analysis indicators recommend sell or buy immediately after the beginning of the monitored period, i.e. around November 2019. The hypothesis was also rejected, since the first indicator emerged around February 2020.

The third hypothesis was formulated as follows: RSI indicators will recommend sell or buy, which will prove to be potentially correct in the next period. RSI indicators recommended specifically two positions. The first of them appeared at the beginning of the year, when the ideal reaction seemed to be selling the position. This turned out to be crucial in the future, since during the first wave of the coronavirus pandemic, a sharp drop in stock prices was recorded. The second indicator emerged in mid-March 2020, when it was recommended buying the positions. As further development showed, this decision would be right since in the next period, stock prices grew. This hypothesis can thus be confirmed.

6 Conclusion

The objective of the paper was to use determined criteria of value screening to select a relevant company for the analysis of its time series using selected methods of technical analysis.

In the introduction, the issue of stock time series was generally defined and general information on its evaluation was provided. Literary research presented the views of other authors to the issue of stock series and technical analysis. In the methodology part, the most common methods used for the evaluation of stock series were described, specifically moving averages and RSI indicators. Moreover, value screening was specified and its parameters were set. The Results part describes the graphs created and explains the results presented. In the chapter dealing with the discussion of the results, specific indicators are described, which emerged in the analysis of the time series and recommendations for traders are made. The conclusion part summarizes the whole structure and process.

References

1. P. Růčková. *Finanční analýza.* Praha: Grada Publishing a. s. ISBN 987-80-271-2028-4 (2019)
2. B. Graham. *Inteligentní investor.* Praha: Grada Publishing a. s. ISBN 978-80-247-1792-0 (2007)
3. D. Gladiš. *Akciové investice.* Praha: Grada Publishing a. s. ISBN 978-80-247-5375-1 (2015)
4. D. Gladiš. *Naučte se investovat.* Praha: Grada Publishing a. s. ISBN 80-247-1205-9 (2005)
5. Rejnuš, O., 2014. *Finanční trhy.* Praha: Grada Publishing a. s. ISBN 987-80-247-3671-6 (2014)
6. P. Syrový. *Investování pro začátečníky.* Praha: Grada Publishing a. s. ISBN 978-80-271-0092-7 (2016)
7. D. Štýbr. *Začínáme investovat a obchodovat na kapitálových trzích.* Praha: Grada Publishing a. s. ISBN 987-80-247-3648-8 (2011)
8. P. Podhajský, T. Nesnídal. *Obchodování na komoditních trzích: průvodce spekulanta.* Praha: Grada Publishing a. s. ISBN 80-247-1851-0 (2006)
9. O. Hartman. *Začínáme na burze.* Praha: Grada Publishing a. s. ISBN 987-80-265-0780-2 (2018)
10. J. Václavíček. *E – Milionář.* Praha: Grada Publishing a. s. ISBN 987-80-7252-220-0 (2008)
11. A. Picasso, S. Merello, Y. Ma, L. Oneto, E Cambria. Technical analysis and sentiment embeddings for market trend prediction. *Expert Systems with Applications.* **135**, 60–70 doi:10.1016/j.eswa.2019.06.014 (2019)
12. D. P. Gandhmal, K. Kumar. Systematic analysis and review of stock market prediction techniques. *Computer Science Review.* **34** doi:10.1016/j.cosrev.2019.08.001 (2019)

13. D. Shah, H. Isah, F. Zulkernine, F., 2019. Stock market analysis: A review and taxonomy of prediction techniques. *International Journal of Financial Studies.* **7**(2) doi:10.3390/ijfs7020026 (2019)

14. B. Weng, L. Lu, X. Wang, F. M. Megahed, W. Martinez. Predicting short-term stock prices using ensemble methods and online data sources. *Expert Systems with Applications.* **112**, 258–273 doi:10.1016/j.eswa.2018.06.016 (2018)

15. G. Kumar, S. Jain, U. P. Singh. Stock Market Forecasting Using Computational Intelligence: A Survey. *Archives of Computational Methods in Engineering*. **28**(6) doi:10.1007/s11831-020-09413-5 (2020)

16. W. Wang, L. Weizi, Z. Ning, L. Kecheng. Portfolio formation with preselection using deep learning from long-term financial data. *Expert Systems with Applications.* **143** doi:10.1016/j.eswa.2019.113042 (2019)

17. K. Li, P. Mohanram. Fundamental Analysis: Combining the Search for Quality with the Search for Value. *Contemporary Accounting Research.* **36**, 1263–1298 doi:10.1111/1911-3846.12466 (2019)

18. K. S. K. Lam, L. Dong, B. Yu. Value premium and technical analysis: Evidence from the China stock market. *Economies.* **7**(3) doi:10.3390/economies7030092 (2019)

19. O. Bustos, A. Pomares-Quimbaya. Stock market movement forecast: A Systematic review. *Expert Systems with Applications*. **156** doi:10.1016/j.eswa.2020.113464 (2020)

20. J.-N. Wang, H.-CH. Liu, J. Du, Y.-T. Hsu. Economic benefits of technical analysis in portfolio management: Evidence from global stock markets. International Journal of *Finance and Economics.* **24**(2), 890–902 doi:10.1002/ijfe.1697 (2019)

Human Resource Management Strategy as a Small Business Competitive Advantage Tool

Jan Urban[1,*]

[1]Institute of Technology and Business, Faculty of Corporate Strategy, Okružní 517/10, 37001 České Budějovice, Czech Republic

Abstract. The goal of a company's HR management (HRM) strategy is to define its medium-term objectives in HRM and development as well as to specify the main management tools needed for their achievement. The HRM goals set by this strategy should be linked to the company's strategic business goals and respond in advance to its medium term HRM requirements dictated by both internal and external factors. Corresponding HRM measures and/or policies should be effective in terms of achieving HRM objectives, as well as efficient in terms of their costs and benefits relations. The purpose of this paper based on a qualitative analysis of interviews with small business owners focused on the importance of their HRM strategy for the competitive advantage of their emerging companies, as well as on relevant secondary sources, is to outline a particular methodology and the main characteristics of successful HRM strategies. The main attention is drawn to the linkage of the HRM strategy to the overall companies´ business strategies as well as their appropriate contents and structure.
Keywords: business strategy; HR management strategy; HR strategy goals and tools; HRM SWOT analysis

1 Introduction

A properly conceived HRM strategy allows a timely response to the future HRM needs of companies [1,2]. Such a response, leaning usually on a companies' HRM SWOT analysis [3,4], can support the strengths of their human resources as well as the strengths of their HRM policy measures, eliminate their weaknesses, face in advance the potential external HRM threats and respond to new HRM external opportunities [5].

The task of the HRM strategy is therefore to clearly define a company's medium-term HRM goals focusing, among other things, on HR development, and to set appropriate tools and measures to achieve them, primarily those whose impact will take some time. The HRM strategy is especially important when HRM can become an important tool for creating the company´s competitive advantage [6]. This quite often concerns small businesses.

An important part of the HRM strategy includes a plan of costs and benefits associated with alternative HRM measures [3,7]. A common practical weakness of HR strategies, on the other hand, is their detachment from the overall business strategy and their rather formal and/or vague nature as well their missing or insufficiently specific measures to achieve their HR goals [2]. In these cases, the HR strategy lacks its efficiency and effectiveness [1,8].

The aim of the paper is to point out the importance and requirements of an effective and efficient HRM strategy and to establish an overall as well as sufficiently detailed methodology on which creating the effective and efficient HRM strategy is based. The main focus is related to the HRM strategy of emerging and small businesses as a tool of their competitive advantage.

2 Methodology

Methodologically, the paper is based on a qualitatively oriented, multiple-case study. Its two mutually interrelated research questions were "*What type of HRM strategies characterize successful emerging and/or small-businesses and which HR strategic management skills do small-business owners need to succeed/survive in their business?*"

Answering these questions was based on the experience and common views of entrepreneurs who successfully survived five years in business. The sample comprised 16 small business owners residing in Central Bohemia and operating small retail shops, personal services and professional service shops and outlets who sustained their business beyond 5 years, employed 10 to 50 employees and mana managed a growing and viable/profitable business entity.

* Corresponding author: 21537@mail.vstecb.cz

Data gathering was based on face-to-face semi-structured interviews providing data for answering the above research questions. Data saturation occurred when no new or relevant themes or supporting information emerged during the data collection process (in relation to the methodology suggested by [9,10]).

The main questions presented in these interviews were to find out whether or not the emerging businesses had an HRM strategy, what were the main goals and measures of their HRM strategy, what methods were used in conceiving their HRM strategy, which point, in retrospect, would the entrepreneurs put more stress on when conceiving their HRM strategy, what, in their view, were the main points on which their business, in terms of their HRM strategy, was based, what were the main differences, in their understanding, of their HRM strategy compared with their less successful competitors, what were the main HRM skills they lacked when they started their business, what main HRM skills they gained when managing their business, and if their HRM strategy changed during and after the first years in business.

At the same time, to confront the results gained form the structured interviews, these results (entrepreneurs´ opinions) were compared with the results and recommendations of secondary sources [11,12] dealing with an HRM strategy creation [13,14] and implementation [15] in general and in small companies [16]. Regarding, however, the questions raised in this paper, specifically the question of entrepreneurial skills and steps for generating a successful HRM strategy of emerging and/or small companies, such sources are relatively scarce [16]. Most of the recommendations in this paper, if not stated otherwise (by a reference to literature), are thus original recommendations which raised from the results gained.

An important part of the method used was a distinction made between an effective and an efficient HRM strategy. This issue was raised by the interviewed entrepreneurs and can also be found in relevant literature [15,16]. Those HRM strategies that focus on HR goals and measures supporting competitiveness (current or future competitive advantages of companies are seen as effective HR strategies, whereas those trying to identify HR policies, which at the same time analyse costs and benefits of alternative HR measures financially, are considered as efficient. Evidently, efficient strategies are crucially important for start-ups and small companies [16]

3 Results and Discussion

According to the interviews, the success of a newly established company depends on its ability to conceive a relevant and realistic HRM strategy. The main characteristics of this strategy contributing to the company´s success, especially in early stages of business development, can be summarized as follows.

(1) The starting point of a successful HRM strategy involves its medium-term business goals. Linkage to the company´s strategy is a key prerequisite for a properly defined HRM strategy; on the contrary, its detachment from the company's overall strategy is one of its most serious weaknesses.

(2) An HRM strategy should set out key HR assumptions on which the fulfilment of the company´s objectives will be based, identify corresponding medium-term HRM objectives (sorted usually by individual HR processes) and propose HRM tools or measures (" HR policies “) leading to their achievement.

(3) A successful HRM strategy structure should follow the above mentioned guidelines.

3.1 Structure of HR strategy

The basic scheme of HRM, corresponding usually to the main steps of its creation, consists of five main areas [15], namely:

- a long-term HR vision of the company containing its basic values or credo. This part of the HR strategy is primarily of motivational and communication importance,
- medium-term HR prerequisites necessary for the fulfilment of the company's strategic business goals,
- characteristics (or a critical analysis) of the current HR status of the organization, its strengths and weaknesses and development trends. This step is used to assess which of the HR prerequisites for successful future development the organization should focus on,
- specific HRM goals of the company and measures, HR policies, or their variants, on the basis of which these goals will be achieved. This set of objectives should also include ways to evaluate them,
- a tactical plan, setting out sub-tasks, their deadlines and responsibilities for their fulfilment. The main HR goals are assigned to individual departments or persons, and the resources that will be released to achieve the goals and the way in which the fulfilment of the goals will be controlled are determined.

The core of the HRM strategy includes the second, the third and the fourth points, which can be seen as an HRM strategic plan. It is also the starting point for continuously updated annual staffing plans.

The strategic HRM plan of the organization should ideally [17] be divided into five to six areas, corresponding to the main sections of HRM. They consist of a medium-term plan of HR costs, new personnel need and recruitment requirements (or, as the case of may be, dismissal of employees), motivation and remuneration, skills development, performance management (setting personal performance goals and their regular evaluation), or a plan of corporate values and corporate culture [18].

3.2 Medium-term HRM assumptions of corporate goals

The starting point for the medium-term HRM strategy encompasses the HR assumptions/prerequisites necessary for the fulfilment of the company's strategic business goals and/or its future market success. The basic tool for their definition is the organization´s strategic SWOT analysis [17] identifying the main factors of future competitive advantages of the company. Examples of these can involve, for example, high quality and reliability of production, low price, an ability to adopt rapid technological or product innovations, an individual/personal approach to customers, etc.

In order to correctly determine the HRM prerequisites for the company´s successful future development, the entrepreneurs that participated in the interviews mentioned the necessity to ask questions such as: "*What future HRM prerequisites are crucial for the company's business strategy", "What HRM requirements will be needed to achieve its competitive advantages*?" or "*What future employee skills and motivation as well as their management and organization methods will be necessary to implement the corporate strategy goals*?

More specifically, the medium-term HR assumptions of the company should relate to all basic HRM areas, i.e. the number of employees, their structure, skills and motivation, their costs, how these employees will need to be managed (for example, what powers will need to be delegated), what organizational structure will be most suitable for the company, what elements of corporate culture will be relied on, etc.

Determining the HR prerequisites that will be needed to fulfil the corporate strategy is, according to most of the entrepreneurs questioned, not only the initial, but also the most important part of the HRM strategy. It is therefore important that the people who create the HRM strategy (if it is not created by the entrepreneurs themselves) understand the critical factors of the future success of the company well.

3.3 Differences in HRM assumptions of different business strategies

HRM assumptions of different business strategies can differ significantly and therefore cannot be mechanically taken over from other companies. For example, HRM assumptions corresponding to the strategy of the lowest possible costs and/or prices will be significantly different from the assumptions required by strategies based on an individual approach to customers or rapid innovation capacity [17].

This may be illustrated by specific examples which the research, or enquiry, revealed. For an organization whose competitiveness is based on technological innovations and their rapid introduction to the market, i.e. a company operating in the field of new technologies, the main HRM prerequisites include the ability to acquire, retain and use top talent and ensure their further development.

In a company of this type, the long-term need to acquire and retain these employees may take precedence over the (short-term) effort to save on their costs. Financing their development should therefore not decrease even in periods of reducing business costs. Another HRM assumption may relate to their remuneration: the basic and starting salaries of individual employees working in equally qualified positions in these conditions may differ significantly depending on their personal abilities or contribution to the creation and introduction of new products.

On the other hand, HRM prerequisites of a company with a decisive emphasis on reducing and controlling operating costs (for example, companies facing strong price competition) will be significantly different. The determination of basic and starting salaries should be based primarily on the requirements of internal equality in remuneration and group motivation. The reason is that teamwork and group performance are usually more important here than the individual contribution or performance of each employee. Differences in the level of basic salaries therefore lack economic justification (individuals with a higher price do not pay) and may also be a threat to teamwork [8].

3.4 Analysis of the current HR status

The second basic step in creating an HRM strategy is to compare the HR prerequisites necessary to fulfil the company's business strategy with the current or near future HRM situation of the company and/or its development tendencies. The purpose of this step is to analyse the main weaknesses or limitations of the current HRM situation of the company, especially those that need to be addressed well in advance [19].

This analysis should, according to the successful entrepreneurs surveyed, ask questions such as: "*How do our current HR situation and development meet our future HR as well as HRM requirements and what are its main weaknesses?*", "*What limitations in terms of future requirements are related the abilities of our employees or the methods they develop?*", "*From the viewpoint of important HRM preconditions for future development, what limitations does our way of managing, motivating or rewarding employees bring?*", "*What limitations can our organization, corporate culture, etc. place*?".

Answers to these questions reveal HR weaknesses/bottlenecks in the given company (both in terms of its current state or development trends), which the HRM strategy should try to eliminate.

3.5 HR goals and measures

The third main step in creating an HRM strategy is setting medium-term HRM goals. They are designed to overcome the existing HRM weaknesses and respond in a timely manner to adverse external HRM trends concerning the organization. The measures necessary to achieve these goals should correspond to them.

Particular HR goals, according to the successful entrepreneurs surveyed, should be as specific as possible: if possible, they should be set in the form of specific values of HR indicators, or other ways in which the achievement of the set goals can be monitored. The specificity of the objectives and the proposed measures as well as the method of their evaluation is usually the most important prerequisite for this step, as pointed out by the successful entrepreneurs.

For example, if the identified weakness involves increased HR costs associated with a higher turnover of new employees, the HR strategy may include a specific HRM objective *"to reduce the turnover rate in the first year of work to 5%"* and indicate measures to achieve it. Other examples include improving employee selection, their adaptation, etc.

3.6 Variant (alternative) HRM measures

An important part of this step is to consider and assess alternative tools or measures that will enable to eliminate the identified HRM weaknesses. An example one of the entrepreneurs stated is the company's effort to respond to employee dissatisfaction with the way they are evaluated (its subjectively perceived injustice), which may lead to a proposal to train managers in the impartial evaluation of subordinates. However, managers may be very aware of the bias of their evaluation, but may not be sufficiently motivated to eliminate it. Therefore, their training is not an adequate solution.

Variant HR tools or measures should be assessed both in terms of their expected benefits and their (direct and indirect) costs. As in other areas of company management, the evaluation of alternative procedures should be based on an analysis of the expected impact of the proposed measures on the costs and revenues of the company, i.e. cost-benefit analysis, successful entrepreneurs suggest.

An example could be a measure consisting in adjusting the method of evaluating candidates in a selection procedure. The impact of this measure on costs may be either a reduction or an increase in the direct costs of the selection process (due to a reduction or an extension of its existing activities, a shortening or a lengthening of its duration, etc.) or an impact on its indirect costs (due to a higher or a lower probability of a wrong HR decision).

Part of this step of the HRM strategy should be the identification of factors that affect the probability of success or failure of the proposed measures. Factors that increase the likelihood of success of a proposed measure, according to the successful entrepreneurs surveyed, may include, for example, employees requiring similar measures, the measure facilitating the work of managers, having influential supporters, or similar measures having been successfully implemented in the past. On the contrary, risk factors may include the fact that managers or employees were not involved in related training, the company does not have sufficient know-how for implementing the measure or its time schedule is not realistic, etc.

3.7 Evaluation of HR measures implemented

The way in which results of the proposed measures will be evaluated must, according to the successful entrepreneurs surveyed, be determined in advance. In simpler cases, this assessment can be based on way how ta particular proposed measure, such as a change in the way of how higher performance is remunerated, has reflected in changes in employee behaviour or satisfaction. In the case of more complex or costly measures, their impact assessment should be based on performance indicators. Examples may include savings in HR costs, increased productivity, reduced scrap, etc. Evaluations of this type may be considered, for example, when assessing the results of a longer-term program aimed at increasing employee motivation. This experience of the successful entrepreneurs surveyed corresponds to certain recommendation in literature [20].

For long-term measures important for achieving the company's strategic HR goals (for example, increasing customer satisfaction, reducing absenteeism or turnover, etc.), evaluating of HR measures should also be based on estimating their financial return, using ROI, i.e. return on investment indicator. This requires to fully define the expected costs and benefits of the planned measure in financial terms and comparing them with one another [20].

In addition to setting out the ways in which the success of the proposed measures will be assessed, this section of the HRM strategy should also include setting deadlines for the main strategic objectives to be critically reviewed in order to assess their relevance [21].

4 Conclusion

The importance of any HRM strategy resides in the fact that it allows to respond correctly and well in advance to future, both internal and external, HR and HRM requirements of the company.

The majority of surveys´ participants claimed that the main HRM challenges for their business development concerned HR related issues, especially *"a high rate of employee turnover and/or hiring the right people"*, typical issues demanding an appropriate strategy approach. At the same time, most of the respondents noted the challenge of keeping staff motivated and happy, as well as keeping an open dialog and clear communications of expectations which they regarded as critical. Though already offering motivational rewards, most companies stressed the necessity of more effective and efficient remuneration strategy. In a similar context, a lot of them expressed a strategic need to improve their interview process to make better hiring decisions.

A large number of the participants stated that, similar to any successful business strategy, the related HRM strategy requires an ability to critically analyse the current HR situation of the organization, both in terms of staff quality and its management and organization, and an ability to bring, if necessary, new or alternative solutions.

Contrary to some simple views, formulating the company's HRM strategy in their opinion means not only determining what the number of its employees should be in the future or what their qualification structure should be, but also how they should be managed, motivated and rewarded, what new skills the company should develop, what methods of development should be used, how it might be necessary to adapt the company's organizational structure, redefine some jobs, or which measures should be taken to develop the company culture..

The main and, as the author believes, original findings of this paper can be summarized in the following sentences. The main conclusion, which the successful entrepreneurs approached in the survey agreed on, is related to the quality of the HRM strategy, i.e. its effectiveness and cost-efficiency, depending on its connection to the business strategy. The precondition for this is the correct determination of HRM assumptions needed for keeping and further development of current and future competitive advantages of the company, concretization of HRM goals as well as a clear and sufficiently specific determination of longer-term measures or HRM policies enabling their fulfilment. The cost-effectiveness of a particular HRM strategy, they claimed, usually also presupposes an assessment of alternative HRM measures and policies, including an estimate of their costs and expected benefits.

Recommendations for further research include focusing on different sizes of small business in different industries and varying geographical locations. Future researchers should also consider studying the issue the extant research has not effectively explained, i.e. why differences in employee-based resources persist across firms that have adopted the same firm-level HR strategy. Moreover, focusing on each of the specific parts of HRM strategy identified within this paper would provide more detailed areas for additional research.

References

1. M. Armstrong, D. Brown. *Strategic Human Resource Management: Back to the future?* Brighton: Institute for Employment Studies (2019)
2. J. Storey, P. M. Wright, D. Ulrich. *Strategic Human Resource Management: A Research Overview*. London: Routledge. ISBN: 9780429490217 (2019)
3. S. A. Alghamdi. The Impact of HRM strategy on Organizational Effectiveness. *American Journal of Theoretical and Applied Business*. **6**(4), 91-97 (2020)
4. S. Hillebrecht. *Die Personalstrategie und Personalplanung des Unternehmens, in: Perspektivenorientierte Personalwirtschaft.* Wiesbaden: Springer Fachmedien Wiesbaden. ISBN: 9783658320935 (2020)
5. F. R. David. Strategic Management: A Competitive Advantage Approach, Concepts & Cases. 15th Ed. New Jersey: Prentice Hall. ISBN: 9780133444797 (2019)
6. P. Boxall. HRM strategy and competitive advantage in the service sector. *Human Resource Management Journal*. **13**(3), 5-20 (2003)
7. C. Chadwick, C. Flinchbaugh. Searching for competitive advantage in the HRM-firm performance relationship. *Academy of Management Perspectives*. **35**(2) (2021)
8. C. Sheehan, H. De Cieri, B. Cooper, T. Shea. Strategic implications of HR role management in a dynamic environment. *Personnel Review*. **45**(2), 353-373 (2016)
9. N. K. Denzin, Y. S. Lincoln. *The Sage handbook of qualitative research*. 4th ed. Thousand Oaks: Sage Publications. ISBN: 9781412974172 (2012)
10. A. Trost. *Neue Personalstrategien zwischen Stabilität und Agilität*. Wiesbaden: Springer Fachmedien Wiesbaden. ISBN: 9783662574072 (2018)

11. E. Knezović, S. Bušatlić, O. Riđić. Strategic human resource management in small and medium enterprises. *International Journal of Human Resources Development and Management.* **20**(2), 114-139 (2020)
12. T. Bagga, S. Srivastava. SHRM: alignment of HR function with business strategy. *Strategic HR Review.* **13**(4,5) (2014)
13. L. Kempová. The Creating of HR Strategy and its Implementation to Organization XYZ. *Acta academica karviniensia.* **3**(2), 33-41 doi:10.25142/aak.2013.022 (2013)
14. E. Lawler, J. Boudreau. *Effective HR Management: A Global Analysis, Stanford: Stanford University Press.* ISBN: 9780804776875 (2012)
15. J. Urban. Jak propojit personální strategii s celkovou strategií podniku. *Práce a mzda.* (2020)
16. J. Urban. *40 manažerských mýtů*. Praha: Grada Publishing a. s. ISBN: 9788027105717 (2018)
17. A. Trost. *Die Entwicklung einer Personalstrategie*. Wiesbaden: Springer Fachmedien Wiesbaden. doi:10.1007/978-3-662-57407-2_3 (2018)
18. J. E. Delery, D. Roumpi. Strategic human resource management, human capital and competitive advantage: is the field going in circles? *Human Resource Management Journal.* **27**(1), 1-21 doi:org/10.1111/1748-8583.12137 (2017)
19. A. Upamali, J. Weerawardena, M.-L. Verreynne. Learning capabilities, human resource management innovation and competitive advantage. *The International Journal of Human Resource Management.* 29(10), 1736-1766 (2018)
20. P. Wright, A. J. Nyberg, R. E. Ployhart. A research revolution in SHRM: New challenges and new research directions. *Research in personnel and human resources management.* **36**, 141-161 (2018)
21. M. Brussel, S. Stella. *Praxishandbuch Personalmanagement für Handwerksunternehmen*. Wiesbaden: Springer Fachmedien Wiesbaden. ISBN: 3662583151 (2019)

Difficulties and Challenges of Strategic Management of Start-up Enterprises in the New Era

Yixue Zhao[1,*]

[1]School of Economics and Management, Lanzhou University of Technology, Lanzhou, Gansu Province, 730050, China

Abstract. Strategic management is a short- to long-term development plan concerning the goals of an enterprise. Formulating appropriate strategic management is of great and far-reaching significance for the current development and sustainable development of the enterprise There is no doubt that the formulation of strategic management is very important. In the new era, under the condition of rapid economic development, the way enterprises formulate strategies and the development of strategic formulation have very important research value. This paper thus examines the characteristics of strategic management and the process of strategic management, the current situation of the development of entrepreneurial enterprises in the new era, and then uses quantitative analysis to analyse the impact of the Midea group on the strategic management of enterprises, puts forward several issues and decisive factors that should be considered for strategic management, and puts forward important implications for the strategic management of entrepreneurial enterprises.

Keywords: strategic management; development planning; social responsibility; entrepreneurial enterprises; the new era

1 Introduction

In the period of rapid economic development, the development of enterprises has been actively leading the transformation and development from the beginning. In other words, uncertain factors such as the international situation of economic development have brought higher requirements for enterprises. Prior to this, the development of enterprises may be to create culture. He Faith attracts capital and registered company stakeholders to produce products or service sales to earn profits. This is the basis of enterprise development. However, as the industry has great momentum and capital inflows become more frequent, many companies have miraculously sprung up. Consumers are becoming more and more strict in their choices. This is definitely a big challenge for companies, and it seems to meet the requirements of the consumer market [1]. The direction of enterprise development is a general trend, but this is only a small part of the factors. The strategic choice environment of enterprises is becoming more and more complex, so it needs a comprehensive and coordinated development to improve their own capabilities, and enhance their own competitiveness. Corporate strategic management summarizes the development background and current situation of venture companies, the historical development of Chinese corporate strategic management, and then analyses the strategic management of Midea Group, as well as the problems of corporate and strategic management, and finally proposes the influence of diversified strategic development. Under the macro environment, enterprises can still maintain their core competitiveness and achieve the conclusion of sustainable development [2].

2 Characteristics and processes of enterprise strategic management

2.1 Characteristics of strategy

"War" refers to battle and war, "strategy" refers to strategy and plan. War in the sense of "strategy" has existed since antiquity and has multiple meanings depending on different application environments. From the perspective of corporate strategy, American experts published a famous theoretical article "Corporate strategy" in 1965. Since then, the word "strategy" has been included in the policy of enterprises. Sustainable development and profitability are the main development direction for the enterprise. Therefore, for an enterprise, strategy means both a strategy and a model, that is, according to the development trend of the enterprise, the strategy is a plan of what development the company wants toto achieve and how to achieve this development. For the process of enterprise development, strategy is a model. Enterprises need to plan what model to choose and what development path to take to ensure the survival of the enterprise in a difficult

* Corresponding author: mzhao320@163.com

environment. The typical features of corporate strategy are integrity, long-term nature, stability, feasibility and hierarchy. The so-called overall situation refers to the comprehensive consideration of all internal and external factors when an enterprise formulates its strategy not just for one department or one aspect. Strategy plays an important leading role in corporate organization and system. It is the main factor of the development of the entire system. The formulation of goals requires comprehensive consideration, and the achievement of such goals requires complex cooperation. Therefore, only based on a comprehensive perspective, we can promote the healthy development of enterprises. Formulating strategies according to a certain stage or a certain part of the goal will inevitably restrict the overall sound development of the enterprise. Long-term nature refers to the long-term development of the enterprise when an enterprise formulates its strategy. Strategy is divided into long-term strategy, medium-term strategy and short-term strategy. Long-term strategy is the goal that an enterprise must achieve within 5-10 years. The medium-term strategy is the direction formulated for 1-5 years, and the short-term strategy refers to the results that need to be achieved within one year. For enterprises, clear mid- and short-term goals serve as long-term goals. Therefore, various factors must be considered comprehensively, and the long-term goals must be established first; afterward, the goals are decomposed into annual plans. Once the stability requirement of the corporate strategy is established, it cannot be changed at will. No matter how the business environment changes, only companies that focus on resource development and sustainability can establish their own professionalism and obtain permanent and stable returns. Feasibility means that the strategy formulated by the company is in line with the development of the company rather than its imagination. Only in line with the development path of the enterprise, it can be a goal that can be sustained. In addition, the company's strategy formulation should conform to reality, proceed from reality, and be based on the company's internal resources rather than lagging behind closed doors. The hierarchy is related to identity. Strategy can be divided into many parts. Each part represents a different part of the strategy, there is no level, only high and low points. Together they form and influence the same strategy. In a hierarchical structure, it generally means that the upper layer guides the next layer, and the next layer supports the upper layer.

2.2 The process of enterprise strategic management

2.2.1 Strategic management analysis

The first and the most important step of strategic management is to assess the state of the enterprise according to the joint action of internal factors and external environment [3]. It is also an important step for an enterprise to understand itself, its industry, and its external environment. It uses a reasonable and feasible evaluation system to assess the environment the enterprise is faced with, adopts targeted and localized means to analyse it, and then mobilizes the superior resources of the enterprise to make strategic decisions.

2.2.2 Strategic management options

The strategic choice is based on the results of the strategic management analysis. The importance of strategic management analysis also stems from its role as a "cornerstone" and strategic choice is directed towards the formulation of strategic plans based on the results of strategic analysis [4]. Within the enterprise, the resource constraints, the feasibility of strategies between departments, and the goals and mission of enterprise development must be taken into account. Secondly, it is necessary to evaluate the scheme. The research and discussion method can better collect the strengths of hundreds of enterprises, discuss and complement the selected strategic scheme, and find the scheme that is more suitable for the development of the enterprise. Finally, the final choice of strategic management is made. The choice of strategic management is only completed at the end of this process.

2.2.3 Implementation of strategic management

The implementation of strategic management is the ultimate embodiment of all theoretical foundations. Strategic analysis and strategic choice ultimately serve to implement the strategy. Each point of the strategic proposal should be reflected in strategic management, including an effective organizational structure, the establishment of a staff appraisal system, an echo of strategic direction and corporate culture. Focus on key competitiveness and sustainable business development. At the same time, provided that the long-term strategy does not change arbitrarily, discuss the new strategy and direction of development in a timely manner according to the various factors they face in the short term.

2.2.4 Strategic management evaluation

The evaluation of strategic management is based on the results of the implementation of strategic management and has the function of "looking ahead and looking back". However, many enterprises ignore strategic management evaluation because of the desire to shorten the process. In fact, it rather corresponds to the summary of the work in this stage and the direction of the development of the work in the next stage, which is conducive to promoting the further development of the enterprise.

3 Development background and current situation of strategic management

3.1 Development background

At present, the background of strategic management development is becoming more and more complex due to frequent economic activities and rapidly improving economic base, which are among the macro factors. We cannot change the adverse effects of macro factors, but we can take positive countermeasures to weaken their adverse effects on us. The integration of the current background faced by enterprises and the in-depth analysis of the background of strategic management development will help to realize further research on the strategic management of start-ups in the new era.

3.1.1 Globalization background

Market globalization, rapid development of enterprises has brought opportunities to developed countries and also expanded overseas markets. These enterprises have been affected by internationalization. Domestic development strategies gradually evolved into multinational strategies and these evolved into globalization strategies and factory separation. It was possible to establish a global arrangement with institutions, promote the coordinated integration of mutual resource utilization, promote the development of the local economy while the enterprises boomed, and ensure the favorable position of state finance to benefit from the global market and global enterprises. Governments of various countries have relaxed restrictions on economic activities in the region and gradually removed antagonism and alienation in politics and culture. The government has also played an active management role. In the case of the intermingling of government and enterprise, the business environment of the enterprise has gradually changed [3], and this has led to the adjustment and reform of the strategic management of the enterprise.

3.1.2 Competitive increase

The most obvious situation brought about by the global economy is the fierce competition among enterprises, so the measures taken by enterprises to respond quickly to this competition and the interpretation of innovation requirements have become an important direction of the organization of strategic enterprise management. Especially in the pandemic, it is more necessary for enterprises to continuously conduct business operations and complete innovation work in order to accelerate the response to specific problems [5]. As mentioned above, the rapid economic changes brought by the global market are followed by rapid environmental changes. The impact of macro factors is undoubtedly severe for businesses, and the speed of awareness of these impacts determines the speed of strategic adaptation of the enterprises. The second reason is the actions that need to be taken, whether to continue to expand scale, increase market share, or adopt a conservative strategy, which will affect the later direction of the business.

3.1.3 Dynamic environmental factors

The global economy, strong competition and technological development have created dynamic environmental factors for businesses [6]. "Dynamic" is relative to "static". Under the combined effect of the above three factors, "opening up" and "transformation" have become the main themes and focus of enterprise development strategy, and these factors will continue to promote social and economic transformation. This leads to a dynamic change in the social environment in which these enterprises find themselves. With the continuous improvement of openness, the "barriers" between countries are gradually breaking down, the boundaries of enterprise development are gradually blurring, international interaction is also improving, the possibility and practicality of political, economic and cultural exchanges between countries is increasing, and some advanced technologies can be used more widely. The business environment of enterprises around the world is therefore changing dynamically.

3.1.4 Promotion of science and technology

The promotion of science and technology has been integrated into all areas of our lives. Thanks to technological progress, transport and information communication are constantly developing [6]. Technological advances have shortened the "distance" between countries and businesses, and the concept of the "global village" has come to fruition. Although the distance between two countries is still very short geographically, with the advancement of communication technology, communication between two countries is no longer as difficult as before, and people's offices are not limited to office space. Using mobile phones, it is possible to surf the Internet and work anytime and anywhere, receive work messages, process office documents, etc. For businesses, this fast office mode poses a challenge to their own strategic management: competitors in the same sector are trying to do better, so what should be our strategic deployment? What is the positive role of science and technology for the development of our businesses today? How can we use it to connect with the trend of the times? All these factors need to be taken into account in strategic management.

3.1.5 Need for social responsibility

As already mentioned, the rapid development of start-ups is intrinsically linked to the support of national policies. With the rapid rise and development of start-ups in line with the trend of the times, the relationship between government and enterprises is becoming increasingly close. The role of government must be naturally accepted by enterprises in order to promote social justice, ensure employment and accelerate economic construction. Therefore, the government can achieve some cooperation with enterprises [7], for example, by issuing relevant regulations. When enterprises help share the burden of "jobs", they can enjoy certain tax incentives. Therefore, businesses may have the problem of "increasing costs due to recruitment beyond their own development needs". Although it has been argued that a simple recruitment mechanism is more beneficial to corporate strategy, the surplus of labour and the need for high social responsibility will bring some difficulties in formulating corporate strategy and financial pressure, especially when social responsibility is mandatory, enterprises may need to consider how to bear social responsibility in strategy formulation [8].

3.2 Development status

Due to the impact of global economy, the intensity of competition has increased. The situation enterprises face in the dynamic environment is complex and fast-moving. The combination of science, technology and social responsibility also promotes the increase of factors that enterprises need to consider when formulating strategies. Under the new situation, corporate strategic management is facing the following status.

3.2.1 The strategic direction tends to internationalization

In formulating a strategy, the focus should be on the transnational or global economic market rather than just the domestic market. The speed of regional economic integration is increasing rapidly. More and more companies are penetrating foreign markets to gain greater market share and economic benefits. This behaviour of expanding its own market must take full account of the fact that the enterprise is formulating its own internationalisation strategy. For example, the enterprise must consider the branches, subsidiaries or offices to be established in different regions of the international market, and the "recruitment and management of employees" strategy, "risk management and control" strategy [9], "cost and control" strategy caused by this problem. We must focus on the trend of nationalization, understand the research of internationalization as the background, formulate appropriate strategies, and smoothly implement the development of enterprise internationalization.

3.2.2 Strategic subjects tend to be diversified

The diversification strategy refers to the strategy of selectively developing other businesses so that the business does not get into trouble because of the development of a single industry. Diversification strategy is related to specialisation. Specialisation is the focus of an enterprise on a particular product, which is an inevitable way to lay the foundations at an early stage of enterprise development. As the development of enterprises, products enter the maturity period, it is necessary to further explore the market. Therefore, the development of new products and the search for new markets becomes the only way. It can be seen that diversification is a possible way of operation, but it cannot be temporary but a comprehensive and long-term strategic behaviour of the enterprise. The main components of a diversification strategy are product diversification, market diversification, investment area diversification and capital diversification. Product diversification means that an enterprise operates two or more products or services simultaneously. Market diversification means that the enterprise's products are dispersed in multiple markets such as domestic market and international market. Regional diversification of investment means that the enterprise's investment activities are in different regions. Capital diversification means the diversity of enterprise capital, such as cash, machinery and equipment, factories, securities, intellectual property rights and trademarks.

3.2.3 The implementation of the strategy tends to be strict

Strictness is embedded in the strategy implementation process. After strategic analysis and decision-making, the subsequent strategic management is the key content of the entire implementation of the company's strategy. Therefore, how to ensure successful strategy implementation, how to link enterprise strategy to development, and how to support the realisation of enterprise objectives in strategic decision-making must be included in the scope that should be considered in the actual strategy implementation situation at this stage. Therefore, enterprises will take strategy implementation as an important part of the evaluation when evaluating the development performance of enterprises, and establish a top-down evaluation standard to include enterprise employees in the evaluation process, supervise employees, and motivate employees to promote strict strategy development through strict strategy implementation procedures. The company's overall strategy is big and broad, but the specific strategy broken down by overall strategy is the responsibility of the manager or department. [1]

4 The development process of strategic management of chinese start-up enterprises

Since the reform and opening-up, China's economy has adapted to the trend of the times, continuously reformed its own development path, and finally achieved dazzling achievements. At present, China's economic development is characterized by diversified development, intelligence and innovation. There is no doubt that the continuous progress and development of enterprises have played a solid role in the process of rapid growth and can make solid progress in the fast-growing economic development. It can be said that the strategic management of Chinese enterprises plays an important role. Strategic management pays more attention to the competitive advantage, development status and value creation of enterprises in the industry, so it has become an important influencing factor and the way to realize the development of business enterprises in the new era.

4.1 Germination and production

The strategic management of Chinese enterprises is beginning to emerge since the reform and opening-up period. The process of its development is an important embodiment of China's reform and opening-up. Prior to reform and opening-up, the planned economic system severely constrained the development of China's economy. At this time, enterprises did not have the right to operate independently, so they did not have the basic conditions for formulating strategic enterprise management, and it was impossible for enterprises to formulate, implement and evaluate strategies. It was only in 1978 that a series of important conferences were held, which gradually loosened the "autonomy" of enterprises, providing a background for the practice of strategic management of Chinese enterprises. With the issuance of a series of standards, industrial enterprises first explored the reform of the enterprise system, began to comply with relevant laws and regulations, and began to enjoy the autonomy of operation and management under the unified guidance and plan of the state, At this time, enterprises also began to explore the "management of their own enterprises"[10]. In 1984, the General Assembly proposed to "expand the autonomy of enterprises" and "enhance the viability of enterprises". By promulgating a series of legislation, it took the improvement of "expanding enterprise autonomy and promoting enterprise development" as its main goal. At this time, a series of researches on management methods for enterprise development began to appear. Many enterprises, such as Qingdao Refrigerator Factory (now Haier Group) and Sichuan Changhong (now Changhong Group), took the lead in introducing advanced management methods such as total quality management, realizing the improvement of enterprise management level and strategic enterprise management, which first entered people's consciousness at this time. At this stage, China's enterprise system exploration is not only the embryonic stage of strategic enterprise management, but also the exploration stage. It has brought some experience for the later practice of enterprise self-management and has a certain role in its promotion.

4.2 Development and determination

After the reform and opening up, strategic corporate governance achieved some development, but at this time the status of a market economy had not yet been fully established. At the initial stage of knowledge of strategic enterprise management, compared with Western countries, there is still a large gap in practice and research, and the relevant management system, strategy formulation, organizational framework and talent building are still immature. However, the introduction of enterprise autonomy still provides a great impetus for strategic enterprise management, provides a solid foundation for strategic management in the development of the enterprise in the new era, and makes an excellent contribution to the establishment of a modern enterprise system. In enterprise practice, a number of new pilot projects have emerged to promote the reform of state-owned enterprises, and enterprise forms such as alliance, merger and leasing have also appeared for the first time. In the process of reform, development needs have promoted the change of the internal organizational structure of enterprises. From the early 1990s, Chinese enterprises began to set up "strategic management" or "strategic planning" departments, such as the "strategic corporate development and investment committee" established by Tsingtao Beer in 1999 [1]. The existing practice of strategic management of Chinese enterprises has begun to penetrate into other enterprises. Businesses have realized the importance of strategic management and its indispensability to the development and planning of the economy. The practice of enterprise strategic management has also stimulated a huge demand for theory and talents, and the evaluation of employees is also based on the introduction of the idea of enterprise strategic management. It also directly encourages the cultivation of talents. Major universities have adopted "strategic management" as a separate discipline for the professional training of students.

4.3 Growth and present

After the establishment of market economy status and after long and continuous practice and examination, the research on the management of Chinese enterprises in the Warring States period has made great progress. Against the background of globalization, the introduction of some Western ideas has had an obvious illuminating and illuminating effect on the enterprise strategy of Chinese countries, and the strategic management of Chinese enterprises has entered a new period of growth. Under the influence of a series of documents and policies, such as management methods and development strategies, strategic enterprise management has started the peak of rapid growth and development. Chinese strategic enterprise management has gradually evolved from domestic development to international direction. Under the challenge and support of policies, Chinese enterprises gradually strengthened their understanding of "strategy", began to formulate strategies in accordance with their own development according to their own actual situation, and widely engaged in more intense, complex and changeable international competition. In addition, entrepreneurs actively participate in various forums where they exchange experiences and knowledge on corporate strategy formulation and strategic practice. By learning from the advanced models of other businesses, they broaden their own thinking and vision. A large number of Chinese enterprises have gradually "set sail", entered the wider global market and gradually become an important force and part of the international market and international competition. In continuous practice, we have acquired advanced technology, resources and management experience.

5 Strategic analysis of midea group co., ltd

In this chapter, the author uses quantitative analysis method to analyse the strategy of Midea Group, makes vertical comparison of financial statements of different years, analyses the influence of corporate strategy on Midea Group, and evaluates its strategic model.

5.1 Midea enterprise and strategy

The Midea Group entered the air conditioning industry in 1985 and began to explore new categories in the home appliance sector. In 2013, Midea was listed on the Shenzhen Stock Exchange. In 2020, Midea Group's stock price reached a record high of 94.76 yuan, up 2453% from the opening price of 3.7 yuan on September 18, 2013; in 2020, it ranked 288th on the Fortune Global 500; on December 24, Midea acquired Hitachi's compressor factory in Thailand and disrupted the global industry. Faced with the development of a new era, Midea has adopted a diversification strategy, and now it has obvious effects.Midea has been pursuing a diversified strategy for more than a decade. In 2018, Midea's market share was 27%, rising to 29.5% in 2019, with an outbreak at the end of 2019. Therefore, in 2020, most enterprises faced the problem that their products could not be sold and most of the goods were piled up in warehouses However, the market share of the Midea Group at the end of 2020 not only did not decline, but increased to 34.5%, which is 8.5% more than in 2019, indicating that the multi-year diversification strategy of the Midea Group has borne fruit in early 2020. Therefore, the Midea Group can maintain its market share in 2020 even if the overall environment is affected by the epidemic These figures show that a diversification strategy can bring economic benefits to businesses (Data source: Midea Group Financial Statements 2018-2020)

5.2 Financial analysis of Midea Group

5.2.1 Analysis of profitability

Table 1. Profitability data (Unit: Million yuan)

	2018	2019	2020
Gross revenue	261,820	279,381	285,710
Operating profit	25,564	29,683	31,493
Net margin	21,650	25,277	27,507
Year-end cash and cash equivalents	17,952	30,442	23,549

Source: Midea Group annual Report 2018-2020.

The profit data of Midea Group from 2018 to 2020 are shown in Table 1. From 2018 to 2020, the total operating income, operating profit and net profit of Midea Group kept growing continuously. The year-end cash and cash equivalents balance both fluctuated slightly in 2020, but still kept growing on the whole.

Table 2. Profitability indicator data (Unit: Percent)

	2018	2019	2020
Gross profit margin	27.54	28.86	25.11
Net operating interest rate	8.34	9.09	9.68
Returns on gross assets	8.2102	8.3712	7.6326

Net profit margin on sales	8.3378	9.0854	9.6779
Return on equity	24.35	23.81	23.17

Source: Midea Group annual Report 2018-2020.

Table 2 shows the profitability index data of Midea Group from 2018 to 2020. From the perspective of profitability index, net operating profit margin on sales increased steadily during this period, indicating that the business situation of the enterprise was good, and the gross profit margin on sales also maintained a stable level during the pandemic. Although the profit margin on total assets declined in 2020, the fluctuation range was relatively small, indicating that the operation status of the enterprise was relatively stable. The fluctuation of return on net assets was relatively small, but the level was relatively high, achieving about three times the profit margin on total assets, indicating that the enterprise was highly profitable

5.2.2 Financial risk analysis

Table 3. Risk Financial Data (Unit: Percent)

	2018	2019	2020
Current ratio	1.4028	1.5	1.3123
Cash Ratio	21.4145	49.1391	44.1001
Debt to Asset Ratio	64.9397	64.4	65.5263
Equity ratio	175.5701	172.6731	182.6973

Source: Midea Group annual Report 2018-2020.

Data on the solvency and capital structure of the Midea Group are presented in Table 3. If the liquidity ratio is higher than 2, the assets of the enterprise can be converted into cash of the year need to pay twice the liabilities of the Midea group due to the nature of the production of goods, the current ratio is less than 2, but in recent years the change in amplitude is small, less affected by the impact of the epidemic. In addition, Midea Group's cash ratio, equity ratio, asset-liability ratio have been steadily increasing, indicating that Midea Group's financial risk level is relatively low and can maintain stability despite the impact of the pandemic.

5.2.3 Operational capability analysis

Table 4. Operational capability analysis. Unit: Percent

	2018	2019	2020
Accounts receivable turnover	14.0668	14.6222	13.6506
Inventory turnover	6.3688	6.3725	6.7015
Total assets turnover	1.0147	0.9837	0.8582

Source: Midea Group annual Report 2018-2020.

Table 4 shows that the Midea Group's accounts receivable turnover has decreased in 2020, indicating that the company's recovery rate has slowed down, which is related to the pandemic outbreak in 2020. The Midea Group's customers and suppliers are in a non-saleable phase and do not have sufficient funds to repay their liabilities. From the perspective of inventory turnover rate, the Midea Group's inventory turnover rate is still higher from 2018 to 2020, indicating that the enterprise has made some achievements in inventory management and improved performance management. From the perspective of fixed asset turnover rate analysis, the Midea Group's fixed asset turnover rate has been decreasing year by year, but it is still at a stable level, indicating that the utilization of fixed assets is relatively appropriate.

5.3 The influence of enterprise strategy on enterprise operation

The above analysis shows that the implementation of appropriate strategies contributes to the long-term development of businesses, which can maintain their key competitiveness even in the face of a widespread economic downturn [6]. The diversification strategy adopted by the Midea Group as an outstanding enterprise in the home appliance sector can help its long-term development. However, many enterprises have not realized that they have adopted an adequate strategy in the early stage of development, or there is no corporate strategic awareness, and most enterprises still have some problems in strategic management in the new era.

6 Problems in strategic management of start-up enterprises in the new era

Although start-up enterprises have the advantages of obvious political support and strong innovation ability, in the face of the rapidly developing and changing economic system and the industrial constraints of large enterprises, small enterprises still need to comprehensively assess the internal and external environment, and then constantly improve their strategic system in combination with their own advantages and disadvantages. When seeking the complete construction of the strategic system, we should take into account the favourable development of the enterprise.

6.1 Ignoring the overall situation of strategic management, strategic objectives and strategic planning are unrealistic

When formulating strategic management of an enterprise, most enterprises cannot look at the future development of enterprises from a long-term perspective. This phenomenon is especially common in start-ups, mainly because start-ups do not have their own positioning, vision and goals established in the early stage of development [11]. All their business development is focused on one inspiration and have not improved their other contents such as lack of organizational structure framework. Slow scale development caused by lack of capital support limits the overall grasp of strategic management by start-ups. In addition, start-ups are busy gaining a competitive position at an early stage of development, so they often ignore the content of 'strategy'. What entrepreneurs need is only a quick entry of capital, not the formulation of a specific enterprise goal. Therefore, even management is faced with the phenomenon of coping with the formulation of strategic management. It only displays relevant content in the report to accomplish a certain task, that's all. Even some enterprises have not introduced the idea and concept of "strategy".

6.2 Pay no attention to human security, post setting and other factors

As mentioned above, CSR commitment will lead to increased corporate costs. However, an equally important issue is how to ensure that employees can get full protection of their rights and get the pay they should get for their work. When reducing costs, companies usually consider: reducing employee wages, cutting holidays, reducing the number of jobs, etc.; however, this does not necessarily lead to strategic corporate governance. On the contrary, in the short term, costs are reduced, but in the long term, employees are in a low-motivation work environment and their performance is not high. Even the general principles of the company are no longer followed. In addition, when recruiting employees, we should also match the quality of employees with the needs of the jobs. This demand-driven recruitment strategy is often more effective.

6.3 Wrong judgment of the economic environment

Either an overestimation of the economic environment or a poor assessment of it makes it difficult for companies to make a proper analysis of the objective situation. In the early stage of business, enterprises often have great ambitions, but do not take into account the speed or complexity of today's economic development, easily get lost in the flow of the economy, misjudge the market environment, and thus it is difficult to develop scientific strategic objectives and planning. In the wake of the pandemic, it is all the more important that businesses rethink the steps they can take to help themselves in times of economic uncertainty [11].

6.4 Untimely strategic adjustment

Although a company's strategy is comprehensive and long-term, if the company faces drastic changes and the company's strategy no longer suits the current development of the company, it is necessary to integrate various factors and adjust the company's strategy. Generally speaking, the formulation, implementation and evaluation of an enterprise strategy is a long-term and time-consuming process. During this time, enterprise managers should conduct a comprehensive analysis of the current situation of the enterprise [12]. If there are differences before and after the analysis, they should make strategic adjustment in time. However, if this factor is ignored, it will lead to the wrong formulation of enterprise strategy and various problems in its implementation. On the other hand, the enterprise strategy is formulated. If the results of strategic evaluation are not corrected and adjusted in time, the development direction of the enterprise will gradually move away from the strategic goal.

7 Conclusion

A diversification strategy can improve the ability of enterprises to withstand risks and can help enterprises to develop quickly after a risk and to recover previous levels of development. Therefore, enterprises should consider the following factors when carrying out strategic management and judge whether the enterprise is suitable for diversification strategy according to its own situation.

7.1 Considering the overall situation of strategic management, formulation of practical strategic objectives and strategic planning

With the rapid economic development, the problem that every business faces is whether to continue to stick to the strategy it has set or to change it according to the market situation. However, the answer is not the same for all businesses. Therefore, enterprises should comprehensively consider the changes and development trend of their industry. For example, in the context of intelligence, if the home appliance industry still sticks to the industry as before without changing and upgrading the embedded core technology, it will only lose market share under the pressure of the industry. Therefore, when formulating business strategy, enterprises should focus on the development trend of the future world and formulate strategies in accordance with their own development according to their own industry. Comprehensively consider the organizational framework, performance incentives and other factors to establish their own core competitiveness. In addition, when formulating strategic planning, we should also consider long-term strategic goals, and not only see the immediate interests, because in the era of rapid development, the risks of enterprises often appear in the form of "they were out before they were ready for change" [3]. Therefore, it is important for enterprises to accurately assess the direction of their industry when formulating strategic management.

7.2 Pay attention to human security, post setting and other factors

Due to the lack of their own financial capacity, start-ups often ignore some potential influencing factors. For example, when formulating strategic management, we do not pay attention to talent recruitment and consistent post setting, which will lead to unclear division of internal responsibilities and unclear definition of responsibility scope. When formulating enterprise strategic management, we should have the idea of "recruiting talents", train professional talents, place them in suitable posts, and give them a salary guarantee system matching their ability to ensure that they can concentrate on serving the company. In addition, the need for social responsibility should be taken into account. In the initial stage, start-ups often have close cooperation with the government. Therefore, while undertaking social responsibilities, they will bring various benefits such as tax relief and promotion, and can improve the corporate image and expand the popularity of enterprises to a certain extent. The form of cooperation with the government will also be deeply recognized by consumers, which can increase their competitiveness.

7.3 Fully understand their own economic environment

The economic environment in which a company is located is one of the important factors in formulating a company's strategy. However, if it does not timely interpret the internal and external environment of the corporate environment and identify the opportunities brought by the new era, it will miss the opportunity. Therefore, when formulating strategies, managers should timely understand the data brought by the economic market, analyse the risks and challenges brought by the current economic environment with the data, and then decide whether to use expansionary or conservative strategies and formulate their own development strategies based on the data [13,14]. Now in the complex economic background, there is a need not only to specialize in a particular product or service, and develop another product line to increase its competitiveness. The diversification strategy of the enterprise widely used, and from a certain degree, it is not only about the effective use of surplus capital, it is about the enterprise's own competitiveness.

7.4 Timely strategic assessment and strategic adjustment

First, the current strategy needs to be adapted to the results of the last strategic assessment. The result of the last strategic assessment is feedback on the implementation of the company's strategy in the previous phase, thanks to which the current process and direction of the company's development can be clearly understood. Therefore, it is an important element that the enterprise must pay attention to. In addition to the overall evaluation of strategy implementation, there is also a strategic evaluation of a specific stage or department, which should also be given special attention. If a start-up wants to achieve sustainable development, all aspects of strategic management should cover the entire development of the enterprise. It should always be based on the 'strategy', evaluating the direction of the stage and monitoring whether it deviates from the overall strategic development objective or still meets the requirements of current economic development. In addition, it is necessary to establish a special Strategic Management Department. In establishing it, it is necessary to select personnel with work experience and good understanding of the enterprise, always pay attention to the changes in the form of the new period, and evaluate and formulate the strategic management of the enterprise.

References

1. A. Tripathi, T. Bagga, S. Sharma, S. K. Vishnoi. Big Data-Driven Marketing enabled Business Performance: A Conceptual Framework of Information, Strategy and Customer Lifetime Value. *11th International Conference on Cloud Computing, Data Science & Engineering (Confluence)*. **11**, pp. 315-320 doi: 10.1109/Confluence51648.2021.9377156 (2021)

2. R. J. Baumgartner, R. Rauter. Strategic perspectives of corporate sustainability management to develop a sustainable organization. *Journal of Cleaner Production*. **140**, 81–92 (2017)

3. N. Pröllochs, S. Feuerriegel. Business Analytics for Strategic Management: Identifying and Assessing Corporate Challenges via Topic Modeling. *Information & Management*. **57**(1) (2018)

4. Z. Zhong, W. Haoran, W. Junsheng. Analysis of Enterprise Strategic Management Issues and Coping Strategies Based on Big Data Analysis. *E3S Web of Conferences*. **214**, 01017 (2020)

5. M. Meng, S. Saide, S. Ratna, M. Muflih. Business Continuity Innovation in Disruption Time: Sociotechnical Systems, Business Analytics. *Virtual Business, and Mediating Role of Knowledge Absorptive Capacity. IEEE Transactions on Engineering Management*. 1–12 (2021)

6. M. H. Shekhar, M. Raihan, R. Sunanda, M. Sabyasachi. Effective Business Strategy Decision Making Method Using Fuzzy Logic. IEEE *2018 9th International Conference on Computing, Communication and Networking Technologies (ICCCNT)*. 1–4 (2018)

7. Y. Liu, L. Peng, T. Lin, Ee. Dong. Research on Ctrip Corporate Social Responsibility Strategy: An Empirical Case. *2021 7th International Conference on Information Management (ICIM)*. 172–76 (2021)

8. E.-M. Vătămănescu, D.-C. Dabija, P. Gazzola, J. G. Cegarra-Navarro, T. Buzzi. Before and after the Outbreak of Covid-19: Linking Fashion Companies' Corporate Social Responsibility Approach to Consumers' Demand for Sustainable Products. *Journal of Cleaner Production*. **327**, 129465 (2021)

9. J. H. de Oliveira, C. M. de Lima, N. F. Almeida, A. M. Mariano, A. C. Bittencourt Reis, J. M. da Silva. Costs influence and mediating effect of corporate risk management on organizational performance: a study applied to the Brazilian Public Service," *2021 16th Iberian Conference on Information Systems and Technologies (CISTI)*. 1-6 doi: 10.23919/CISTI52073.2021.9476392 (2021)

10. X. Erming, L. Weiguang. Strategic Management of Chinese enterprises in the past 40 years (1978-2018): Review, Summary and Prospect [J]. *Economic and management research*. **39**(09), 3-16 (2018)

11. T. Malviya. A multi-dimensional business strategy model for effective decision making: A holonic representation of enterprises. *2017 IEEE International Systems Engineering Symposium (ISSE)*. 1-6 doi: 10.1109/SysEng.2017.8088297 (2017)

12. Z. Ping. Manager Interaction and the Formation of Strategic Consensus: The Moderation Effect of Manager Role. *Science Innovation*. **7**(4), 115-119 doi: 10.11648/j.si.20190704.12 (2019)

13. M. Borges, J. Bernardino, I. Pedrosa. Data-driven decision making strategies applied to marketing. *2021 16th Iberian Conference on Information Systems and Technologies (CISTI)*. 1-7 doi: 10.23919/CISTI52073.2021.9476506 (2021)

14. A. Najdawi, S. K. Patkuri. Modeling Business Intelligence Process: Toward Smart Data-Driven Strategies. *2021 International Conference on Computational Intelligence and Knowledge Economy (ICCIKE)*. 198-202 doi: 10.1109/ICCIKE51210.9410804 (2021)

The Impact of Work-related Stress on Job Satisfaction and Organizational Trust during COVID-19 Pandemic

Altanchimeg Zanabazar[1,*], *Sarantuya* Jigjiddor[1], *Tsolmon* Jambal[2]

[1]National University of Mongolia, Business School, Management department, Mongolia
[2]Institute of Technology and Business, Faculty of Corporate Strategy, Okružní 517/10, 37001 České Budějovice, Czech Republic

Abstract. One of the industries significantly affected by the Covid-19 pandemic is tourism. As a result of lockdowns, tourism came to a halt, which has caused severe damages not only to tourism-related business but also aviation, hospitality sector, which resulted in unemployment, reduced work hours with minimum salary or temporary leave and even closure the businesses.
As a result of the pandemic, 99 travel agencies liquidated their business activities and the revenues of 88 percent of the whole tourism sector reduced and 56 percent of employees were laid off.
This paper analyses the impact of Work-related stress on the job satisfaction and organizational trust in the tourism sector during the unprecedented changes related to the Covid-19 pandemic. The survey was conducted online in April-May, 2021 and involved 275 employees working for different travel agencies. Numerous tests are used to verify the proposed hypothesis, such as factor analysis, reliability analysis, correlation analysis and PLS analysis.
The results of the survey show that the Work-related stress caused by Covid-19 pandemic has a negative impact on job satisfaction and organizational trust among surveyed employees.
Keywords: work-related stress; job satisfaction; organizational trust; COVID-19 pandemic

1 Introduction

The virus, first identified in Wuhan city, the People's Republic of China in December 2019, quickly spread around the world, was declared as a pandemic by the World Health Organization in March 2020 [1].

According to World Travel and Tourism Council, more than 50 million jobs are at risk due to the emerging situation [2].

The National Security Council of Mongolia issued Directive No. 01/01 "Temporary border shutdown" on January 30, 2020 and introduced lockdown on February 01, 2020. The measures taken by the government made the businesses completely suspend their activities. This applied also to businesses operating in tourism and hospitality sectors. Many of them were even forced to file for bankruptcy as well as well as laying off their employees [3].

The study conducted by Mongolian Chamber of Commerce and Industry shows that 90 percent of businesses in tourism sector have been affected by those unprecedented circumstances [4]. As a result, employee well-being is threatened significantly due to loss of income affecting their lifestyle, standard and quality of living and mental health conditions.

Numerous studies have been conducted among health professionals to evaluate the impact of the pandemic related to job stress, depression, anxiety and insomnia; however, there are not many studies addressing the level of employee fear, irritation and stress, and depression related to uncertainty among the employees in tourism and hospitality sector [5-6].

Tourism is definitely one of the sectors seriously affected by the Covid-19 pandemic. Due to the pandemic, 99 travel agencies suspended or closed down their business, 88 percent of them faced a 91-100% income shortage and 56 percent of them reduced the number of their employees [7].

This study is aimed at analysing employee Work-related stress and job satisfaction of local tourism travel agencies´ employees and their impact on the organizational trust.

Job stress and Job Satisfaction

Lazarus and Folkman describe stress as a physiological response to a potential threat to human well-being. It is a physical and mental reaction to the changes that people face in their daily life or needs that come up in life. Job stress is characterized by discrepancies between the demand of workplace and personal skills of the person who strives to meet

* Corresponding author: altanchimegz@num.edu.mn

the needs [8-9]. On the other hand, job satisfaction also plays an essential role in preventing and handling employee stress [10].

When people feel and perceive less support or cooperation from their co-workers and managers, it can lead to stress, which negatively contributes to the loss of self-confidence, decrease in job satisfaction and performance [11-12]. Many studies have been conducted to identify possible negative impacts of Work-related stress on employee work teams as well as organization as such. Work-related stress.

Recently, a new term appeared - pandemic stress originating from the spread of Covid-19. People are now spending a significant portion of their daily life working online due to lockdowns introduced and have easier access to large quantities of information on the pandemic, and their increased fear over uncertainty of their employment and stability of the future income often leads to stress and depression [13].

One of the crucial factors in achieving organizational objectives is employee job satisfaction [14]. Employee job satistifaction refers to the employee attitude to their job or subjective experience gained throughout their professional life [15].

According to Herzberg, performance outcomes is a source of both job satisfaction or dissatisfaction that occur due to certain workplace circumstances and hygiene factors. The issue of job satisfaction needs to be addressed as it is closely related to numerous organizational problem such as absenteeism, staff turnover, commitment and productivity [16].

Many researchers have confirmed the negative impact of Work-related stress on mental health and decrease in employee job satisfaction [10, 17-19].

During the Covid-19 pandemic, job security is a main stress-inducing factor, which leads to a decrease in job satisfaction and contributes to mental health concerns [20].

Work-related stress and Organizational trust

In the present, business success is heavily dependent on the organizational trust, even more than it used to be. Trust plays a crucial role in interpersonal communication and collaboration. Trust is considered as a success factor for any organization and its management since zero trust in the organization hinders maintaining open and sincere communication between managers and employees, which often leads to mistrust [21].

Organizational trust is an essential factor maintaining communication and collaboration among the employees of the organization and demonstrates the individuals' expectations regarding organizational communication and behaviour [22]. In simple terms, organizational trust shows employee expectations and trust concerning the belief in the organizational activities as well as its employees. Employees who show high levels of trust in the organizational management and co-workers tend to be more loyal.

Studies show that employees who suffer from stress often lose trust in organizational commitment and job security [23-24]. Job related stress reduces organizational trust and directly or indirectly affects staff turnover [25].

Job Satisfaction and Organizational Trust

Job satisfaction and organizational trust are factors that contribute to achieving the goals of a given organization and gain a competitive advantage over competitors. Organizational trust brings various positive effects both for employees and the organization. Higher level of organizational trust among its employees increases employee loyalty to the organization and higher employee job satisfaction [26].

There is a positive correlation between organizational trust and job satisfaction; high level of organizational trust results in high level of employee job satisfaction [27-29].

Research design: Based on literature review, the following research model is proposed to achieve the objectives of the study:

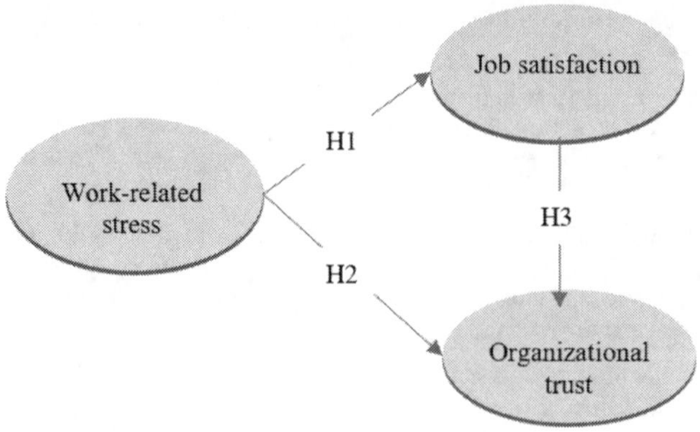

H1: Work-related stress influences has a negative impact on employee job satisfaction.

H2: Job stress negatively affects the organizational trust.

H3: Job satisfaction influences the organizational trust.

Figure 1. Word cloud of publication keywords on the topic of employee engagement

Source: Author.

2 Methodology

The objective of the study is to investigate the impact of Work-related stress on employee job satisfaction and organizational trust in the tourism sector in the wake of the unprecedented changes caused by the Covid-19 pandemic.

A total of fifteen travel agencies participated in the survey. SPSS 23.0 and Smart PLS 3.0 software, which are widely used in the social sciences research, are used for the data analysis. Numerous tests are performed to confirm or reject the formulated hypothesis, such as factor analysis, reliability test and regression analysis.

Data collection and data source

When selecting the sample, 95% confidence level with 5% confidence interval was considered. There were selected 275 out the total of 618 employees working in 15 travel agencies. We used the data of 275 participants, which is a sufficiently representative sample. The survey was conducted online between April 12 and May 20, 2020. The survey questionnaire consisted of 18 questions related to organizational trust, job satisfaction and stress; each question was rated on a 5-point Likert scale. In the study, the changes in demographic factors including age, sex, marital status, education, salary and employment status were considered.

Table 1. Background information on the Respondents

Age			Length of employment in the company			Changes in employment		
	n	%		n	%		n	%
20-25	16	5.9	Up to 1	32	11.6	Paid holiday	13	4.7
26-30	78	28.3	1-3 years	144	52.4	Decrease in worked hours	80	29.1
31-35	73	26.5	4-10 years	51	18.5	No salary	114	41.5
36-40	80	29.1	11-15 years	36	13.2	Lost job	68	24.7
41-50	22	8.1	16 years and more	12	4.3	No change	-	-
51 and older	6	2.1						

Source: Author.

Of the 275 participants, 58.9% were female and 41.1% male; as for age, they were relatively young employees at their twenties - fourtiesged 20-40's (89.8%). The majority of the participants (64 %) worked 1-3 years for their organization; 18.5 % of them worked in the sector for 4-10 years, and remaining 17.5% of them for more than 16 years. The results show that in terms of the pandemic-related consequences, 24.7% of the participants experienced unemployment, 41.5% unpaid leave, 29.1% reduced working hours, and only 4.7% had a paid leave. This indicates that employees in the tourism and hospitality sectors have faced the biggest risks associated with loss of income and loss of job.

Reliability analysis of variable

Cronbach α test is performed to test the reliability of the questionnaire. The test showed a value greater than 0.7 [30] and was thus evaluated as capable of measuring the proposed variables. Moreover, Composite Reliability coefficient showed a value of 0.5 [31]; the questions included in the questionnaire are reliable for the purposes of the study. Reliability test results are summarized in Table 2 below.

Table 2. Questionnaire reliability analysis

Set of items	Number of questions	Cronbach's α	AVE	CR
Work-related stress - WS	6	.892	.630	.910
Job Satisfaction - JS	6	.913	.679	.927
Organizational trust - OT	6	.869	.597	.899

Source: Author.

Correlation analysis

To estimate the correlations of the factors' correlations, Pearson Correlation analysis is carried out and the results are summarized in Table 3.

Table 3. Pearson Correlation analysis

Correlations				
		WS	JS	OT
WS	Pearson Correlation	1		
	Sig. (2-tailed)			
JS	Pearson Correlation	-.580**	1	
	Sig. (2-tailed)	.000		
OT	Pearson Correlation	-.655**	.593**	1
	Sig. (2-tailed)	.000	.000	
**. Correlation is significant at the 0.01 level (2-tailed).				

Source: Author.

The results of the analysis indicate notable negative correlations [r=-.580, $p<0.01$] of work-related stress, job satisfaction and organizational trust in the case of the surveyed participants show Job satisfaction and organizational trust among the participants show significant positive correlations [r=593, $p<0.01$].

Structural equation modelling

To ensure the reliability of the survey data and evaluate the survey model, R-Square is estimated. We used PLS analysis to measure the proposed model's predictive power. Factors related to the independent and dependent variables are carefully checked step by step. The results of the analysis show that all hypothesis formulated for the purposes of the study can be confirmed; regression analysis show a significant p-value ($p <0.05$). The results of the analysis are summarized in Table 4.

Table 4. Research result PLS-SEM

Hypothesis	Path	Standardized Beta	t value	P value	Remarks
H1	WS→JS	-0.492	6.492	.000	Supported
H2	WS→OT	-0.610	7.459	.000	Supported
H5	JS→OT	0.330	5.779	.000	Supported

Source: Author.

According to the results, participants' work stress of participants impacted slightly on job satisfaction[β=-0.492] while it strongly affected to the organizational trust [β=-0.610; $p<0.01$]. Job satisfaction has a moderate impact on organizational trust β=0.330; $p<0.01$]. Normative estimate is presented in below Figure 1.

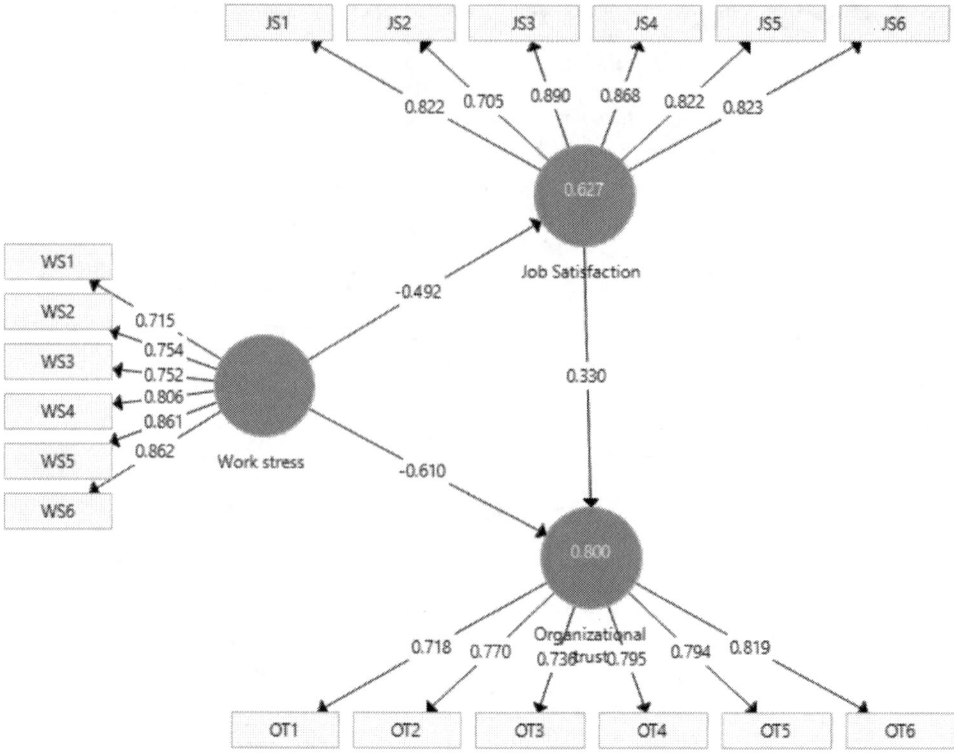

Figure 2. Path Model and PLS-SEM Estimate

Source: Author.

The effects of all moderating variables turned out to be positive (Beta coefficient) and statistically significant ($p < 0.05$). It can thus be concluded that the hypotheses formulated are valid.

3 Conclusion

The uncertainty arising from the pandemic is the main factor that causes the stress of employees. High stress level affects productivity, reduces employee job satisfaction and organizational trust, which may result in the failure of the entire organization [32]. Therefore, organizations need to tackle the problem through identifying employee stress level, its causes and finding solutions to resolve the situations [17].

In the case of tourism and hospitality sectors, employees have faced numerous challenges including job security, layoffs, income decrease, work break, work and life balance and livelihood, which are the main sources of stress reducing employee job satisfaction as well as organizational trust.

The objective of the paper was to investigate the impact of work-related stress on employee job satisfaction and organizational trust in the case of travel agencies in the wake of unprecedented changes caused by the pandemic.

According to the results of the study, work-related stress can be a significant factor reducing job satisfaction and organizational trust. The results of the survey are consistent with the results of earlier studies conducted in the area, such as research by Feather & Rauter [24], Bhatti & et al. [17], Chao & et al. [18], Hoboubi & et al. [10], Hsu & et al. [19], and Said & El-Shafei [25].

The results show that an increase in work-related stress by one unit results in a decrease in job satisfaction by 0.492 (Beta) unit and a decrease in the organizational trust by 0.610 (Beta) unit.

Moreover, job satisfaction has a significant impact on organizational trust among employees. Organizational trust is a psychological commitment of the employee to the organization by accepting its philosophy and confidence in its' future that is nurtured overtime.

An increase in job satisfaction by one unit results in an increase in organizational trust by 0.330 (Beta) unit, means higher job satisfaction level tend to make employees be loyal and trust their organization, which make them deepen their commitment and show positive attitudes towards the organization.

The results were also consistent with the previously conducted studies by numerous researchers, such as Perry & Mankin [27], Fard & Karimi [28], Artar & Erdil [29] and Aygün [26].

It can thus be concluded that there is a positive correlation between organizational trust and job satisfaction, which means that in organizations where the trust level remains high employees show higher job satisfaction.

As we decided to conduct the current survey due to the increasing public discussion on ambiguity of the tourism sector, we have no possibility to compare the previous state of work-related stress level and its impact on organizational trust. Further research could focus on comparing the changes in work-related stress.

Therefore, it is essential for any organization to deal with work-related stress and take preventive measures reducing the negative impacts that may result from the stress and other negative job practices to maintain healthy work environment where employees tend to remain loyal and positive towards not only to their jobs but also to their organizations.

References

1. World Health Organization. *WHO director-General's opening remarks at the media briefing on COVID19* [online]. Available at: https://www.who.int/dg/speeches/detail/who-director-general-s-opening-remarks-at-the-media-briefing-on-covid-19---11-march-2020 (2020)
2. World Travel & Tourism Council. *Coronavirus Puts Up to 50 Million Travel and Tourism Jobs at Risk Says WTTC* [online]. Available at: https://wttc.org/News-Article/Coronavirus-puts-up-to-50-million-Travel-and-Tourism-jobs-at-risk-says-WTTC (2020)
3. L. Sarangerel et al. *COVID-19 Impact on small businesses*. Ulaanbaatar: Edmarket. (2020)
4. MNCCI. *Research report of employment impact of Covid-19.* Ulaanbaatar: Mongolian National Chamber of Commerce and Industry. (2020)
5. J. Lai, S. Ma, Y. Wang, et al. Factors Associated With Mental Health Outcomes Among Health Care Workers Exposed to Coronavirus Disease 2019. *JAMA Network Open.* **3**(3):e203976, 1-12. doi:org/10.1001/jamanetworkopen.2020.3976 (2020)
6. S. Pappa, V. Ntella, T. Giannakas, V. G. Giannakoulis, E. Papoutsi, P. Katsaounou (2020). Prevalence of depression, anxiety, and insomnia among healthcare workers during the COVID-19 pandemic: A systematic review and meta-analysis. *Brain, Behavior, and Immunity.* **88**, 901-907 doi:org/10.1016/j.bbi.2020.05.026 (2020)
7. Y. Badyelgaji. *Covid-induced challenging losses to a national economy of East Asia: The 2020 economic case of the tourism sector in the Ulaanbaatar area of Mongolia*. Ulaanbaatar: Mongolian National University (2020)
8. T. W. Colligan, E. M. Higgins. Workplace stress: Etiology and consequences. *Journal of Workplace Behavioral Health.* **21**(2), 89-97. doi:org/10.1300/J490v21n02_07 (2006)
9. M. Beheshtifar, R. Nazarian, Role of occupational stress in organizations. *Interdisciplinary, Journal of Contemporary Research in Business.* **4**(9), 648-658 (2013)
10. N. Hoboubi, A. Choobineh, F. K. Ghanavati, S. Keshavarzi, A. A. Hosseini. The Impact of Work-related stress and Job Satisfaction on Workforce Productivity in an Iranian Petrochemical Industry. *Safety and Health at Work.* **8**(1), 67-71. doi:org/10.1016/j.shaw.2016.07.002 (2017)
11. A. Y. Poor, E. M. Ahmed. F. M. Ansari, M. Movahedi. Investigating the relationship between Work-related stress and organizational trust in Sirjan Municipality. *World Applied Sciences Journal.* **32**(10), 2042-2047 (2014)
12. H. C. Yang, Y. H. Ju, Y. C. Lee, Y. C. Effects of Work-related stress on self-esteem, job satisfaction, and turnover intention. *Journal of Transnational Management.* **21**(1), 29-39. doi:org/10.1080/15475778.2016.1120613 (2016)
13. A. R. Ahmad, H. R. Murad. The Impact of Social Media on Panic During the COVID-19 Pandemic in Iraqi Kurdistan: Online Questionnaire Study. *The Journal of Medical Internet Research.* **22**(5) doi:org/10.2196/19556 (2020)
14. M. L. Azic. The Impact of Hotel Employee Satisfaction on Hospitability Performance. *Journal of Tourism and Hospitality Management.* **23**(1), 105-117 (2017)
15. M. Miembazi, Y. Qian. Research on the impact of organizational commitment on job satisfaction between China and Congo-Brazzaville. *Journal of Research in Business, Economics and Management.* **8**(3), 1444–1450 (2017)
16. R. Kreitner, A. Kinicki, A. *Organizational behavior.* Boston: Irwin/McGraw-Hill (2001)
17. N. Bhatti, M. A. Hashmi, S. A. Raza, F. M. Shaikh, K. Shafique. Empirical Analysis of Work-related stress on Job Satisfaction among University Teachers in Pakistan. *International Business Research.* **4**(3), 264-270 (2011)

18. M.-C. Chao, R.-CH. Jou, C.-Ch. Liao. Workplace Stress, Job Satisfaction, Job Performance, and Turnover Intention of Health Care Workers in Rural Taiwan. *Asia-Pacific Journal of Public Health.* **27**(2), 1827–1836 doi:org/10.1177/1010539513506604 (2015)

19. Y.-Y. Hsu, Ch.-H. Bai, Ch.-M. Yang, Y.-CH. Huang, T.-T. Lin, Ch.-H. Lin. Long Hours' Effects on Work-Life Balance and Satisfaction. *BioMed Research International.* **2019**(1), 1-8. doi:org/10.1155/2019/5046934 (2019)

20. T. Nyanga, A. Chindanya. Covid 19 Pandemic Shifting The Job Satisfaction Landscape Among Employees. *Business Excellence and Management, Faculty of Management, Academy of Economic Studies, Bucharest, Romania.* **10**(5), 168-176 (2020)

21. E. M. Ahmed, A. Y. Poor, F. M. Ansari, M. Movahedi. Investigating the Relationship Between Work-related stress and Organizational Trust in Sirjan Municipality. *World Applied Sciences Journal.* **32**(10), 2042-2047 (2014)

22. P. S. Shockley-Zalabak, S. Morreale, M. Hackman. *Building the high trust organization: Strategies for supporting five key dimensions of trust.* San Francisco, CA: Jossey-Bass (2010)

23. M. Sverke, J. Hellgren, K. Näswall. No security: A meta-analysis and review of job insecurity and its consequences. *Journal of Occupational Health Psychology.* **7**(3), 242–264 doi:org/10.1037/1076-8998.7.3.242 (2002)

24. N. T. Feather, K. A. Rauter. Organizational citizenship behaviors in relation to job status, job insecurity, organizational commitment and identification, job satisfaction and work values. *Journal of Occupational Psychology.* **77**(1), 81–94 doi:org/10.1348/096317904322915928 (2004)

25. R. M. Said, D. A. El-Shafei. Occupational stress, job satisfaction, and intent to leave: nurses working on front lines during COVID-19 pandemic in Zagazig City, Egypt. *Environmental Science and Pollution Research.* **28**, 8791–8801 doi:org/10.1007/s11356-020-11235-8 (2021)

26. M. Aygün. Analysis of the relationship between job satisfaction and organizational trust of physical education and sports teachers. *Journal of Pedagogical Research.* **5**(1), 203-215 doi:org/10.33902/JPR.2021167583 (2021)

27. R. W. Perry, L. D. Mankin. Organizational Trust, Trust in the Chief Executive and Work Satisfaction. *Public Personnel Management.* **36**(2), 165-179 doi:org/10.1177/009102600703600205 (2007)

28. P. G. Fard, F. Karimi. The Relationship between Organizational Trust and Organizational Silence with Job Satisfaction and Organizational Commitment of the Employees of University. *International Education Studies.* **8**(11), 219-227 doi:org/10.5539/ies.v8n11p219 (2015)

29. M. Artar, O. Erdil. Relationship Between Job Satisfaction, Organizational Trust And Work Alienation. *13th International Strategic Management Conference, The European Proceedings of Social & Behavioural Sciences.* 194-205 doi:org/10.15405/epsbs.2017.12.02.17 (2017)

30. J. Nunnally. *Authors should use the forms shown in Table 3 in the final reference list.* New York: McGraw-Hil (1978)

31. J. Hair, W. Black, B. Babin, R. Anderson, R. Tatham. *Multivariate Data Analysis (6th ed.).* Upper Saddle River, NJ: Pearson Prentice Hall (2006)

32. S. H. Haynes, M. C. Leone, L. D. Keena, D. C. May, R. Ricciardelli, E. G. Lambert. The association between different forms of organizational trust and correctional staff job stress. *Journal of Crime and Justice.* **43**(5), 623-639 doi:org/10.1080/0735648X.2020.1734056 (2020)

Formation of Human Resources within an Enterprise under Conditions of Globalization Changes

Kateryna Zaika[1,*], *Andriy* Pochtovyuk[1], *Kateryna* Pryakhina[1], *Oleksandr* Kushch[2]

[1]Kremenchuk Mykhailo Ostrohradskyi National University, Faculty of Economics and Management, Management Department, Pershotravneva Street 20, 396 00, Kremenchuk, Ukraine
[2]Kremenchuk Mykhailo Ostrohradskyi National University, Faculty of Mechanics, Transport and Natural Sciences, Human Health Department, Pershotravneva Street 20, 396 00, Kremenchuk, Ukraine

Abstract. Strengthening competition in world markets leads to qualitative changes in the system of organization and management of businesses which must adapt to constantly changing conditions in the intensified competition at the international level. It is necessary to improve SMEs regulation system, including facilitating the establishment of inter-firm contacts and raising the qualifications of specialists and administrative staff. In this article, the authors research a formation process of new qualitative characteristics of enterprise human resources under globalization changes. The concentration of production is being replaced by the concept of intellectual-intensive production with a high level of education and information-technological equipment of the personnel. These peculiarities of the globalization process must be taken into account when solving issues related to the development of small businesses in Ukraine. It is revealed that this concept focuses attention not only on quantitative, but also on qualitative characteristics of personnel, including the level of competencies, professionalism, and a lack of focus on the development of intellectual potential. Competency as a socio-economic category becomes a system-forming basis that determines the formation of new processes for the preparation of human resources. The research reviewed the current state of SMEs in Ukraine under negative effects of the COVID-19 pandemic. The results of statistical analysis indicate a steady downward trend in the number of individual entrepreneurs. Changes in the manufacturing sector have also brought about changes in relation to labor resources. Ideas of self-development, lifelong education, emphasizing informal and non-formal learning are becoming popular. All of the above indicates that the provision of organizations with competent personnel is becoming an increasingly urgent problem and one of the main priorities. A specific model of a competency-based approach to the human resource formation system was proposed. It determines the behaviour of people in the performance of professional duties in a simple language for specialists and increases the return on the use of competencies. A high level of professional competencies of personnel is one of the important conditions for the effective work of business entities.
Keywords: human resources; SMEs; globalization; competency

1 Introduction

The value of human resources grows every year, so that for enterprises the decisive factor of success is their quality formation. According to the resource-based theory, heterogeneous enterprises acquire sustained competitive advantages by exploiting their strategic physical, human and organisational capital resources [1]. The process of forming qualitative characteristics is a complex and time-consuming phenomenon. It is an acquisition of parameters, each of which is defined by specific indicators. This process is carried out under the influence of a number of factors and conditions. The most important factors of a human resource strategy are the following: legal requirement, employee engagement, career advancement programs, corporate image, and performance management system [2]. Globalization for the formation of human resources – not very clear is the destruction of the boundaries of enterprise activities in the economy, technology, information, as well as in social society, which requires the creation of special conditions that could combine all resources and direct them towards the achievement of the goals that have been set. Due to globalization, the world as such is changing significantly [3]. This requires creating adaptive corporate structures and training programmes. The increasing prevalence of globalization is driven by a number of factors, including shortage of talent in developing countries, global market forces, technological forces, global cost forces [4]. In the context of economic globalization, it is the knowledge, skills, creativity and entrepreneurial abilities of employees that are considered as the main resources for improving the efficiency and competitive advantages of enterprises. Moreover, the process of forming human resources of a particular enterprise includes the following components:

* Corresponding author: kateorlova.977@gmail.com

- resource planning;
- recruitment of staff, selection of candidates to fill vacancies;
- determination of wages and benefits.

Understanding the human resource formation as a holistic phenomenon contributes to deepening scientific understanding of specific problems. The development of a holistic view of personnel formation among practitioners contributes to the development of a strategic vision, the formation of realistic ideas about the limits of possibilities of targeted management impact on human resources, a scientifically based understanding of probabilities and risks of implementing a particular strategic plan in the field of management. The essence of human resource management is to provide employees with an opportunity to acquire and enhance education, staff rotation and delegation of authority, career planning and development. In the context of globalization changes, the human resource formation reflects such negative trends as a significant reduction in natural population growth due to declining birth rate and rising mortality, unfavorable dynamics of gender and age structure of the population, increasing demographic and economic burden on the working part of population, irrational migration flows, low level of social development and living standards of the population. Some enterprises do not support their employees' desire to gain new knowledge and skills, and this problem needs to be studied further [5]. The increase in the demand for human resources determines the need for such enterprises to establish a system for their development, that is to say, the creation of a mechanism for changing human resources taking into account the purposes and specificities of a given organization's functioning. With the help of the system of human resource development, there should be a qualitative, and in extreme cases, quantitative improvement of employees' abilities by identifying and attracting their available labor potential. In the course of the development, human resources in the organization are transferred to a higher quality level. The goal of the research is to develop a new model of the human resource management system, focused on management mechanisms based on the competency approach and their impact on the long-term efficiency of small and medium-sized enterprises (SMEs).

Transformational changes that are characteristic of the domestic labor market are in line with modern globalization trends and transformations of the global labor market [6]. The spread of globalization processes has affected the development of human resources and contributed to the spread of global trends in domestic and foreign practice. T. Ivanović and M. B. Jovanovska [7] believe that since companies started to extend their operations beyond national borders and employ people originating from different countries, new challenges for human resource management have emerged. At present, there are many different approaches to determining human resources. The classics of economics have traditionally focused their research on determining the role of an individual in the system of economic relations and the use of his/her abilities in the process of labor. The theory of human resource management is reflected in many works of modern authors, such as Y. Qamar Y. U. and T. A. Samad [8], H. Wang et al. [9], R. J. Ward [10], C. S. Long and S. S. Xuan [11], B. E. Becker and Huselid M. A. [12]. Despite the authors' research in this area, there is no unified common definition of concepts of «human resources» and «human resource management» in the scientific literature. It will be appropriate to understand human resources as a set of socio-cultural indicators and personal and psychological qualities of employees, the formation of which occurs at the stage of elementary education.

The terminology in the field of human resource management is quite diverse. In modern practice of scientific developments, it is possible to distinguish the concepts of personnel management system and personnel development system. . The personnel management system is a combination of specific methods, tools, procedures and technologies for working with human resources. I. Smyrnova and K. Rudyka believe that such management is essential for all organizations – large and small, commercial and nonprofit, industrial, and operating [13]. Small and medium-sized enterprises have a number of distinctive properties that determine the features of management. Among them, modern researchers highlight flexibility, rapid response to globalization changes and focus on the regional market, as they are often the basis for the preservation of small and medium-sized cities. Research on the practice of human resource development (HRD) management of small and medium-sized enterprises is significantly limited for many reasons. According to C. T. Nolan and T. N. Garavan [14], research on HRD in SMEs has to date utilised a limited number of theoretical perspectives and insufficiently accounted for the heterogeneity of SMEs.

Regarding this period of economic development, a progressive scenario of intensive development of SMEs and increasing their competitiveness through the formation of a system of effective human resource management are relevant. The capabilities to manage human resources in small and medium-sized enterprises are significantly different in terms of the features of large companies [15]. The type of personnel policy of SMEs is characterized by a low level of staffing requirements, a low level of funding for social programs, a lack of problems with human resources or change of job descriptions, as well as a large number of innovations in personnel management. Put more formally, the contextual dimension of « company size» is related to other dimensions of organizational characteristics [16]. Improving the management of human development of a small enterprise includes the formation of an organizational mechanism for managing human development and the process of transforming knowledge into intellectual assets of the enterprise.

According to Y. Dong, H. Zhang and L. Li [17], the traditional human resource management system is faced with a number of issues, e.g. the quality of staff recruitment cannot be guaranteed, the training performance is not consistent with the actual performance, the enterprise performance appraisal is unfair, and the salary distribution is unfair, which seriously affects the enthusiasm and loyalty of employees and threatens the survival and development of enterprises.

Thus, it is necessary to form and use qualitatively new approaches to human resource management, which allow ensuring compliance of staff with the stringent requirements of the external environment.

The concept of efficiency in relation to human resource development should be considered in an expanded aspect, namely as the effectiveness of the human resource development system of the organization as a whole. However, the limited resources of small businesses do not allow the use of a wide range of management tools that are successfully used in large businesses. The competency approach therefore becomes the basis that allows one to develop a system of labor relations that combines the requirements of businesses working with staff and the tasks of a modern education system. Over the past 15 years, the concept of the competency-based approach has become an integral part of the human resource management model. In any modern organization, its staff should be considered a strategic asset, which should primarily involve certain competencies (personal and professional knowledge, skills and patterns of behavior necessary for the effective performance of their functions). According to G. Özçelik and M. Ferman [18], the competency approach is an effective measurement tool that helps employees agree on a common language and comprehend what is understood by superior performance. Competency as such is a system-forming basis for the formation of new processes of staff training, changing the supply and demand, the development of relations between employers and employees on the evaluation and payment of labor, the transformation of social, labor and professional relations at the internal and state levels. This necessitates a comprehensive understanding and systematization of scientific knowledge in the field of the competency approach.

2 Methodology

Within the research framework, the essential characteristics of the competency approach in the modern concept of formation and development of human resource within enterprises in the conditions of globalization changes are considered. Substantiation of the relevance of its application is based on analyzing fundamental concepts and scientific hypotheses presented in the works of classics of economics, specialists and practitioners in the field of labor economics, personnel management and globalization. The information and empirical basis of the study included analytical materials of the *opendatabot* resource, consensus forecasts of the Ministry of Economy of Ukraine, survey results of the Union of Ukrainian Entrepreneurs (SUP), survey results of the Resource Efficient and Cleaner Production Centre (RECP), global report of the International Trade Center, publications in scientific periodicals, materials of scientific conferences, as well as information from the State Statistics Service of Ukraine. The research was conducted within the framework of systemic, complex, as well as scientific methods of dialectical cognition: analysis and synthesis, induction and deduction, comparison and analogy, statistical and mathematical modeling.

3 Results and Discusssion

In the global perspective, the main actors in economic life are not large corporations, but small and medium-sized businesses, which account for up to 90 % of all global companies and provide 70 % of jobs. Small and medium businesses in Ukraine bring 55 % of the GDP to the country's economy, in particular, the share of small businesses is 16 % of the GDP. Globalization conditions create interrelated trends between small and medium-sized businesses. Their value as a stabilizer in crisis conditions is growing due to flexibility, dynamism and high adaptive properties. The globalization of the world economy creates a problem for further survival and development of small businesses in Ukraine. The process of their development has been slowed down due to a number of significant reasons, which include the negative effects of the COVID-19 pandemic. A report by the US National Bureau of Economic Research (NBER) states that the pandemic will not have serious negative consequences for globalization [19]. However, its pace has slowed recently compared to previous decades.

The economic trends of the current crisis are serious. A wide-scale downturn in the economy is forcing one to adapt to new, more complex realities. The COVID-19 pandemic has affected firms of all sizes, but SMEs can be particularly vulnerable for several reasons [20]. These include limited access to high-quality resources, a lack of financial resources, and difficult access to information and scientific advances. Overall, 75 % of small and medium-sized enterprises suffered serious losses as a result of the pandemic. More than 80 % of SMEs have optimized costs due to reduced profits, of which 28 % have reduced wages, 21 % staff and 22 % sent staff on unpaid leave. Furthermore, Figure 1 shows an infographic on changes in the number of sole proprietors and other business entities.

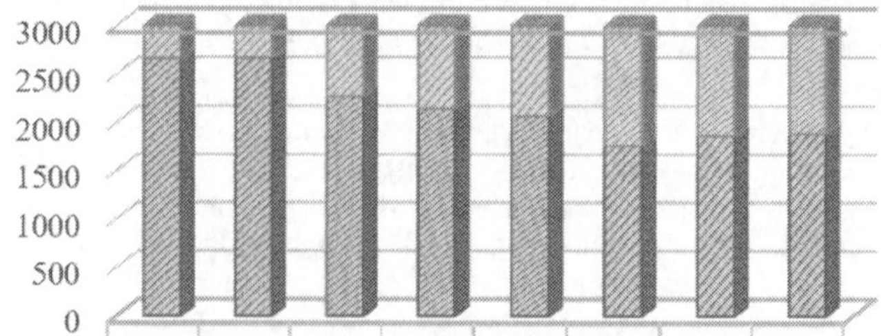

	2013	2014	2015	2016	2017	2018	2019	2020
legal person, mln	1281,6	1317,1	1278,6	1121,3	1185,1	1235	1298,4	1350,6
private entrepreneur, mln	2664,3	2670,3	2267,9	2143,5	2069,7	1763,3	1866,1	1885,9

Figure 1. Infographics of the number of individual entrepreneurs and legal in Ukraine (2013 – 2020)

Source: Authors based on State Statistics Service of Ukraine data [21].

The results of statistical analysis indicate a steady downward trend in the number of individual entrepreneurs. Since 2013, it has decreased by 779 thousand, while the number of legal entities has increased by 69 thousand. In 2014, the number of private entrepreneurs grew by 0.2 %. The private entrepreneurs closed for the next four years. – not very clear. In 2015, their number decreased by 15 % to 2.267 million, and in 2016, it decreased by another 5.5 % (2.143 million). In 2017, the decrease in the number of private entrepreneurs was 3.4 %, and in 2018, the situation from 2015 was repeated, as the number of entrepreneurs decreased by 16 %. The revival of small businesses began in 2019, when they registered 7.5 % more private entrepreneurs than the previous year. Modest growth rates were shown by 2020, when 1,885 million individual entrepreneurs were registered (+ 1.1 %). In 2020, 232 thousand individual entrepreneurs ceased to exist against 199.8 thousand in 2019. The new business requirements that have emerged in connection with the COVID-19 pandemic encompass the need for rapid access to data on employees, including their competencies. The current understanding of the concept of competencies is very wide [22]. In a general sense, competencies comprise characteristics of an individual, including motivational and psychophysiological characteristics, knowledge, skills and abilities, as well as human values that contribute to improving work efficiency. More specifically, they are the abilities needed to solve labor problems and obtain the necessary results. The competency-based assets in the field of human resources are related to absorbing knowledge, experience, or artificial goods through the competency-based processes [23].

In the practice of human resource management, the concept of professional competencies is given in a slightly different meaning. It is considered in terms of the suitability of an employee for a specific position he/she holds in a particular organization. A recognized tool for assessing professional competencies is the competencies' model.

Table 1. Model of the competency-based approach to the human resource formation system

Criterion	Characteristic
Mission	Increasing competencies of the workforce as a strategic potential, a key competitive advantage
Purpose	Systematization of knowledge, skills and abilities, development of staff competencies in relation to sustainable development
Tasks	Formation of competencies of specialists capable of implementing positive changes. Assessment of competencies leading to an increase in the importance of the staff in achieving the target parameters of the given organization´s economic efficiency.
Object	Competencies of the company's personnel
Personnel categories	The entire workforce (human resources) of the enterprise
Subject areas	Assessment, training, promotion, motivation
Result	Improving the staff competencies in order to achieve better results for individual and team work

Source: Authors.

A distinctive feature of the competency-based approach in the activities of the organization's personnel is the reorientation of its management goals from solving operational personnel tasks (e.g. the timeliness of retraining of employees, advanced training or replacement of managerial positions) to systematic problems of higher strategic order that go beyond the usual responsibility of the personnel management service. The goals of human resource management in the context of this approach are designed in such a way as to show that these processes contribute to the improvement of the enterprise and its divisions. The competency-based approach describes not only the knowledge and skills of a person, but also a model of professional behavior, through which a direct connection of the applied knowledge, skills and abilities is established with the specific results of the employee's labor activity. The dynamism of professional competencies is well understood at the level of organizations and is the basis for building such procedures for human resource management as adaptation, assessment, training, and career advancement. According to human resources professionals, the competency model is a portrait of the «ideal employee», integrating all the abilities and qualities necessary for successful work in a particular workplace in this organization. Within enterprises, the competency model is used as a basis for comparison in the process of making key management decisions regarding personnel.

The technology for competency model development includes a detailed analysis of professional activities: identifying the characteristics of successful and unsuccessful behavior by expert methods, meaningful grouping of primary characteristics, defining competencies and their indicators. This study substantiates structuring of professional competencies in the areas of their manifestation, established in accordance with the level of job requirements. It is proposed to distinguish three areas of professional competencies: social and personal, corporate and subject-practical, which meet three levels of requirements for the position (general, organizational and special). Such a view allows one to consider professional competencies as a systemic compliance of employee personality to the requirements of society, organization and a specific workplace. Solving the problem of developing competencies in the human resource management system, it is necessary to create conditions for its formation, taking into account the influence of external and internal factors of competency management (organizational or functional). The organizational competencies involve a set of staff competencies together with the given organization's ability to systematically achieve specific results. The organizational competencies can be: the ability to master the production of new products with minimal losses, the ability to innovate, the competency in creating foreign representative offices and branches, competencies in taking into account the needs of customers when developing new products, etc. The specified technology for creating the competencies is a unique model, the versatility of which lies in the fact that it allows one to link the personnel management system with the strategic goals of the enterprise (Fig. 2). This connection is built through the definition of professional and personal qualities of managers and specialists, that is, the key company competencies.

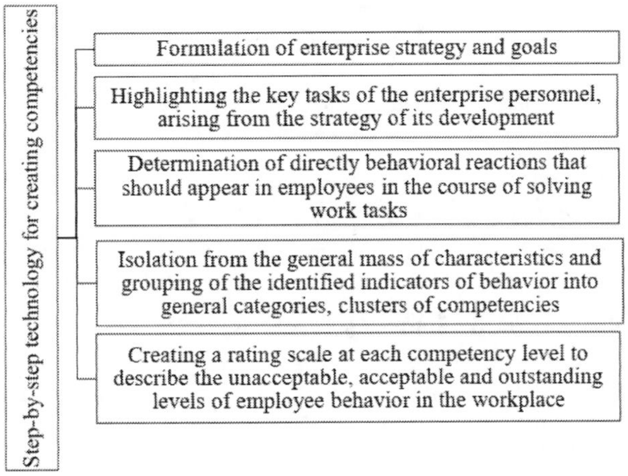

Figure 2. Compontents of the step-by-step technology of human resource competencies formation

Source: Authors.

The proposed model determines the real behavior of people in the performance of professional duties in a simple and accessible language for managers and specialists, and increases the return on the use of competencies. The model underlies the work with human resources, facilitating the recruitment process, developing standards for successful performance of work, increasing the effectiveness of training, retraining and developing employees, developing criteria for assessing performance, ensuring personnel development and career planning.

Thus, it expands the range of analyzed issues, forms the prospects for further research in the direction of the competency-based approach to human resource management. Considering an enterprise as a hierarchical organizational system for making managerial decisions, it is necessary to determine that for each managerial level it is possible to define its own competencies that ensure the emergence or strengthening of the enterprise's competitive advantages. This allows one to assert that for the successful formation and implementation of the long-term development of a company, it is necessary to recognize the presence of competencies of certain levels, compliance with strategic goals and the ability to manage their dynamics.

4 Conclusion

It may be argued that the control of globalization becomes the most important challenge, starting from worldwide labour force resources usage and finishing by recruiting, training and development and salaries forming, and orienting to other EU countries and knowledge employees [24]. For all organizations, large and small, production as well as their operating in the field of human resources service is of paramount importance. Although there has been some research conducted and there are a number of organizations considering SMEs as the main generator of economic development, some factors should not be overlooked, e.g. attracting, motivating, developing and retaining long-term employees have been recognized as important challenges that are encountered by SMEs [25]. Human resource management is recognized as one of the most important areas of enterprise life, which can increase its efficiency many times over. In the context of the development of management science, the analysis of the concept of « competencies» was carried out. It has been established that the provision of organizations with competent personnel is becoming an increasingly urgent problem and one of the main priorities of socio-economic development and increasing the competitiveness of the national economy. Competency models are seen as an important way to achieve high organizational efficiency by focusing on individual abilities and capabilities of employees in various projects. The competency model proposed in the study can be used to coordinate and harmonize human resource management processes. The list of studies devoted to the competency-based approach indicates that this topic is extremely relevant and scientifically in demand. In domestic and foreign science, a certain theoretical and empirical basis has been created to substantiate the role and importance of the approach in personnel management. However, its socio-economic aspects have not been highlighted. In addition, it should be noted that competencies have not been systematically considered as a scientific problem in the field of labor economics, which is related to the system of human resource management.

The scientific novelty of the study lies in the creation of basic provisions for working with personnel based on the competency-based approach in terms of the diversity of existing trends in social and labor relations and the complexity of economic development. Further research can be aimed at the scientific and practical development of the systemic use of the approach, in particular, to create a synthesis of its methodology and strategy for human resource management, which is a fundamental basis for understanding the development strategy of the labor economy in the context of globalization changes.

References

1. V. H. Nam, H. N. Luu. How do Human Resource management practices affect innovation of small and medium sized enterprises in a transition economy. *Journal of Interdisciplinary Economics.* 1-22 (2021)
2. B. D. Stewart, D. B. Webster, S. Ahmad, J. O. Matson. Mathematical models for developing a flexible workforce. *International Journal of Production Economics*. **36**(3), 243-254 (1994)
3. Z. Stofkova, V. Sukalova. Sustainable Development of Human Resources in Globalization Period. *Sustainability.* **12**, 7681 (2020)
4. R. Saranya, A. Asok. Globalisation and Its Impact on Human Resource Management. *Journal of Research in Humanities and Social Science.* **7**(4) 21-24 (2019)
5. N. Hunko. Factors Influencing the Formation of Human Resources. *Economics & Sociology*. **6** (2), 65-72 (2013)
6. I. Trunina, D. Zagirniak, K. Pryakhina, T. Bezugla. Diagnostics of the enterprise personnel sustainability. *Problems and Perspectives in Management.* **18**(2), 382-395 (2020)
7. T. Ivanović, M. B. Jovanovska. Individual and social aspects of global human resources staffing. *Dynamic Relationships Management Journal*. **8**(1), 55-64 (2019)
8. Y. Qamar, T. A. Samad. Human resource analytics: a review and bibliometric analysis. *Personnel Review.* (2021)
9. H. Wang, D. Tan, W. Liu. Electronic Human Resource Management Survey Enabled by Big Data Analysis. *IEEE 15th International Conference on e-Business Engineering, ICEBE 2018.* 340-345 (2018)
10. R. J. Ward, The challenge of development: Theory and practice in human resource management. 1st ed. Routledge, (2017).
11. C.S. Long, S. S. Xuan. Human resources development practices and employees' job satisfaction. *Social Issues in the Workplace: Breakthroughs in Research and Practice (2 Volumes).* 1726-1738 (2017)
12. B. E. Becker, M. A. Huselid. Strategic human resources management: Where do we go from here? *Journal of Management.* **32**(6), 898-925 (2006)
13. I. Smyrnova, K. Rudyka. Improving the personnel management system of an industrial enterprise. *The Economic bulletin of Donbass.* **4**(58), 179-185 (2019)
14. C. T. Nolan, T. N. Garavan. Human Resource Development in SMEs: a systematic review of the literature. *International Journal of Management Reviews*. **00**, 1-54 (2015)
15. B. Z. Poljasevic, S. Petković. Human resource management in small and medium-sized enterprises: conceptual framework. *Facta universitatis. Series: Economics and Organization.* **10**(3), 301-315 (2013)
16. J. D. Kok. *Human Resource Management within small and medium-sized enterprises*. Thela Thesis. ISBN 90-5170-705-3 (2014)

17. Y. Dong, H. Zhang, L. Li. Mechanism construction of human resource management based on blockchain technology. *Journal of Systems Science and Information.* **9**(3), 310-320 (2021)

18. G. Özçelik, M. Ferman. Competency Approach to Human Resources Management: Outcomes and Contributions in a Turkish Cultural Context. Human Resource Development Review. **5**(1), 72-91 (2006)

19. National bureau of economic research. *Acess to Macroeconomic outcomes and COVID-19: a progress report* [online]. Available at: https://www.nber.org/system/files/working_papers/w28004/w28004.pdf (2020)

20. World Bank Group. *Acess to Small and medium enterprises in the pandemic: impact, responses and the role of development finance* [online]. Available at: https://openknowledge.worldbank.org/bitstream/handle/10986/34552/Small-and-Medium-Enterprises-in-the-Pandemic-Impact-Responses-and-the-Role-of-Development-Finance.pdf?sequence=1&isAllowed=y (2020)

21. State Statistics Service of Ukraine [online]. Available at: http://www.ukrstat.gov.ua/ (2021)

22. I. Prabawati, Meirinawati, T. AOktariyanda. Competency-based training model for human resource management and development in public sector. *The 2nd International Joint Conference on Science and Technology (IJCST)* 2017. **953** (2018)

23. P. Rastgoo. The role of human resources competency in improving the manager performance. *Acta universitatis agriculturae et silviculturae mendelianae brunensis*. **1**, 341-350 (2016)

24. D. J. Kelly. *Acess to Human Resource Development: For Enterprise and Human Development* [online]. Available at: https://ro.uow.edu.au/artspapers/114/ (2016)

25. T. Beqiri, B. Aziri. Human resources management development in micro and small enterprises. *Human Research in Rehabilitation.* **11**(2), 143-150 (2021)

Approach to Corporate Education of Employees in Companies in the Czech Republic, Russia, Slovakia and Ukraine: A Comparative Analysis

Zdeněk Caha[1*], *Iva* Klementová[1], *Michaela* Procházková[1], *Mario* Bogdanovic[2], *Michaela* Ruso[1]

[1]Institute of Technology and Business, Faculty of Corporate Strategy, Okružní 517/10, 37001 Ceske Budejovice, Czech Republic
[2]Istrian University of Applied Sciences in Pula, Croatia

Abstract: Effective and systematic corporate education is a key factor for competitiveness the current turbulent economy. The results presented in this paper are based on a questionnaire survey. The research was carried out on a sample set of Czech, Russian, Slovak and Ukrainian companies. The aim of the survey was to determine to what extent companies in four post-communist countries apply a strategic approach to education, which, among other things, is linked to the presence of long-term and annual education plans. The results of the research showed that micro-companies and small companies in particular do not do so. The research also revealed that some companies, even though they declare they have an LTEP/AEP in place, are not adhering to their plans.
Key words: Czech Republic; Russia; Slovakia; Ukraine; strategic approach; corporate education; plan

1 Introduction

Effective and systematic corporate education, as a part of overall company strategy, has been shown to be a key factor in maintaining a competitive advantage in the current knowledge economy [1,2]. According to Armstrong [3], strategic education and development, like strategic HR management, plays a key role in the achievement of an organisation´s success. Within this context, employees are, without a doubt, one of the most important resources a company possesses when it comes to generating potential competitive advantage [4]. It is therefore important to educate and develop employees systematically and effectively [5,6]. The education of an organisation´s employees is the most effective if it has a cyclical character, is based on the education strategy and policy of the particular organisation, and is in compliance with the overall strategy of the organisation [7]. One reason for the increasing interest in entrepreneurship education is the positive impact of entrepreneurship on sustainable economic growth. The basic cycle for the systematic education of employees of an organisation consists of: the identification of the need for education; the planning of the education; the implementation of the education process; and the assessment of the results and effectiveness of the education programme. The systematic education of employees and the conceptual planning and management of their careers supports the achievement of an organisation´s strategic goals and contributes to the long-term competitiveness of a company [8]. Bencsik and Sólyom [9] also claim that the more a company is able to maintain, develop and utilise the knowledge of its employees, the more it is able to achieve business success, as knowledge becomes a resource of long-term competitiveness and an advantage from the business strategy point of view. Rauch and Hatak [10] underlines the importance of the strategic planning of corporate education by claiming that the cost of education should represent a profitable investment, which is why education planning should be based on an analysis of the present situation, as well as the future needs and demands of individual work positions. According to Moifrrnkuzzx et al. [11], the presence of development and education specialists is an indication of a strategic approach, particularly in large and medium-sized companies.

According to the results of an international survey conducted in 2010 [12], less than a third of companies reported having written education plans or similar education programmes in place, and less than half reported having an allocated budget for such activities. Estimations made by the International Labour Organisation [13] show that a 1% increase in the number of days spent on education leads to a productivity increase of 3% in a company. Good quality training enables people to develop their capacities and to exploit social opportunities, as well as improves the productivity of employees and companies. These factors combined contribute to the future strengthening of innovation and development and encourages domestic as well as international investment. This in turn, creates jobs, reduces unemployment, leads to higher levels of pay and reduces social inequality. Collier, et al. [14] claim on the basis of their own research, that companies that systematically educate their employees have a 13% higher probability that they will survive. They go on to say that more research is required into the long-term impact of education or human capital on a company´s overall performance. Kitching and Blackburn [15] confirm the positive impact of a strategic approach to education. Korenková [16] is convinced that it is the people who work for a company that help it distinguish itself from others and who determine

* Corresponding author: caha@mail.vstecb.cz

whether a company will be successful or not through their active and creative approach to innovation and to customers themselves. Support for education is therefore crucial for maintaining competitiveness. Staff education and development should therefore be a part of overall company strategy and policy, and should be a coherent and systematic process. Korenková goes on to state that large companies prefer an active and systematic approach to education, whereas a random approach prevails in medium-sized companies and a passive approach in small companies. These different approaches are largely determined by the availability of funds, whereby the majority of small companies prefer to invest into innovation and improvements in technologies and leave education to the employees themselves. Tej and Slávikovà and Hrvolová [17] divided companies into three categories according to their approach to education, namely companies without an education strategy, companies with a non-systematic strategy and companies with a positive systematic strategy. They hold the view that companies achieve competitive advantages through their employees and that, although the effectiveness of the money invested into education is often difficult to enumerate, education is an investment that certainly pays off for companies. Cagáňová, et al. [18] characterise the orientation of companies active in the field of education and employee development in Slovakia and present the results of research conducted in 340 companies in 2012. Their working assumption was that education and a comprehensive approach to education are the characteristics of an innovative organisation. They concluded that organisations should become so-called "learning companies" with the aim of achieving permanent education focused on the improvement of performance not only on a short-term basis, but also on a long-term one.

2 Methodology

The aim of the research was to determine to what extent companies in four post-communist countries apply a strategic approach to education, which, among other things, is linked to the presence of long-term (LTEP) and annual education plans (AEP). In relation to the research goal, the following three research questions were formulated:

RQ1: To what extent do companies in the four post-communist countries (the Czech Republic - CR, Russia - RUS, Slovakia -SLO, and Ukraine - UKR) apply a strategic approach to corporate learning (existence of long-term training plans, annual training plans).

RQ2: If there are LTTP or ATP, to what extent they are followed?

RQ3: To what extent does the existence of LTTP or ATP depend on the company size and the scope of its operations in the individual countries?

The research was conducted from mid-2016 to January 2018. The data was obtained by means of a questionnaire survey distributed among selected companies in the Czech Republic, Russia, Slovakia and Ukraine. In addition to establishing basic identification data on the companies (size, number of employees, scope of activities), the questionnaire also sought to determine data on the long-term education plans (LTEP) and annual education plans (AEP) of the respondents. The questionnaire was addressed to personnel managers or other competent managers or owners of companies. The respondents could choose between a printed or an electronic version of the questionnaire by means of an electronic link within the Google Disc web platform. The data obtained this way was entered into Excel, checked and subsequently adjusted for use in the R programme. The basic methods of inductive and descriptive statistics particularly contingency tables and graphs, Chi-square tests of independence, two-selection asymptotic U-tests and regression methods with artificial variables - logit regression model were applied to the data analysis [19].

The results are presented by means of tables and graphs. The statistical analysis of the existence of an LTEP and the existence of an AEP in companies in the Czech Republic (CR), Russia (RUS), Slovakia (SLO) and Ukraine (UKR) were processed in relation to company size, number of employees, and scope of company activity.

The following part of this paper includes a written and statistical analysis and evaluation of the existence of an LTEP (AEP) in the selected companies in the individual countries. In cases where the information on corporate education was missing, it was assumed that such an education plan did not exist. The differences in terms of the existence of an LTEP in Czech, Russian and Slovak companies was not great. The proportion among Ukrainian companies was less than half. As previously stated, the significance of these differences was also assessed by means of statistical tests. In addition, the adherence to an LTEP was assessed by means of contingency tables and graphs. Relative frequencies were calculated for the different numbers of companies in the individual countries.

The dependence of the existence of an LTEP (AEP) on company size and scope of company activity in the individual countries was also analysed by means of contingency tables and graphs. The hypothesis that the proportion of companies with an LTEP (AEP) grows in relation to company size was also tested. The dependence of the median proportion of companies with an LTEP (AEP) on the number of employees and on country was quantitatively examined and compared by means of logistic regression.

3 Results and discussion

The first step of the analysis was to look at the existence of an LTEP in companies according to country. The results are presented in the contingency tables (1, 2)

Table 1. Numbers of companies with and without an LTEP

	Yes	No	Total
Czech Republic	290	319	609
Russia	22	28	50
Slovakia	45	40	85
Ukraine	29	116	145
Total	386	503	889

Source: Authors

Table 2. Proportion of companies with and without an LTEP (in %)

	Yes	No	Total
Czech Republic	47.6%	52.4%	100.0%
Russia	44.0%	56.0%	100.0%
Slovakia	52.9%	47.1%	100.0%
Ukraine	20.0%	80.0%	100.0%
Total	43.4%	56.6%	100.0%

Source: Authors

It is clear that the country with the highest number of companies with an LTEP is Slovakia (52.9%), followed by the Czech Republic (47.6%) and Russia (44.0%). The country with the lowest number (20.0%) is Ukraine. The significance of the dependence of the proportion of companies with an LTEP on the country of origin was proven by a Chi-square test of independence. The results (see Table 2) were as follows: X-squared = 39.89; df = 3; p-value = $1.113.10^{-08}$. This means that the statistical significance of this dependence was confirmed at the significance level of 5%. However, it should be noted that the results may be influenced by the low number of Russian companies in the sample set.

Two-selection asymptotic U-tests were conducted on the population proportions to determine the significance of the difference in the proportions of companies with an LTEP in the individual countries. To do this, the population proportions of companies with an LTEP in the individual countries were identified by the symbols πCR, πRUS, πSLO, πUKR, the corresponding selection proportions identified by the symbols pCR, pRUS, pSLO, pUKR, and the corresponding total numbers of companies identified by the symbols nCR, nRUS, nSLO, nUKR.

Firstly, the companies with an LTEP in the Czech Republic and in Russia were compared. The data presented in Table 1 shows that the existence of an LTEP is lower in Russia, with only 22 out of the 50 Russian companies (44%) having one. Of the 609 Czech companies that participated in the research, 290 have an LTEP in place (47.62%) – see Table 1 . A zero hypothesis H0: πCR = πRUS was subsequently tested against an alternative hypothesis H1: πCR > πRUS. The test statistic was

$$U = \frac{p_{\check{C}R} - p_{RUS}}{\sqrt{p^*(1-p^*)(\frac{1}{n_{\check{C}R}} + \frac{1}{n_{RUS}})}}, \quad \text{where} \quad p^* = \frac{p_{\check{C}R} n_{\check{C}R} + p_{RUS} n_{RUS}}{n_{\check{C}R} + n_{RUS}}$$

On the basis of the data in Tables 1 and 2

$$p^* = \frac{p_{\check{C}R} n_{\check{C}R} + p_{RUS} n_{RUS}}{n_{\check{C}R} + n_{RUS}} = \frac{0{,}4762\,.609 + 0{,}44.50}{609 + 50} = 0{,}4552$$

and

$$U = \frac{0{,}4762 - 0{,}44}{\sqrt{0{,}4552\,(1 - 0{,}45852\,(\frac{1}{609} + \frac{1}{50})}} = 0{,}494$$

and the p-value = 0.314. The zero hypothesis (H0) cannot therefore be refuted at the significance level of 5%. This means that the proportion of companies with an LTEP in the Czech Republic is not statistically significantly higher than that in Russia. In addition, the validity of the alternative hypothesis was not proven (the difference is not statistically significant).

Similar comparisons were subsequently made for different combinations of countries. For the comparison between the Czech Republic and Slovakia (45 out of 85 Slovak companies had an LTEP in place) the results were as follows: U = -0.925 and the p-value = 0.823. On this basis, the same conclusion can be drawn. For the comparison between the Czech Republic and Ukraine (29 out of 145 Ukrainian companies had an LTEP in place) the results were: U = 6.050 and the p-value = $7.24.10^{-10}$. On this basis, the opposite conclusion can be drawn. The test confirmed a statistically significant difference in the existence of an LTEP between the Czech Republic and Ukraine. For the comparison between Russia and Slovakia the results were: U = -1.36 and the p-value = 0.913. Once again, the zero hypothesis could not be refuted. There is no statistically significant difference in the existence of an LTEP between them. For the comparison between Russia and Ukraine the results were: U = 10.53 and the p-value = 0. In this case, the zero hypothesis was refuted. The test confirmed at the significance level of 5% that the proportion of Russian companies with an LTEP was significantly higher than the proportion of Ukrainian companies with an LTEP. For the comparison between Slovakia and Ukraine the results were: U = 17.35 and the p-value = 0. On this basis, the same conclusion can be drawn as in the previous case. The test confirmed that the number of companies with an LTEP in Slovakia was significantly higher than that in Ukraine.

The second step of the analysis was to look at to what extent an LTEP, where it exists, is or is not adhered to. The relevant data is presented in the following tables (3, 4).

Table 3. (Non-) existence of an LTEP and (non-) adherence to it (in terms of numbers of companies)

	Yes	Yes/Not adhered to	No	No response	Total
Czech Republic	247	43	316	3	609
Russia	9	13	28		50
Slovakia	40	5	40		85
Ukraine	21	8	114	2	145
Total	317	69	498	5	889

Source: Authors

Table 4. (Non-) existence of an LTEP and (non-) adherence to it (in terms of the proportion of companies (in %))

	Yes	Yes/Not adhered to	No	No response	Total
Czech Republic	40.6%	7.1%	51.9%	0.5%	100.0%
Russia	18.0%	26.0%	56.0%	0.0%	100.0%
Slovakia	47.1%	5.9%	47.1%	0.0%	100.0%
Ukraine	14.5%	5.5%	78.6%	1.4%	100.0%
Total	35.7%	7.8%	56.0%	0.6%	100.0%

Source: Authors

An existing LTEP is most often *not adhered* to in Russia (26.0%) and most often *adhered* to in Ukraine (5.5%). However, in Ukraine, the non-existence of an LTEP is the highest (78.6%) among the four countries, or garnered no response at all (1.4 %).

The third step of the analysis was to look at the existence of an AEP in companies according to country. The results are presented in the contingency tables (5, 6).

Table 5. Numbers of companies with and without an AEP

	Yes	No	Total
Czech Republic	293	316	609
Russia	8	42	50
Slovakia	46	39	85
Ukraine	29	116	145
Total	376	513	889

Source: Authors

Table 6. Proportion of companies with and without an AEP (in %)

	Yes	No	Total
Czech Republic	48.1%	51.9%	100.0%
Russia	16.0%	84.0%	100.0%
Slovakia	54.1%	45.9%	100.0%
Ukraine	20.0%	80.0%	100.0%
Total	42.3%	57.7%	100.0%

Source: Authors

It is clear that the country with the highest number of companies with an AEP is Slovakia (54.1%), followed in succession by the Czech Republic (48.1%), Ukraine (20.0%) and Russia (16.0%). The significance of the dependence of the proportion of companies with an AEP on the country of origin was proven by a Chi-square test of independence. The results (see Table 7) were as follows: X-squared = 57.01; df = 3; p-value = $2.56.10^{-12}$. This means that the statistical significance of this dependence was confirmed at the significance level of 5%. However, it should be noted that the results may be influenced by the low number of Russian companies in the sample set.

As before, two-selection asymptotic U-tests were conducted on the population proportions to compare the existence of an AEP in the four countries. The results are summarised in table 7 (the compared countries, selective proportions, numbers of measurements, test statistic, the p-value of the test):

Table 7. Results of U-tests - comparison of the existence of an AEP in the four countries

	p1	p2	n1	n2	u	p-value
CR_RUS	0.481	0.16	609	50	4.380	$5.940.10^{-06}$
CR_SLO	0.481	0.541	609	85	-1.043	0.852
CR_UKR	0.481	0.2	609	145	6.150	$3.867.10^{-10}$
RUS_SLO	0.16	0.541	50	85	-9.058	1
RUS_UKR	0.16	0.2	50	145	-1.755	0.960
SLO_UKR	0.541	0.2	85	145	17.978	0

Source: Authors

On the basis of the results presented in Table 7, it can be concluded that the zero hypothesis on the existence of an AEP can only be refuted for the following comparisons: Czech Republic - Russia; Czech Republic - Ukraine; and Slovakia - Ukraine. This means that the proportion of companies within a country that have an AEP is significantly higher in the Czech Republic than in Russia and Ukraine and is significantly higher in Slovakia than in Ukraine. The differences are not significant between the remaining pairs of countries.

The dependence of the existence of an LTEP on country and company size was subsequently analysed.

The results are presented hierarchically in the contingency tables (8, 9) that follow.

Table 8. Number of companies with an LTEP according to country and company size

	Yes	No	Total
Czech Republic	**290**	**319**	**609**
micro-company	35	107	142
small company	77	103	180
medium-sized company	89	75	164
large company	89	34	123
Russia	**22**	**28**	**50**
micro-company	3	10	13
small company	17	17	34
medium-sized company	2	1	3
Slovakia	**45**	**40**	**85**

micro-company	5	8	13
small company	11	19	30
medium-sized company	15	11	26
large company	14	2	16
Ukraine	**29**	**116**	**145**
micro-company	3	25	28
small company	5	26	31
medium-sized company	11	44	55
large company	10	21	31
Total	**386**	**503**	**889**

Source: Authors

Table 9. Proportion of companies with an LTEP according to country and company size (in %)

	Yes	No	Total
Czech Republic	**47.6%**	**52.4%**	**100.0%**
micro-company	24.6%	75.4%	100.0%
small company	42.8%	57.2%	100.0%
medium-sized company	54.3%	45.7%	100.0%
large company	72.4%	27.6%	100.0%
Russia	**44.0%**	**56.0%**	**100.0%**
micro-company	23.1%	76.9%	100.0%
small company	50.0%	50.0%	100.0%
medium-sized company	66.7%	33.3%	100.0%
Slovakia	**52.9%**	**47.1%**	**100.0%**
micro-company	38.5%	61.5%	100.0%
small company	36.7%	63.3%	100.0%
medium-sized company	57.7%	42.3%	100.0%
large company	87.5%	12.5%	100.0%
Ukraine	**20.0%**	**80.0%**	**100.0%**
micro-company	10.7%	89.3%	100.0%
small company	16.1%	83.9%	100.0%
medium-sized company	20.0%	80.0%	100.0%
large company	32.3%	67.7%	100.0%
Total	**43.4%**	**56.6%**	**100.0%**

Source: Authors

The results reveal that the proportion of companies with an LTEP grows significantly with company size in the Czech Republic and in Russia, whilst this growth is less significant in Slovakia and Ukraine.

The dependence of the proportion of companies with an LTEP on country and the scope of company activity was subsequently analysed. The results are presented in the contingency tables (10, 11) that follow.

Table 10. Numbers of companies with a LTEP according to country scope of company activity

	Yes	No	Total
Czech Republic	**290**	**319**	**609**
multinational	108	79	187
national	82	96	178
regional	99	143	242
(vacant)	1	1	2

Russia	**22**	**28**	**50**
multinational	5	2	7
national	5	6	11
regional	12	20	32
Slovakia	**45**	**40**	**85**
multinational	14	16	30
national	16	10	26
regional	15	14	29
Ukraine	**29**	**116**	**145**
multinational	10	20	30
regional	19	96	115
Total	**386**	**503**	**889**

Source: Authors

Table 11. Proportion of companies with an LTEP according to country and scope of company activity

	Yes	No	Total
Czech Republic	**47.6%**	**52.4%**	**100.0%**
multinational	57.8%	42.2%	100.0%
national	46.1%	53.9%	100.0%
regional	40.9%	59.1%	100.0%
(vacant)	50.0%	50.0%	100.0%
Russia	**44.0%**	**56.0%**	**100.0%**
multinational	71.4%	28.6%	100.0%
national	45.5%	54.5%	100.0%
regional	37.5%	62.5%	100.0%
Slovakia	**52.9%**	**47.1%**	**100.0%**
multinational	46.7%	53.3%	100.0%
national	61.5%	38.5%	100.0%
regional	51.7%	48.3%	100.0%
Ukraine	**20.0%**	**80.0%**	**100.0%**
multinational	33.3%	66.7%	100.0%
regional	16.5%	83.5%	100.0%
Total	**43.4%**	**56.6%**	**100.0%**

Source: Authors

In the majority of countries (with the exception of Slovakia), the highest proportion of companies with an LTEP were multinationals, whereby the highest proportion was in Russia (71.4%), and the lowest proportion of companies with an LTEP were regional companies, whereby the lowest proportion was in Ukraine (16.5%).

The dependence of the median proportion of companies with an LTEP on the number of employees was examined by means of a regression analysis. An artificial variable y, which achieves the value 1 for a company with an LTEP and the value 0 for a company without an LTEP, was used as the alternative LTEP variable. The number of employees (x) of a company was used as the explanation variable. Because a linear regression model does not have good properties, a logit model was used instead. In the logit model, the dependence was modelled on the basis of a logistic regression function. The logit model took the following form:

$$y_i = \frac{1}{1+e^{-(\beta_0+\beta_1 x_i+\varepsilon_i)}},$$

where εi, i = 1,…n, are random components. The estimated form of the model was

$$\hat{y}_i = p_i = \frac{1}{1+e^{-(b_0+b_1 x_i)}},$$

where p_i are the probability estimations, P(Y = 1 | xi), i = 1,…,n a b0, b1 are estimations of parameters β0, β1. The expression on the right side of the equation represents the distribution function of a logistic random value. The model was subsequently transformed by means of logarithmic transformation into the following linear form:

$$\hat{L}_i = \ln \frac{p_i}{1 - p_i} = b_0 + b_1 x_i .$$

The left sides of this equation $\hat{L}_i$ are so-called estimated logits. It is a linear model with variables L_i and x_i. The logits are the logarithms of the chances of condition Y = 1 (an LTEP exists in the company) and are linear functions with regression parameters. The interpretation of the coefficient b_1 is as follows: if the variable x changes by one unit, then the logit changes by b_1 units. This means that the chance (i.e. the proportion p/(1-p)) of condition Y = 1 will be exp(b_1) times greater. The absolute member b_0 is equal to the chance of condition Y = 1, if the explanation variable is zero. As there is heteroskedasticity in the model, the regression coefficients in the R programme were estimated by means of the maximum likelihood (MLE) method. The maximum likelihood estimation was obtained by maximising the combined distribution (or in fact its natural logarithm) of a selection of the explanatory variables.

If the artificial variables i are also for the categorical variable state, the logit model can be used to express the dependence of the proportion of companies with an LTEP on the number of employees and on country. The result is dependence expressed in the form of the following logistic functions for each country:

Czech Republic	p = 1/(1+exp-(-0.479+0.00212x);
Russia	p = 1/(1+exp-(-1.325+0.0635x);
Slovakia	p = 1/(1+exp-(-0.295+0.00286x);
Ukraine	p = 1/(1+exp-(-1.717+0.00101x);

as presented in the graph below.

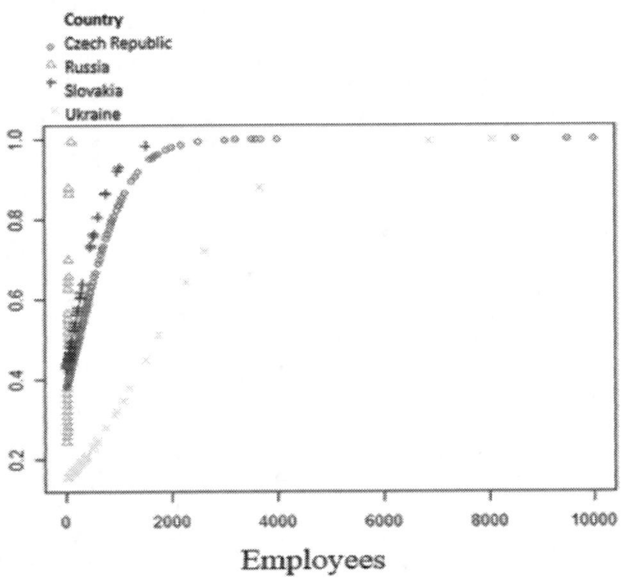

Graph 1. Dependence of the proportion of companies p with an LTEP on the number of employees x and on country
Source: Authors

As is evident in Figure 1, the proportion of companies with an LTEP grows significantly. With the exception of Slovakia, the estimations of the regression coefficients are statistically significant.

In a similar way, the logit model was subsequently used to estimate the dependence of the proportion of companies with an AEP on the number of employees and on country. The result is dependence expressed in the form of the following logistic functions for each country:

Czech Republic	p = 1/(1+exp-(-0,682+0,00441x);
Russia	p = 1/(1+exp-(-2,486+0,0404x);
Slovakia	p = 1/(1+exp-(-0,441+0,00484x);
Ukraine	p = 1/(1+exp-(-2,100+0,00281x);

as presented in the graph below.

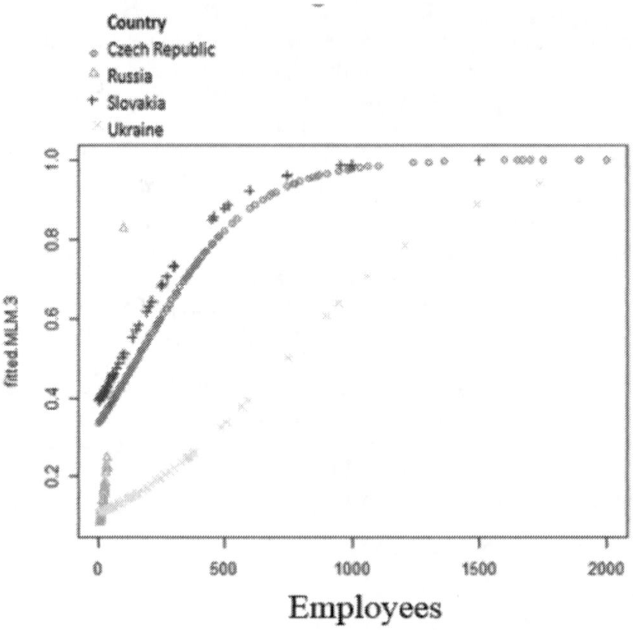

Graph 2. Dependence of the proportion of companies p with an AEP on the number of employees x and on country
Source: Authors

The conclusions with regards to the dependence are identical to the conclusions from the previous model.

The basic precondition for successful corporate education is a strategic approach and planning. The research survey conducted in four post-communist countries showed that micro-companies and small companies in particular do not do so. This finding is consistent with the results of an international survey in which less than a third of companies reported having written education plans or similar education programmes in place [12].

The research also showed that some companies, even though they declare they have an LTEP/AEP in place, are not adhering to their plans.

4 Conclusion

The proportion of companies in which an LTEP exists in the Czech Republic is 47.6%, in Russia 44%, in Slovakia 52.9% and in Ukraine 20.0%. Tests on the proportions revealed that only the differences between Ukraine and the other countries were statistically significant and between the other countries insignificant. By means of a Chi-square test of independence it was shown that the existence of an LTEP and an AEP in a company statistically significantly depends on the country. The proportion of companies which have an LTEP in Slovakia was 54.1%, in the Czech Republic 48.1%, in Ukraine 20.0%, and in Russia only 16.0%. Tests revealed that the percentage difference with regards to the presence of an LTEP was only statistically significant between the Czech Republic and Russia, between the Czech Republic and Ukraine, and between Slovakia and Ukraine.

The adherence to an LTEP by companies in the individual countries was also analysed. In contrast to the other countries, the proportion of companies with an LTEP in the Czech Republic statistically significantly grows with the size of the company (from 24.6% among micro-companies to 72.4% among large companies). The case is similar with regards to the scope of a company´s activities, i.e. the proportion of companies with an LTEP in the Czech Republic also statistically significantly grows with the scope of a company´s activities, unlike in the other countries.

The quantitatively expressed dependence of the median proportion of companies with an LTEP and/or AEP on the number of employees was obtained by means of logistic regression. The dependence for each country was expressed by a formula (used for prediction purposes) and illustrated graphically. As the graphs indicate, a significant growth in the existence of an LTEP as well as an AEP was proven for the companies in all the countries, with exception to Slovakia.

The key precondition for effective corporate education is a strategic approach and planning. The research survey showed that micro-companies and small companies in particular do not do so.

The research also showed that some companies, even though they declare they have an LTEP/AEP in place, are not adhering to their plans.

Acknowledgement: This paper is a partial output of the project no. SV202102R: Analysis of corporate education on a selected sample of companies of various size categories in the Czech Republic and of the project IVSUPS 003 "Current Trends in Human Resource Management".

References

1. Y.-H. Wang, S.-B. Zhou, Research on Managers' Training Needs of Small Medium-sized Enterprises in Mainland - A Case Study of Chongqing. *Proceedings of the International Conference on Human Resource, Organizational Behavior and Leadership,* London, United Kingdom of Great Britain and Northern Ireland, 157-160 (2009)
2. S. Akrofi, Evaluating the effects of executive learning and development on organisational performance: implications for developing senior manager and executive capabilities. *International Journal of Training and Development,* **20**(3), 177-199 (2016)
3. M. Armstrong, *Armstrong's handbook of human resource management practice*. Philadelphia: Kogan Page Ltd, (2014)
4. A. R. Masalimova, M. Usak, A. R. Shaidullina, Advantages and Disadvantages of National and International Corporate Training Techniques in Adult Education. *Current Science,* **111**(9), 1480-1485 (2016)
5. R. Blahuš, Experiential Education as a Part of Human Development in Czech Republic Business Companies. *Proceedings of the 7th European Conference on Management Leadership and Governance*, Antibes, France, 467-476 (2011)
6. J. Kirkpatrick, W. K. Kirkpatrick, *The Kirkpatrick Four Levels: A Fresh Look After 50 Years 1959 – 2009.* [s. l.]: Kirkpatrick Partners, LLC (2009)
7. Chernykh, S.I., Parshikov, V.I. Corporate education abroad: Social and philosophical analysis. International Review of Management and Marketing [online]. Econjournals, volume 6, Issue 6, 2016, pp. 272-277 (2016)
8. JI, X. Study on Corporate Training Quality Monitoring. In: *Proceedings of the 2015 Conference on Informatization in Education, Management and Business*. Paris: ATLANTIS PRESS, 2015, pp. 356–359 (2015).
9. A. Bencsik, A. Sólyom, Education and Training Practice Strategies in Small and Medium Sized Enterprises. *Proceedings of the 6th International Conference on Management and Artificial Intellige*nce, Bali, Indonesie, 65-69 (2011)
10. A. Rauch, A., Hatak, I.. A meta-analysis of different HR-enhancing practices and performance of small and medium sized firms. Journal of Business Venturing, 31(5), 485–504 (2016)
11. U. Moideenkutty, U., Murthy, Y. S. R., Al-Lamky, A. Localization HRM practices and financial performance: Evidence from the Sultanate of Oman. Review of International Business and Strategy, 26(3), 431–442 (2016)
12. Czech Statistical Office, Further Vocational Education of Employed Persons in 2010 (CVTS). Prague: Czech Statistical Office (2013)
13. International Labour Organization, *A Skilled Workforce for Strong, Sustainable and Balanced Growth: A G20 Training Strategy*. Geneva: International Labour Organization, (2011)
14. W. Collier, F. Green, Y.-B. Kim, J. Peirson, Education, Training and Economic Performance: Evidence from Establishment Survival Data. *Journal of Labor Research*, **32**(4), 336-361 (2011)
15. J. Kitching, R. Blackburn, *The Nature of Training and Motivation to Training in Small Firms*. Kingston upon Thames: Small Business Research Centre at Kingston University, (2002)
16. M. Korenková, Vplyv regionálnych rozdielností na možnosť vzdelávania sa zamestnancov v podniku. *Proceedings of the 17th International Colloquium on Regional sciences,* Brno, Czech Republic, 264-269 (2014)
17. J. Tej, G. Sláviková, M. Hrvolová, *Vybrané kapitoly z manažmentu ľudských zdrojov*. Dubnice nad váhom: Dubnický technologický inštitút v Dubnici nad Váhom, (2010)
18. D. Cagáňová, M. Čambál, K. Stachová, Z. Stacho, Education of Employees in Organisations in Slovakia. *Electronic Journal of Knowledge Management*, **12**(4), 210-220 (2014)
19. D. N. Gujarati, C. P. Dawn, *Basic Econometrics*. Fifth ed. New York: Mc Grow-Hill, (2008)"
20. Z. Caha and J. Urban, Human Resource Management in SMEs in the Czech Republic – specifics and key requirements. *SHS Web of Conferences* (2020)

Financial Health of Enterprises as a Tool of Business Potential of an Area

Petra Pártlová[1*]

[1] Institute of Technology and Business in České Budějovice, Faculty of Corporate Strategy, Okružní 517/10, České Budějovice 37001, Czech Republic

Abstract. The paper discusses the evaluation of the financial stability of an enterprise, depending on the business potential of a municipality. The aim of the research is to prove a relation of the prediction models and their ability to predict the financial ability of an enterprise depending on sample clusters of the business potential in a municipality. In the first part of the research, the municipalities are clustered using the cluster analysis and the municipalities are classified into four clusters - economic, social, infrastructural, and environmental. Subsequently, financial data on business entities located in the area are used and the prediction models are calculated using the Czech indices of the Neumaiers. The results are further graphically evaluated in terms of size and sectors of the enterprise. The ANOVA method is verified to assess the independence of the creditworthiness models on sample clusters. Regarding the method and the analytical tool, it is possible to state their suitable choice; the outputs are relevant, valid, and usable in practice. The influence of the creditworthiness on the environmental cluster in the area is proved.
Keywords: IN Index, cluster, ANOVA method, creditworthiness

1 Introduction

The enterprises of proper financial health have been discussed by the researchers for several decades [1; 2; 3; 4; 5]. It is usually related to the enteprises showing growth in the number of employees, turnover, profitability, etc. Such enterprises are increasingly discussed in relation to the economy, and the authors focus on the factors affecting the financial stability of the enterprise and their contribution to the economy [6; 7; 8; 9; 10; 11; 12; 13; 14]. The factors influencing the prosperity of an enterprise, and thus their competitiveness is classified into endogenous: size, financial performance, capital structure, liquidity, flexibility in external financing, earnings stability, plan of financial results, audit quality, disclosure, corporate governance, ownership structure, investors' protection, corporate ethics, individual risks, or internal risks, etc. [15; 16; 17; 18] and exogenous: risk free rate decrease, Financial market development, the GDP growth, bank lending to non-financial private sector growth, foreign direct investment growth, money supply growth, unemployment rate growth, the appreciation of domestic currency, politic stability, government expenditures growth, banking system liquidity, corporate tax rate growth, long-term interest rate growth, inflation growth, corruption growth, systematic risks or risks of external environment, etc. [19; 20; 21; 22]. During an economic crisis, general attention is paid to the enterprises that show stable financial health, contributing to economic growth. At the same time, they play a key role in dealing with some important economic issues, such as reducing unemployment and creating jobs. The current discussion therefore focuses on factors contributing to a stable, competitive business environment in different countries [23; 24]. Research into the financial health of an enterprise is related to the research of the benefits of small and medium-sized enterprises. Many SMEs have good financial conditions, and such conditions contribute significantly to the creation of new jobs [25]. Blažek and Uhlíř [26] even identify the SMEs as a key tool for regional development, addressing the unemployment issues, creating good conditions for the creation of innovations in order to reduce negative regional disparities. However, it should be mentioned that the population of small and medium-sized enterprises is rather heterogeneous, with the largest share of the start-ups disappearing within a few years of their establishment [27]. Another part of the enterprises, not nearly as numerous, is stagnant and does not expand significantly. The smallest part of newly established enterprises is growing rapidly, and such enterprises have the largest share in creating new employment [27]. Currently, there are two strong directions of regional development and the promotion of competitiveness: the regional innovation systems and the theory of global production networks. In such theories, the small and medium-sized enterprises play a special role and the specific ways of their cooperation and the agglomeration effects induced by them (economies of scale in conditions of intensive cooperation and created networks of specialized services together with highly flexible response to the changing demand). The small and medium-sized enterprises and their emerging clusters, together with supporting institutions, are therefore considered to be the most important subjects of economic development in the regions

* Corresponding author: partlova@mail.vstecb.cz

as analysed. The benefit of this theory is to stimulate a comprehensive discussion about the interrelations between the competitiveness and socio-cultural environment of the regions and the regional rooting of the enterprises. In such relation, the positive influences are emphasized, especially for large enterprises in the networks of contacts, influencing the possibilities of further development of small and medium-sized enterprises in the respective region to a large extent. [28]. In this context, many regional economists see the regions as the intersection of spatial relations and networks of contacts, known as networking [26]. The location of the enterprise in the region seems to be a major factor that significantly affects the financial stability of the enterprise. Black and Henderson [29] argue that the enterprises located in agglomeration regions show the impact of positive externalities. Such positive externalities primarily mean knowledge transfer and the influence of municipal institutions. The enterprises in agglomeration areas also have better access to some specialized resources (such as educated workforce) [29]. Sleuwaegen, Goedhuy; Krasniqi; and Damborský and Wokoun [30; 31; 32] focus their research on the influence of geographical location on the location of small and medium-sized enterprises, and the proximity of sales markets has a positive effect on development and stability of the SMEs, even more in the services sector. Another important aspect is related to the location of medium-sized enterprises close to the suppliers. The dependence on competitors' proximity is found for the micro-enterprises. Regarding the secondary and tertiary enterprises, the authors discuss the availability of professional services and the overall economic growth of the region as important factors. In the Czech Republic, the SMEs play a crucial role, as small and medium-sized enterprises in the Czech economy make up the vast majority. In the long run, the representation of SMEs has been around 99% [41], with them employing less than 60% of the total number of employees [42]. Based on such figures, it is deduced very clearly that the SMEs both contribute significantly to job creation, and they also have a major impact on the social and economic development of municipalities, cities, and regions, stimulate growing market dynamics and contribute to the stabilization of the economic system. The author of the paper focuses on a little-explored topic so far, in particular the influence of the financial health on the location of the enterprises in the relevant cluster of the examined region. The aim is to find out whether there are clusters (economic, social, infrastructural, and environmental) with a significant share of the enterprises.

2 Methods

The most important data sources were secondary data for most municipalities in the region of South Bohemia, provided by Ekotoxa. The data were focused on the description and evaluation of the state of the territory in four pillars (economic development and business activities, human resources, infrastructure and the environment). Four indicators were selected for the economic pillar, seven indicators for the social pillar, eleven indicators for the infrastructural pillar and seven indicators for the environmental. Emphasis was placed on the easy availability of data, their sufficient explanatory power and the possibility of updating. All the municipalities in the region of South Bohemia were included in the data collection, except for municipalities with more than 25,000 inhabitants.

For the calculation of the prediction models, related to the financial health of the enterprises not located in the territory of the tested municipalities, see the table. The data sample after generating the set from the Albertina database of Bisnode contained a total of 42,592 data rows. Each row contains the following: Enterprise identification: enterprise name, ID number, municipality region, municipality size, enterprise data: NACE, number of employees, NACE5A code, M_NACE, CZ-NACE5A, year of financial statements, financial statements for the year of operation of the enterprise: balance sheet, profit and loss statement, cash flow statement, sample indicators of profitability, activity, liquidity, indebtedness, productivity and more.

Subsequently, calculations, partial indicators, and items for the calculation of prediction models were carried out:

- EBIT / interest payable - the indicator compares EBIT with interest expense. The indicator shows how large such interest is in EBIT;
- Assets / Liabilities - this is an inverse indicator to the debt ratio (by default, the calculation is Liabilities / Assets or debts). The indicator says what part of all the assets (assets), and therefore all the resources (liabilities), is made up of external sources. In practice, therefore, it tells what level of risk and financial interest the owners bear;
- EBIT / Assets - profit, if it is not covered by money, is only a virtual quantity, as it is based on the current accounting principle. The indicator is thus an indicator of the degree of efficiency used by the total assets;
- Revenues / Assets - Revenues include both sales and other revenues (extraordinary). They do not relate purely to the main activity of the enterprise. The indicator is an alternative indicator of activity.
- Current assets / (short-term liabilities + Short-term bank loans) - this is a modified liquidity indicator;
- Overdue liabilities / income;
- Different weights V1-V6;
- MPO – based on the methods of the Neumaiers.!Neumaierová, 2002, 2005)

Table 1. Number of enterprises in the region in 2017

Year	Original data set	Adjusted data set
2017	9708	6983

Source:Author

Methods used

- Regarding the analysis of financial health of the enterprise, the descriptive method is used. Pivot tables, line and other descriptive graphs are created.

  Clustering of the municipalities uses the principle based on the creation of four models – the Economic Cluster, the Social Cluster, the Infrastructure Cluster, and the Environmental Cluster based on four groups of variables (economic, social, infrastructure and environmental area). According to such principle, a municipality is classified into a group according to its highest average ranking in the appropriate group (pillar). This means that the average ranking of the municipality in each pillar is calculated and then the group (the pillar with the best average ranking) is used.

  The clusters (pillars) are analyzed by the Stepwise-backward-regression method in terms of its business potential. In order to analyze the business potential of rural areas in the clusters, the researcher identified it as a decisive index of the progressiveness of the economic structure, determining the socio-economic differentiation of the regions in the Czech Republic. Each cluster is analyzed by a stepwise method. The result is the finding of significant variables that significantly affect the business potential of each cluster [33].

- Regarding the comparison between the business potential of the area and the real economic state of the enterprises, the method of analysis of variance is used, which determines the significant dependence of profitable enterprises (financial health of the enterprises) on four models as created previously: economic cluster 1, social cluster 2, infrastructure cluster 3 and environmental cluster 4.

  Analysis of variance (ANOVA) is a statistical method that allows the comparison of several means of independent random samples. The method was established by R.A. Fisher (1890-1962).

  The aim of the ANOVA analysis is to prove that the values of the qualitative character of X affect the values of the quantitative character of Y. That is, the creditworthiness of the enterprise affects the clusters. The values of the character Y, which belong to the value xi of the factor X, are denoted by yi1, yi2, ..., yini. The principle of the ANOVA method, which we prove the dependence of Y on X, and the dependence of the clusters on the creditworthiness of the enterprise, is based on the fact that the total variability measured by the sum of the squares of deviations from the overall average, is divided into the variability within the selections and the variability between the selections.

  It is assumed that there are k levels (values of xi) of X, with an effect on Y, as expressed by the following relation: $\mu i = \mu + \alpha i$, i = 1, 2, ..., k,

  and:

  - μ i is the mean of Y in the ith group (relevant to the value of xi factor),
  - μ the total mean of Y,
  - αi is the effect of the value of xi factor on the character of Y.

  Then, null hypothesis H0 is formulated, with all the selections coming from the same sample, and that the values of the factor X have no effect on the values of the character Y. Assuming that the values of αi come from the same base set with zero mean E (αi) and constant variance of $\sigma 2$.

  H0: $E(\alpha 1) = E(\alpha 2) = ... = E(\alpha k) = 0$,

  An alternative hypothesis is formulated to the contrary, as H0 will not apply, provided that at least two items, such as i and j, apply:

  H1: $E(\alpha 1) \neq E(\alpha 2)$.

If accepting H0, the independence of the values of Y on X is deduced, and Y factor does not depend on the X factor. Otherwise, if H0 is rejected, Y depends on X and X affects Y.

The following hypotheses are carried out in the research:

H1: Spatial differentiation, based on the analysis of four pillars (economic, social infrastructural and environmental), does not prove their conditionality on the financial health of the enterprise.

H2: In one of the clusters of the area under research, there are an increased number of creditworthy enterprises compared to other clusters.

H3: The ANOVA tool together with the cluster analysis is appropriate to find out the relations of the clusters to the financial health of the enterprises.

3 Results

Analysis of the financial health of an enterprise in the region of South Bohemian region by the size of the enterprise

As reported by figure 1, the categories of the small enterprise, micro-enterprise and middle-sized enterprise are creditworthy in more than 40%. The creditworthiness of large enterprises is approximately 33%. The grey zone is mostly related to large enterprises (50%), followed by middle-sized enterprises (34%) and small enterprises (32%); the micro-enterprises are the least common.

By the analysis of the IN 95 index, the most bankrupt enterprises are tracked in the group of micro-enterprises (35%), followed by the medium-sized enterprises with 25%, the small enterprises (26%) and the large enterprises with 17%.

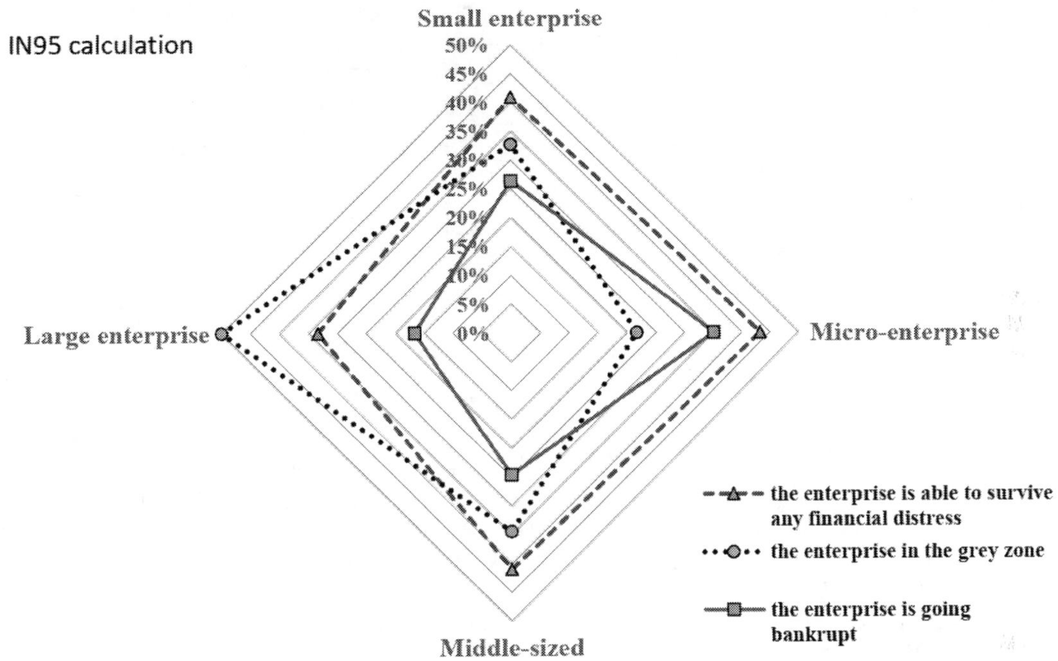

Figure 1. Financial health of South Bohemian enterprises by IN 95, in relation to the size of the enterprise

Source: Author

Relation of the financial health of the enterprises by NACE

The analysis of the impact of the financial health of the enterprises on the sector is performed. IN 95 is also used in the analysis

Regarding IN 95, the percentage share of A (agriculture, forestry and fishing) and C (manufacturing) sections worth mentioning as almost one third of the enterprises is either creditworthy, grey or bankruptcy zone. F section (construction) reports more than 40% of the enterprises as creditworthy, 20% in the grey zone and 35% in the bankruptcy zone. More than 50% of the creditworthy enterprises in in sections G (wholesale and retail trade; repair of motor vehicles and motorcycles) and H (transportation), see figure 2.

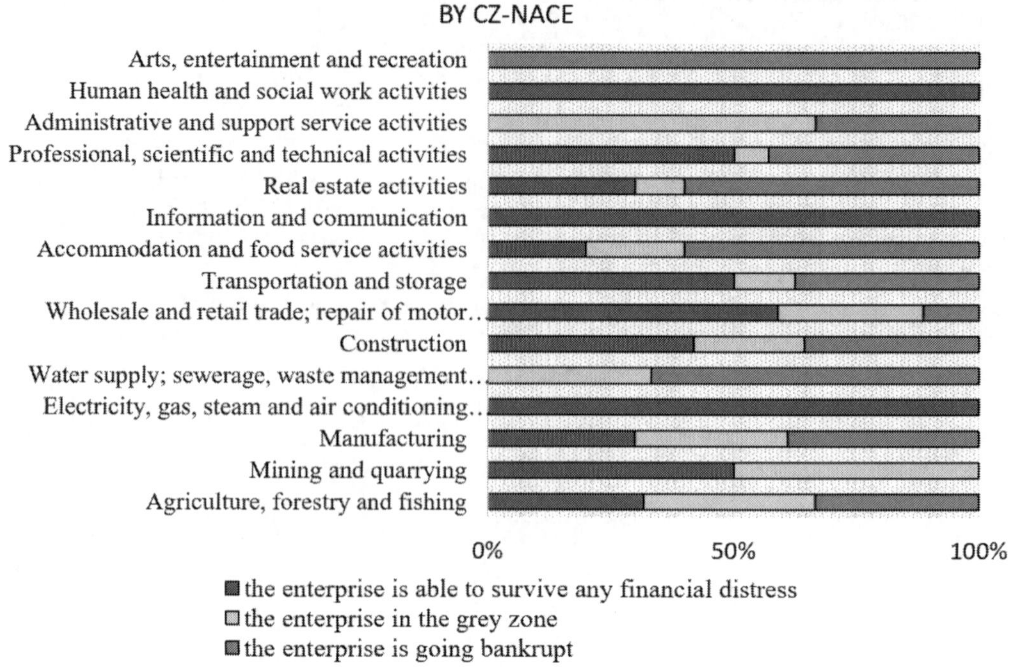

Figure 2. Financial health of South Bohemian enterprises by IN 95, in relation to CZ-NACE
Source: Author

Relation of the financial health of the enterprises by the municipalities

The enterprises of the region of South Bohemia and their financial health are also analysed in relation to the municipalities (type I, type II and type III, based on Act 128/2000 on Municipalities). Type III municipalities (municipality with extended powers) are not included in type II and I. Similarly, type II municipalities (municipality with authorized municipal office) are not included in type I.

By the analysis carried out through the IN95 index, it is identified that the largest number of enterprises in the region of South Bohemia are the creditworthy enterprises, both in municipalities I. II and III. This is followed by a group of the enterprises in the bankruptcy zone; very few enterprises are reported in the grey zone, according to the IN 95 analysis.

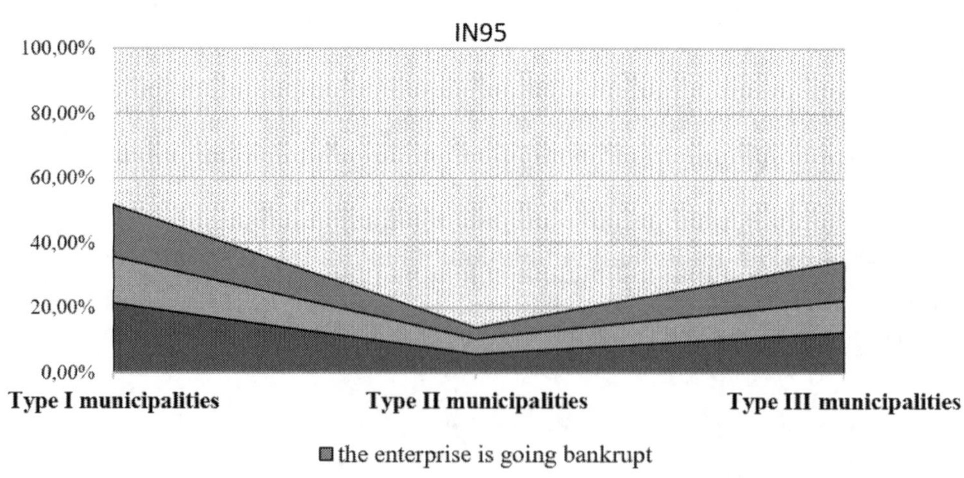

Figure 3. Financial health of South Bohemian enterprises by IN 95, in relation to municipalities
Source: Author

Financial health of the enterprises in the region of South Bohemia, by IN 95

The total comparison of the number of South Bohemian enterprises and their classification into different prediction zones revealed the following: by IN 95 analysis, it is reported that most of the enterprises are creditworthy, followed by the enterprises in the bankruptcy zone. A large number of the enterprises are included in the grey zone. On the other hand, the analysis of IN 01 and 05 shows the opposite trend, most of the enterprises are found to be the bankrupt or the grey zone, a smaller part of the enterprises in the region of South Bohemian is described as creditworthy.

Relating the clusters to the health of the enterprises

The health of the enterprises is measured by Neumeier index IN 95. The enterprises are divided by the sectors and size. Each group is analysed by ANOVA model with grouping variable corresponding to the category to which its location belongs to. The categories are computed above.

$$IN95 = V(1) \cdot A + V(2) \cdot B + V(3) \cdot C + V(4) \cdot D + V(5) \cdot E - (6) \cdot F$$

and:

A = assets / debt

B = EBIT / interest expense

C = EBIT / total assets

D = sales / total assets

E = current assets / current liabilities

F = overdue / sales

Table 2. Relating the clusters to the health of the enterprise according to index 95

IN 95						
	""	"DF"	"SumOfSq"	"MeanSq"	"FRatio"	"PValue"
	"Model"	3	76.5344	25.5115	2.4213	0.0649
{**ANOVA** →	"Error"	653	6880.2151	10.5363	""	""
	"Total"	656	6956.7495	""	""	""
	"All"	1.7039				
	Model[1]	1.4184				
CellMeans →	Model[2]	1.2770				
	Model[3]	1.7025				
	Model[4]	2.5692				
PostTests →	{"Model" → Tukey {""}}}					

Source: Author

The resulting table 2, see below, it is clear, that after the analysis of the index 95, the creditworthiness of the enterprises is reported to be category 4, and the environmental cluster is significantly higher than in other categories, i.e., the economic, the social and the infrastructural clusters. No model identified this difference at the 5% significance level. For all of them, the level of significance is about 6.5%, i.e., the error of the first kind (that is, the probability of a misidentified model is 6.5%). However, due to the consistency and systematic nature of such three indices, it is assumed that the creditworthiness in the 4th category (environmental cluster) is significantly higher than the creditworthiness of the remaining clusters. Similar analyzes with the same result are also carried out regarding the other Neumaier indices IN 01 and 05.

It is also interesting to determine the frequency of the creditworthy enterprises in the clusters, regardless of what point value was calculated for the indices in the range of creditworthy enterprises, figure 4. The analyzed values for 2017, for example for IN 95, ranged from 1.6 to 17.6 points (see the formula for IN 95). Subsequently, all creditworthy enterprises are included in the map of the municipalities according to the clusters. Map 1 shows that most creditworthy enterprises in the area are in cluster 3, with 176 enterprises, followed by a cluster of economically with 37 enterprises, then environmental cluster with 25 enterprises and social cluster with 21 enterprises, as calculated by IN95.

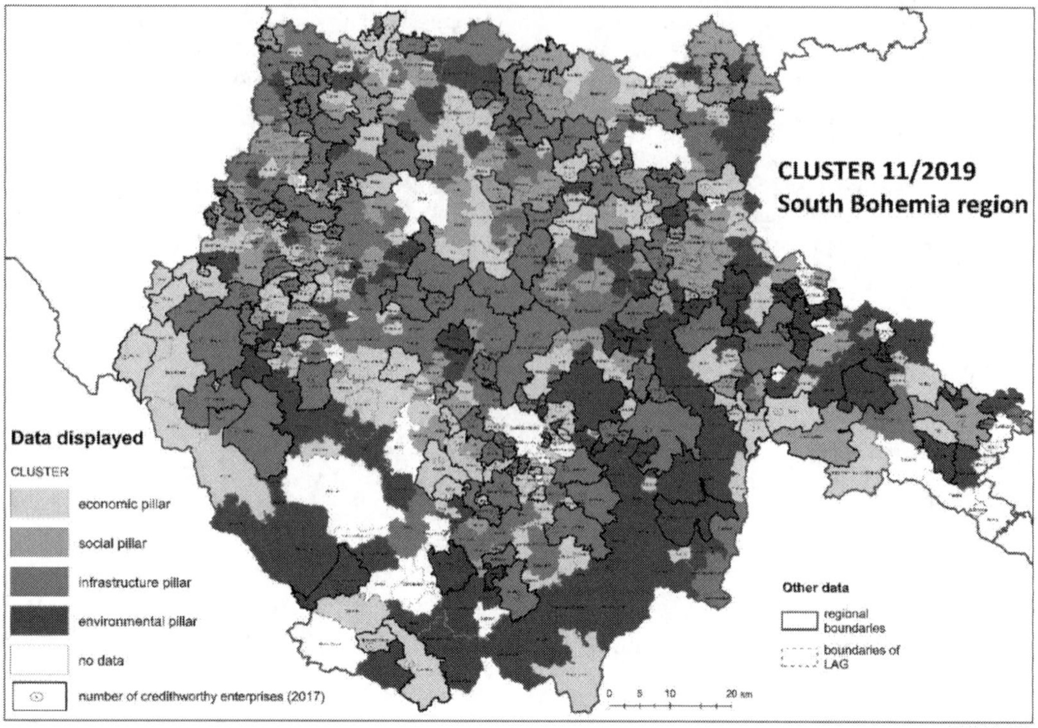

Figure 4. Location of the creditworthy enterprises in the clusters
Source: Author

Suggestions and recommendations

Before the research, several hypotheses were established. The first consisted in the design of an economic model that would allow the determination of the economic potential of the municipalities; the second question was focused on finding and verifying suitable analytical methods for the model. A partial goal was the actual outputs of research, especially in terms of the usability of financial health indices of the enterprises in relation to the business potential of the municipalities. The following summaries and recommendations discuss the fulfilment of the above-mentioned objectives.

The structure of the pillars, including 29 indicators, reflecting the specifics of the regions in the Czech Republic, is considered appropriately designed. However, its final extent should be determined only after wide application, within all the regions in the Czech Republic. However, based on the example of the South Bohemian region, their number proved to be sufficient. Looking ahead, there is a need to precisely define the procedure for collecting relevant secondary and primary data, to specify the minimum scope of the observed period, to define the qualitative aspect of the database and the form of their processing before using it for the analysis.

The analytical tool Cluster Analysis proved to be suitable. Using the cluster analysis of municipalities (Clustering municipalities), four clusters of municipalities were created - economic, social, infrastructural and environmental, according to four groups of variables (economic, social, infrastructure and environmental environment), so that in each cluster, there were the municipalities that excel most in the area. Each municipality was included in a specific cluster according to which group (pillars) the municipality had the highest average ranking. This procedure brought the expected outputs and the structuring of municipalities.

An integral part, in accordance with the above scientific approach, is the analysis of the potential of rural municipalities in relation to the financial health of the enterprises. Here, the use of the ANOVA method proved to be suitable, which proved the relation to one of the clusters.

The potential of municipalities is an integral part of the total potential of the territory; however, it is not a sum, but an intersection of economic and financial indicators monitoring the financial health of the business sector in the territory. The method is based on four pillars (economic, infrastructural, social and environmental), similarly to the first procedure. Based on the cluster analysis, it is then possible to define a summary map of the potential of municipalities for the selected territorial unit with subsequent prediction of the business potential in the given municipality. In the region of South Bohemia, the calculation is based on three monitored indices of the Neumaiers (IN 95, IN 01, IN 05); the paper presents the results only for the IN 95 index) showing that the largest business potential, analyzed for the creditworthy enterprises, with good financial health, is traced to the municipalities of the environmental pillar, which corresponds to the character of the region.

The influence of the financial health of the enterprises in the research was also evaluated from the point of view of sectoral differentiation and size categorization. The results cannot be considered as generally valid and final, yet they are believed to have scientific value. In accordance with the results and findings, the following action is recommended:

- Research into different components of the business environment should continue to ensure the collection of input data from business entities in times of economic recession when other results of the financial health of the business sector in the addressed regions of the Czech Republic can be predicted.
- The follow-up research should be focused on the sub-sectors, from the point of view of sectoral and size differentiation of business entities and find out their influence on the created clusters of municipalities and their potential.

Regarding practical outputs focused on the local and regional development of the Czech Republic, it is possible:

- Create a guidebook for municipal self-government, including recommendations on how and with what tools of the external and internal environment to increase the business potential of the municipality. Such tools include, in general, the creation of an area-based local development strategy for well-defined sub-regional rural areas with a specific business sector, the creation of public-private partnerships at local level, the multi-sectoral design and implementation of a strategy based on the synergies between business actors from different sectors of the local economy, the application of innovative practices, the creation of clusters of local and regional partnerships.

4 Conclusion

Many research institutes, both in the Czech Republic and abroad, address the issue of stabilization and development of the business environment, especially in rural areas. The solution proves to be a limiting factor in ensuring the functionality of the small and medium-sized enterprises in the area, which is a crucial socio-economic accelerator and stabilizer for the countryside. However, a prerequisite for a successful solution for the development of business activities is knowledge of business potential, including good creditworthiness indicators and characteristics of local business entities.

Lundström a Stevenson [34] carry out an extensive research of SMEs, regarding their economic potential in the region, and its development. In particular, they analyse the economic and financial indicators, identified as crucial for the existence and development of the SMEs. Their research activities and the outputs show that the enterprises showing good financial health, and conversely, the enterprises in danger of extinction (bankruptcy), are the key entities in terms of the dynamics of development of the region in which such enterprises carry out their activities.

Similar conclusions are revealed by Parsley and Dreessen [35], also showing the crucial importance of stable and pro-growth SMEs in the context of regional development. They also consider the enterprises that are of good financial health and that are an important source of employment and well-being in the area where they operate to be the pro-growth and stabilizing part of the region. In their conclusions, they similarly state the importance of the enterprises with a pro-growth character, emphasizing the need to know their development limits, business potential, and also the conditions of their stability and development. They identify the knowledge of such pro-growth parameters as a basic precondition for a more efficient and effective setting of the state support. Hill [36] adds that various forms of business support significantly affect the potential and limits of the development at the local and regional level. Shucksmith et al. [37] in their study point to the need for targeted financial and non-financial support to the growth-oriented enterprises in the region. They also notice that the financial support and the subsidy should be oriented both to pro-growth companies, and to those that are located in pro-growth regions, and in the municipalities if there is a prospect of better development conditions.

Similarly, Pělucha [38] carries out the research, the aim of which is to demonstrate the targeting of financial support to the municipalities with a better socio-economic level of development in the context of supporting the dynamics of territorial cohesion. From its results, a statistically significant relation between the subsidies provided for non-agricultural activities and the socio-economic level of the supported municipalities is not fully demonstrated. In terms of the intensity of the subsidy provided, however, rather peripheral and structurally weak municipalities are the most supported. The author adds that the financial share of total public expenditure is marginal compared to the group of agricultural instruments. The research results throughout the Czech Republic show that the current financial subsidies in the agricultural sector is focused exclusively on the transformation of the agricultural enterprises in order to support both their production function and non-productive, environmental function, which is gradually disappearing from the rural area nowadays.

It is a fact that such targeted subsidies contribute, both in extensive areas and especially in high-production areas, to a departure from their original role as mere producers, and a new role for landscape maintenance (sometimes called landscape managers) is emerging. Pělucha (2006, p. 166) further notes that "such new role mostly works to prevent the creation of self-sufficient business entities that could function in the future without any support from public sources." The author's dissertation (Possibilities of Diversification of Agricultural Enterprises in the LFA of the Bohemian Forest) also focuses on this issue isadvantaged areas. In the work, the author also addressed the process of designing a multifunctional management system, specified individual business activities in terms of their effectiveness on the economic results of agricultural businesses. She deals with the definition of production and non-production functions in the rural landscape and at the same time. she discusses the issue of objective setting of the financial instruments for the enterprises operating in naturally disadvantaged areas. In her work, the author also addresses the process of designing a multifunctional management system, specified business activities in terms of their effectiveness on the economic results of the agricultural enterprises. The thesis also includes the case studies focused on the calculation of economic efficiency of agricultural enterprises operating in the conditions of the border region of the Bohemian Forest with and without subsidies from the EU. The author draws similar conclusions as Pělucha. Without targeted subsidies, it is not possible to ensure the functionality of non-productive functions in the landscape. The crucial importance of landscape managers, who are mainly private farmers with a permanent residence in the place of their business, is confirmed. The farmers who bought land for the purpose of allocating subsidies from the EU are not such landscape managers, with few exceptions, they rather contribute to the degradation of the landscape. In conclusion, the author states that from the point of view of future generations and the preservation of the rural landscape, it is necessary and essential for these generations to maintain agricultural and non-agricultural activities in harmony. These conclusions, as reached more than 15 years ago, are followed by the author in this publication, and in the issues presented. The new theoretical definition of rural space in the systemic anthropo-ecological concept fully respects this.

The follow-up construction of the model of the total economic potential, and its four sub-areas (economic, infrastructural, social and environmental) implement such principle methodically into the working experience. The verification of the proposed methodologies on a regional scale and in more than 600 municipalities with valid and practically usable results indicates the correctness of the procedure. At the methodological level, the publication defines the procedure of aggregation of four sub-models into one final indicator; the calculations of four sub-models with commenting on the outputs from the solution are presented. In the next phase of the solution, the integrated model of economic potential is supposed to be verified at the regional level.

In terms of the importance of the relations between the financial health of the enterprises in the area and the structural representation of the pillars, such relationship is clearly demonstrated. The performed analysis revealed a strong relation between the financial health of the enterprises and the environmental parameters of the area where the enterprises work.

It is proved that the enterprises show better economic results in areas where the environmental aspect is emphasized, and the surroundings are not severely degraded. The data (Figure 4) also show that the degrading trend might be caused both by pumping over the natural potential of the territory and by not using it, and both of these extreme limits have a devastating effect on the economic potential of the territory. It can therefore be concluded that if this dependence was confirmed in other regions as well, this would be an important finding in the context of a new direction of regional policy and more effective targeting of financial and non-financial instruments.

In fact, a comprehensive review of the literature shows that that the research with a similar focus has been carried out yet neither in the Czech Republic nor abroad. On the one hand, this naturally limits the possibility of comparison with the outputs of other research organizations, but at the same time the solution contains a charge of new scientific knowledge and new research findings.

References

1. E. B. Deakin. A Discriminant Analysis of Predictors of Business Failure. *Journal of Accounting Research*. **10**(1), 167-179 (1972)
2. E. I. Altman, R. G. Haldeman, R. G., P. Analysis. A new model to identify bankruptcy risk of corporations. *Journal of Banking and Finance*. **1**, 22-54 (1977)
3. M. E. Zmijewski. Methodological issues related to the estimation of financial distress prediction models. *Journal of Accounting Research*. **22**, 59-82 (1984)
4. T. Shumway. Forecasting Bankruptcy More Accurately: A Simple Hazard Model. *Journal of Business*. **74**(1), 101-24 (2001)
5. Y. J. Wang, H. S. Lee. A clustering method to identify representative financial ratios. *Information Sciences*. **178**, 1087-1097 (2008)
6. M. Henrekson, D. Johansson. Gazelles as job creators: a survey and interpretation of the evidence. *Small Business Economics*. **35**(2), 227-244 (2010)
7. C. Brush, D. Ceru, R. Blackburn. Pathways to entrepreneurial growth: The influence of management, marketing, and money. *Business Horizons*. **52**(5), 481-491 (2009)

8. O. Bajo-Rubio et al. Foreign Direct Investment and Regional Growth: An Analysis of the Spanish Case. *Regional Studies*. **44**(3), 373-382 (2010)
9. J. Bradley. Stimulating Sustainable Growth Using Cohesion Policy: Realistic or a Holy Grail? *International Evaluation Conference, Cohesion Policy 2014-2020: Towards Evidence-Based Programming and Evaluation*. Vilnius. (2013)
10. J. Kaczmarek. Construction elements of bankruptcy prediction models in multi-dimensional Early Warning Systems. *Polish Journal of Management Studies*. **5**, 130-143 (2012)
11. J. Du, Y. Temouri. High-growth firms and productivity: evidence from the UK. In Small Business Economy. 44, 123-143 (2015)
12. M. Karas, M. Režňáková. Building A Bankruptcy Prediction Model: Could Information About Past Development Increase Model Accuracy? *Polish Journal of Management Studies*. **17**(1), 116-130 (2018)
13. M. Kovacova, T. Kliestik, P. Kubala, K. Valaskova, M. Radišić, J. Borocki. Bankruptcy Models: Verifying Their Validity As A Predictor Of Corporate Failure. *Polish Journal of Management Studies*. **18**(1), 167-179 (2018)
14. J. Belás, Z. Strnad, B. Gavurová, M. Čepel. Business environment quality factors research-SME management's platform. K.utb.cz [online]. Available at: https://publikace.k.utb.cz/handle/10563/1009535. (2019)
15. W. R. Nelson, R. Perli. Selected indicators of fiacnial stability. In: *4th joint central bank research conference on risk measuremtn and systemic risk*. Frnakfurt am Main: ECB, 1-30 (2005)
16. C. S. Armstrong, J. E. Core, D. J. Taylor, R. E. Verreccia. When does information asymmetry affect the cost of capital? *Journal of accounting research*. **49**(1), 1–40 (2011)
17. O. Barron, X. Sheng, M. Thevenot. Information enviroment and the cost of capital: a new approach, Working papers 2012-12, American University, pp 1-35 (2012)
18. L. Kotek, A. Nosek, A. Fiala, V. Bartos. Risks in industrial management systems. *MM Science Journal*. 12, 1608-1612 (2016)
19. W. A. Allen, G. Wood. Defining and achieving financial stability. *Journal of Financial Stability*. 2, 152-172 (2006)
20. European Commision. Approaches for assessing the impacts of the Rural Development Programmes in the context of multiple intervening factors, Findings of a Thematic Working Group established and coordinated by The European Evaluation Network for Rural Development. [online]. Available at: https://enrd.ec.europa.eu/sites/default/files/eb43a527-c292-f36c-fc51-9ea5b47cedae.pdf (2010)
21. T. C. Artiach, P. M: Clarkson. Disclosure, conservatism and the cost of equity capital: a review of the foundation literature. *Journal Accounting and Finance*. 51, 2-49 (2011)
22. V. Kubíčkova et al. *Dynamické podniky služieb – gazely*. Praha: Wolters Kluwer s.r.o., 152 (2016)
23. OECD. Rural - Urban Partnerships: An integrated approach to economic development. OECD Rural Policy Reviews. ISBN 978-92-64-20481-2 (2013)
24. J. R. Baldwin, D. Beckstead, W. M. Brown, D. L. Rigby. Agglomeration and the Geography of Localization Economies in Canada. *Regional Studies*. 42, 117-132 (2008)
25. D. L. Birch. Job Creation in America. In: J. Blažek, D. Uhlíř: *Teorie regionálního rozvoje: nástin, kritika, implikace*. Praha: Univerzita Karlova (2011)
26. J. Blažek, D. Uhlíř. *Teorie regionálního rozvoje: nástin, kritika, implikace*. Praha: Univerzita Karlova (2011)
27. J. Brüderl, P. Preisendörfer. Fast-growing Businesses: Empirical Evidence from a German Study. *International Journal of Sociology*. **30**(3), 45-70 (2000)
28. T. Mori, K. Nishikimi, E. T. Smith. The Number-Average Size Rule: A New Empirical Relationship Between Industrial Location and City Size. *Journal of Regional Science*. **48**(1), 165-211 (2008)
29. D. Black, V. Henderson. A theory of urban growth. *Journal of Political Economy*. **107**(2), 252-284 (1999)
30. L. Sleuwaegen, M. Goedhuys. Growth of firms in developing countries: Evidence from Cote d'Ivoire. *Journal of Development Economics*. 68, 117-135 (2002)
31. B. A. Krasniqi. Building an Expanded Small Firm Growth Model in Transitional Economy: Evidence on Fast Growing Firms. *Journal of East-West Business*. **18**(3), 231–271 (2012)
32. M. Damborský, R. Wokoun. Lokalizační faktory malého a středního podnikání v podmínkách ekonomiky ČR. *Ekonomie a management*. 2, 32-43 (2010)

33. R. Christensen. *Plane Answers to Complex Questions: The Theory of Linear Models*, Third Edition. Springer-Verlag, New York (2002)

34. A. Lundström, L. A. Stevenson. *Entrepreneurship Policy. Theory and Practice*. New York: Springer, 2005, ISBN 0-387-24140-X (2005)

35. C. Parsley, E. Dreessen. Growth Firm Project: Key Findings, Industry Canada, 2004, [online]. Available at: https://www.ic.gc.ca/eic/site/061.nsf/vwapj/GrowthFirmsProject_KeyFindings_Eng.pdf/$file/GrowthFirmsProject_KeyFindings_Eng.pdf (2012)

36. B. Hill. Understanding the Common Agricultural Policy, Velká Británie: Taylor & Francis Ltd. ISBN 978-1-84407-777-9 (2012)

37. M. Shucksmith, D. L. Brown, S. Shortall, J. Vergunst, M. E. Warner. *Rural transformations and rural policies in the US and UK*. Hoboken: Taylor& Francis. ISBN: 9781136502743 (2012)

38. M. Pělucha. Rozvoj venkova v programovacím období 2007-2013 v kontextu reforem SZP EU. 1. vyd. Praha: IREAS, ISBN 80-86684-42-3 (2006)

39. Neumaierová, I., 2002. Výkonnost a tržní hodnota firmy: metody, ukazatele, využití v praxi, 1.vyd. Praha: Grada, 215. ISBN 80-247-0125-1

40. Neumaierová, I., 2005. Neumaier Ivan Index IN05: Index IN05. In Evropské finanční systémy: Sborník příspěvků z mezinárodní vědecké konference. Brno: Masarykova univerzita v Brně, 143-146

41. V. Koráb, P. Koudelková. What determines innovation in small and medium enteprises? A case study from the Czech Republic. *Transformations in Business and Economics*. 15. 78-88 (2016)

42. Ministry of Industry and Trade. MIT. Report on SME development and their support (Zpráva o vývoji malého a středního podnikání a jeho podpoře). Available at https://www.mpo.cz/cz/podnikani/male-a-stredni-podnikani/studie-a-strategickedokumenty/zprava-o-vyvoji-maleho-a-stredniho-podnikani-a-jeho-podpore-v-roce2016--232792/ (2017)

Pros and Cons of Home Office during the Covid-19 Pandemic

Eliška Obdržálková[1], *Michala* Moravcová[2,*]

[1]Institute of Technology and Business in České Budějovice, Faculty of Corporate Strategy, Okružní 517/10, 370 01 České Budějovice, Czech Republic
[2]University of Ss. Cyril and Methodius in Trnava, Faculty of Social Sciences, Hajdóczyho 1, 917 01, Trnava, Slovakia

Abstract. Due to the Covid-19 pandemic, many changes have taken place, not only in the economic and health sector but also in the labor market. Many workers had to shift to work from home regardless of their previous experience in having a home office, bringing many challenges and possible benefits and drawbacks. The paper aims to analyze the advantages and disadvantages of the home office trend during the Covid-19 pandemic and their differences in terms of location through a literature review method, synthesis, and subsequent comparison. The research has shown that the most common benefits of working from home include time flexibility, no need to commute, and a reduced risk of contracting the Covid-19. Better communication and physical health are among the least mentioned and therefore the least frequent benefits of the home office trend. The results show that social isolation, work-life balance, and lack of equipment are the most common disadvantages. The least common cons include issues with performing tasks, back pain, or long meetings. In terms of location, the most common advantages and disadvantages vary considerably. In some cases, however, the differences in terms of location were slight to almost non-existent. The findings indicate that it is not clear whether the home office trend and the transition during the Covid-19 pandemic are positive or negative, as it has both advantages and disadvantages. They vary in terms of location, which may be due to the different development of countries and cultural differences.
Keywords: home office; working from home; advantages; disadvantages; Covid-19; pandemic

1 Introduction

In March 2020, a global Covid-19 pandemic was announced. It is a highly infectious viral disease caused by a new type of coronavirus, scientifically named: Severe and Acute Respiratory Syndrome Coronavirus 2 - SARS-CoV-2 [1]. The Covid-19 pandemic is still ongoing and continues to claim many casualties. As this highly infectious disease spreads very quickly, many government measures are being announced to try to reduce the likelihood of infection as much as possible [2]. As reported by Kyzlinková et al. [3], these measures have a significant impact on society in many ways. The pandemic thus brought many changes. Not only in health and economics but also work and employment, for example [4]. Companies in these conditions are forced to react flexibly not only to the competition [5], but also to change the whole setting of the existing operation [6,7]. This also includes product [8], marketing activities [9] and human resource management [10-13].

One of the many measures includes the transition to a home office, which is becoming a particular trend. This measure not only seeks to reduce the risk of Covid-19 infection but also helps employers prevent the loss of the workforce due to quarantine [2]. Most of the workers worked from home, as mentioned by Kyzlinková et al. [3]. The home thus became the workplace for many employees. However, even this trend brings with it many challenges and affects both employers and employees. They were forced to adapt to this change very quickly, often regardless of whether they already had had experience with this form of work [3]. Apart from adaptation, however, it is important to consider some of the risks, ie the negatives and also the positives that the transition to a home office brings. A home office can influence the employees in several aspects. The transition to working from home can affect productivity, both mental and physical health, working hours, money savings, and finally concentration of the employees.

This thesis aims to analyze the advantages and disadvantages of the home office trends during the Covid-19 pandemic and their differences in terms of location.

2 Literary research

This is a very topical issue, so the advantage is that the data are more recent. The transition to a home office that can be considered a trend today, and its impact on employees during the Covid-19 pandemic, have been studied by Awada et al. [14]. They obtained the necessary information for the subsequent analysis from the online questionnaire designed and

* Corresponding author: michala.moravcova@ucm.sk

executed by them. After analyzing the responses from a total of 988 respondents from 40 states, most of them from California, Awada et al. [14] discovered that productivity at work during the Covid-19 pandemic is not fundamentally different compared to their productivity in the office. However, they said that the benefits of working from home include, for example, improved mental and physical health, better communication with their co-workers, and the comfort of home. They also point to the increased number of working hours, especially among respondents who stated in the questionnaire that they have school-age children [14]. An increase in the number of working hours can be considered a certain disadvantage, ie a negative.

Ipsen et al. [15] similarly dealt with work from home and with working conditions, who after their study concluded that people have a rather positive experience of working from home. This is mainly because they can appreciate the benefits associated with the home office. The study by Ipsen et al. [15] consisted mainly of examining and analyzing the experiences of people from 29 European countries who shifted to a home office. Using an online survey of 5,748 respondents from Europe, they identified the key advantages and disadvantages of working from home. The most frequently mentioned benefits include, for example, reduced risk of spreading COVID-19 and a reduced risk of possible infection, the flexibility of time in terms of breaks and meals, as well as saving time in commuting. Lack of contact with colleagues or going out, poor physical conditions, or lack of necessary equipment were then assessed as the most common and most important disadvantages of working from home [15]. Using an online questionnaire with thousands of respondents from Italy and subsequent analysis of the answers, Bolisani et al. [16] pointed out that it is not possible to state unequivocally whether working from home is more advantageous or vice versa, mainly due to the balance between positive and negative opinions of workers working from home during the Covid-19 pandemic. They also gave various examples of disadvantages such as the inability to suspend long meetings and isolation from colleagues. On the contrary, better focus on work or the possibility of one 's meals are some of the aspects that favor working from home [16]. One of the unusual aspects that affect employee satisfaction and productivity when working from home during a pandemic is also noise. This factor was addressed by Puglisi et al. [17]. Through an online questionnaire on perceived noise in the workplace, with a total of 1,934 respondents also from Italy, they concluded that noise nuisance, which occurs mainly in employees working in the same room with other members of the household, has a negative effect on concentration, attention, or relaxation [17]. These effects can be considered as disadvantages of working from home and therefore noise is one of the possible negatives of the home office.

Using the same method, i.e., by investigating and subsequently analyzing the responses from the respondents, Grossmann et al. [2] confirm some positive experiences with working from home. They cite, for example, the need for commuting, time flexibility, or, for example, focusing on work as the best-rated benefits [2]. Chen [18] also mentions many identical advantages and disadvantages of working from home during the Covid-19 pandemic in their study of the impact of working from home on employees. Based on other expert studies and surveys, Chen [18] concluded that, among other things, insufficient skills of employees, uncertainty about redundancies, wage cuts, and bankruptcies are some of the significant negatives of working from home for employees. For managers, for example, it is the inability to thoroughly control and supervise subordinates or the lack of interaction with them [18].

Mehta [19] examined whether the work commitment and happiness of employees working from home during the Covid-19 pandemic was at the same level as before the pandemic, and thus in line with previous studies. Thanks to responses from 440 respondents in India, the study shows that working from home during the aforementioned pandemic has a significant impact on employee happiness. Mehta [19] further states that employees feel more involved in their work, mainly because working from home allows them independence and comfort. These aspects have a positive effect on their work commitment, and this leads to happiness [19].

Based on the already available data of current sample surveys, Kyzlinková et al. [13] conclude that employees perceive working from home differently, not only positively. However, it is worth noting that the move to a home office has helped mitigate some of the negative effects of the Covid-19 pandemic, both in terms of economics and health. Other positive impacts include time flexibility, restrictions on commuting to work and the associated time savings, work performance without interruption, or savings in the employer's operating costs [13]. They thus identify with most of the advantages reported by Grossmann et al. [2]. The absence of contact with colleagues, social isolation, problems with the organization of time, and complications in fulfilling the task, then Kyzlinková et al. [13] specify working from home as the main disadvantages. Identical or at least very similar advantages and disadvantages of working from home are confirmed by a study of the Ministry of Labor and Social Affairs of the Czech Republic [20], which examined the experience of employees and employers from the Czech Republic with working from home during the Covid-19 pandemic. Among other things, they also stated that the possibility of working at least partially from home for most respondents has a positive effect on mental health and family life [20]. Pfeifer [21] also examined the mental health and hard work of Czech employees and managers during the Covid-19 pandemic by analyzing a questionnaire with 174 answers. He came to a somewhat different opinion than, for example, the Ministry of Labor and Social Affairs of the Czech Republic. According to the results of the survey, employees and managers fall into mental stress when working from home, which negatively affects their mental health [21]. Pfeifer [21] stated that the reason may be the often-mentioned increased sensitivity to social isolation. It is also a very interesting finding that the mental health and desire to work for employees working from home during the Covid-19 pandemic depends on whether the lockdown is ordered. At the time of the lockdown, the desire to work is up to 20% lower and the sensitivity to social isolation is higher [21].

In addition to the advantages and disadvantages, Okuyan and Begen [22] have discussed recommendations and measures that will help eliminate some of the disadvantages and thus have a positive effect on working from home during the Covid-19 pandemic, based on information from other professional monographs and surveys. These include, for example, setting a border between work and personal life through breaks and pre-determined daily routines, which can prevent obesity or an imbalance between work and personal life [22]. Rieth and Hagemann [23] also agree with the fact that the issue of work-life balance is a disadvantage of working from home. Using an online survey, they obtained answers from 465 employees from Germany and focused on the conflict between family and work, besides other factors, during the pandemic. The results show that people who have children at home in the home office experience significantly more stress, a greater loss of concentration, and thus consider working from home to be rather negative [23]. Among other things, this disadvantage may be related to emotional exhaustion and burnout in employees working from home during the Covid-19 pandemic [4]. Barriga et al. [4] used an online questionnaire with 1,044 respondents from South America, specifically Ecuador, to observe and analyze the influence of conflict between work and family on the burnout syndrome, emotional exhaustion during the home office. The results showed that working from home has a negative effect on work-life balance, so they are also consistent with a study by Irawanto et al. [24], who came to this conclusion based on an analysis of responses from 472 respondents from Indonesia.

The coronavirus crisis and work at this time, especially work from home, were dealt with by López Peláez et al. [25], who came up with unique and interesting advantages of the transition to a home office. Thanks to working from home, there is greater digitalization and innovation in which it is appropriate to emphasize their availability or functionality and can therefore be considered as positives of working from home. Thanks to the data and information from other expert surveys and subsequent analysis, López Peláez et al. [25] discovered that remote work is an indicator of greater digitalization and contributes to improving the economic results of companies. Whether the positives and negatives of working from home during a pandemic differed by gender were examined by Khan and Umer [26] through an online questionnaire with 222 respondents from Japan and the United States. After analyzing the answers, they concluded that the advantages include more men than women, money saved for non-commuting, as well as the comfort of home [26]. According to Khan and Umer [26], this may be because women with children sometimes must take care of family or housework during working hours. A significantly higher proportion of men than women cited fewer business meetings as an advantage. There are no significant gender differences in the negatives, or challenges, that employees face when working from home [26].

Farooq and Sultana [27] and subsequently Guler et al. [28], who have a completely different view on this issue, also examined the transition to a home office and its impact on employees' productivity. Using an online questionnaire with 250 Indian respondents, Farooq and Sultana [27] examined employee productivity at home and aspects that could affect their productivity. They analyzed the responses and concluded that working from home during the Covid-19 pandemic, employee productivity declined, and the results showed that it had a greater impact on productivity for women than for men. Decreasing productivity could be caused, for example, by ineffective communication, which respondents often mentioned [27]. Guler et al. [28], who, using the same method and also based on the analysis of answers from 149 respondents, concluded that employee productivity did not decrease after the transition to a home office, but on the contrary, increased. Their work was of better quality and more efficient [28]. Guler et al. [28] in their study also focused on the health of home workers and reported interesting effects such as weight gain or complaints of back and neck pain. Thus, musculoskeletal pain can also be mentioned as a negative.

Regarding the counterproductive work behavior of employees working from home during the Covid-19 pandemic and whether the home environment has an impact on this factor have been investigated by Nemteanu et al. [29]. The data on which they based their study were collected from 641 employees from Romania through an online questionnaire [29]. According to Nemteanu et al. [29,p.601] "remote work significantly and strongly affects professional isolation and to a lesser extent, but still significantly, counterproductive work behavior and the ability of self-regulation of employees".

3 Methodology

As already mentioned, the home office trend, which spread widely during the COVID-19 pandemic, has many advantages and disadvantages. This thesis aims to analyze the advantages and disadvantages of the home office trends during the Covid-19 pandemic and their differences in terms of location. To achieve the goal, it is necessary to ask the following research questions.

1. What are the advantages and disadvantages of working from home during the Covid-19 pandemic and what are the least common ones?

2. Are there any differences in advantages and disadvantages in terms of location?

The basic source of data will be articles from the citation databases Web of Science and Scopus. It will therefore be a literature review. Using Covid-19 keywords, pandemics, home office, working from home, advantages and disadvantages, possible relevant resources are sought and further analyzed. The sources that analyze the advantages and disadvantages of working from home during the Covid-19 pandemic using empirical research are selected. The respondents to the analyzed studies come mainly from Europe. However, some data are obtained from respondents from around the world.

For example, from Asia, mostly from India, Indonesia and Japan. Or, for example, from respondents from South and North America.

With the help of keywords Covid-19, pandemics, home office, work from home, advantages, and disadvantages, resources are researched from the citation databases Web of Science and Scopus. About 150 professional publications are displayed, from which only a few relevant sources are selected that meet a certain criterion. Resources need to include an analysis of the pros and cons of working from home during the Covid-19 pandemic. Thus, we work with sixteen professional articles that meet this criterion. These sources are then examined for the already mentioned advantages and disadvantages of the home office, frequently mentioned in the results, then in the discussion of the results and the conclusion. Next, a synthesis of selected advantages and disadvantages is done. They are recorded in a spreadsheet editor in Excel and then compared using statistical and logical functions. Subsequently, the results are examined based on absolute and relative frequency. Furthermore, the advantages and disadvantages are divided in terms of location. Of the sixteen sources, only twelve are used, for which the location is given and thus meets the criterion. The resources are divided in terms of location into two parts, namely Europe and the other continents. Therefore, it works with seven sources from Europe and five sources from other continents. Due to the unbalanced distribution, it is calculated based on relative frequency, with seven sources with respondents from Europe representing 100% and five sources with respondents from other continents also representing 100%. The results are displayed in tables, followed by the final comparative research method, which determines the differences in advantages and disadvantages in terms of location.

4 Results

As shown in Figure 1, some benefits of working from home during the Covid-19 pandemic have been mentioned more frequently. The highest percentage achieves the advantage of time flexibility. This is a 20% representation. The advantage of no need to commute and the associated savings is 17%. Another benefit of working from home was health, i.e., a reduced risk of contracting the Covid-19 virus. The advantage of the reduced risk of infection was mentioned by up to five professional articles, which is only two less than the need for commuting and thus reaches up to 13% representation. Better concentration, as an advantage of the home office trend, also reaches 13%. Other analyzed benefits of working from home are the comfort of home and more time spent with the family. These advantages have been mentioned in four academic publications and represent the 10%. The last benefits, which have achieved at least one repetition, and have therefore been mentioned in at least two articles, are better mental health, the ability to prepare food on one's own, and independence reach 5%. The remaining benefits were mentioned in only one of the academic articles used. In chart one, they are referred to as "other benefits" and acquire only 3%, thus reaching the lowest percentage.

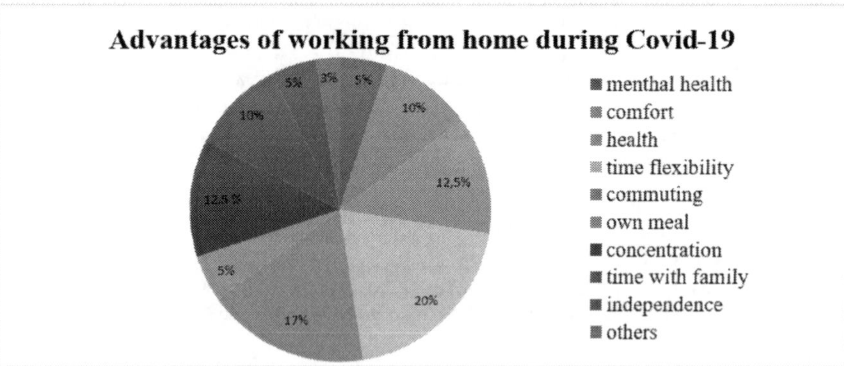

Figure 1. The most common advantages of working from home during Covid-19

Source: Authors.

Figure 2 shows the disadvantages of working from home during the Covid-19 pandemic and their percentage. The highest percentage is achieved by the disadvantage of social isolation, which is 19%. The following disadvantage, the imbalance of personal and professional life, was mentioned in up to five sources used and thus reaches up to 16% of the representation. Three percent less, ie 13%, achieved the disadvantage of lack of equipment, which was mentioned in four selected professional sources. Other disadvantages that appeared in multiple references were the lack of contact with colleagues and obesity or weight gain. Specifically, they were mentioned in three professional articles that meet the criteria and reach up to 10%. Only two sources mentioned more disadvantages, which include an increased number of working hours, possible technical problems, more stress, lack of concentration, and poor communication. These disadvantages amount to 6% and are fewer common disadvantages. The remaining disadvantages were always mentioned in only one of the sources used. In Figure 2, they are marked as "other" and reach only 3%.

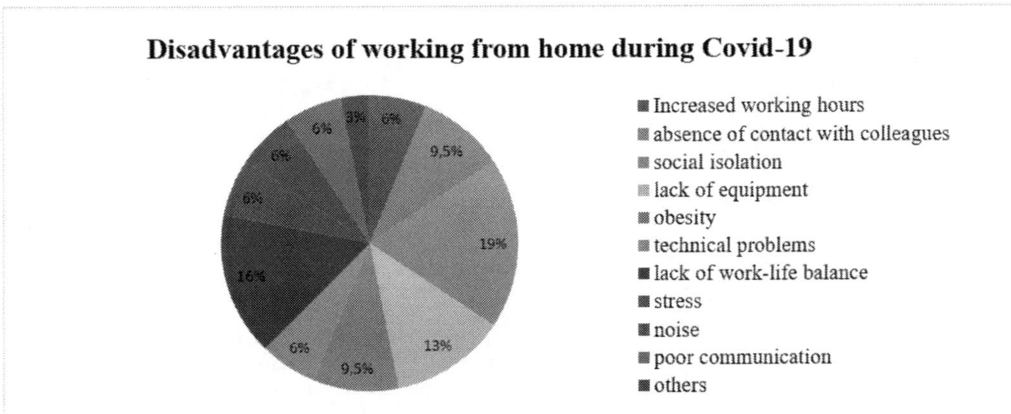

Figure 2. The most common disadvantages of working from home during Covid-19

Source: Authors.

Regarding the differences in advantages and disadvantages of working from home during the Covid-19 pandemic in terms of location, there are certain differences, which are shown in the tables below. Table 1 shows the differences in benefits. Table 1 shows that there are significant differences in the benefits of time flexibility mentioned in five of the seven sources that analyze the responses of European respondents and thus reach up to 71% of the representation. While respondents from other continents have not mentioned this advantage once and it is 0%. The difference in terms of location is significant for this advantage. Significant differences in terms of location are also in the advantage of the lack of commuting, which is more common in Europe than in other parts of the world. In Europe, this advantage is 71%. However, as can be seen from Table 1, the difference in terms of location is 31% smaller than in terms of time flexibility, as it is at least 40% on other continents. Similarly, there are benefits such as the ability to prepare one's meal and better concentration. Both of these benefits were mentioned only by respondents from Europe. However, respondents from other continents did not address these benefits. As can be seen in the table, the difference in terms of location is 29% for the benefit of preparing one's meals and 57% for better concentration. On the contrary, the advantages that were more often mentioned by respondents from other continents are the comfort of home and independence. In other continents, home comfort benefits up to 60%, compared to only 14% in Europe. It is very similar for independence, where the difference in terms of location is 40%. For the remaining advantages, the differences in terms of location are only slight. This is a benefit of the reduced risk of contracting the Covid-19 virus and the more time spent with the family. For both advantages, the difference in terms of location is only 11%.

Table 1. Differences in advantages in terms of location

Advantages	Europe	Other continents
Lower risk of Covid-19 infection	29 %	40 %
Time flexibility	71 %	0 %
Commuting	71 %	40 %
Own meal	29 %	0 %
Concentration	57 %	0 %
Comfort	14 %	60 %
More time with family	29 %	40 %
Independence	0 %	40 %

Source: Authors.

Table 2 shows the disadvantages of working from home during the Covid-19 pandemic and their differences in terms of location. Table 2 shows that there are slightly fewer disadvantages than the advantages listed in Table 1. The disadvantage - social isolation points to many differences. This disadvantage was mentioned in five sources analyzing the responses of European respondents and accounts for 71% of Europeans. Whereas respondents from other parts of the world have been mentioned only once, reaching only 20% across the other continents. In terms of disadvantages, this is the biggest difference in terms of location, namely a 51% difference. The disadvantage, which was also mentioned by more respondents from Europe, was the absence of contact with colleagues. However, for this disadvantage, the difference in terms of location is only 9% and is therefore much smaller than for the disadvantage of social isolation, as can be seen in Table 2. Disadvantages more often mentioned by respondents from other continents are imbalance between personal and professional life and poor communication. The disadvantage of a lack of work-life balance is 40% on other continents, compared to only 14% in Europe. The difference in terms of location is therefore 26%. At the disadvantage of poor communication, the difference is even greater, namely 14%, as it was not mentioned by European respondents at all. The difference in this disadvantage in terms of location is therefore 40%. As can be seen from Table 2, for the remaining disadvantages, such as lack of concentration, technical problems, lack of equipment, and increased working hours, the differences in terms of location are insignificant. The difference is always around 6% or a maximum of 11%.

Table 2. Differences in disadvantages in terms of location

Disadvantages	Europe	Others continents
Social isolation	71 %	20 %
Absence of contact with colleagues	29 %	20 %
Lacko of concentrationt	14 %	20 %
Technical problems	14 %	20 %
Lacko of equipment	29 %	40 %
Lacko of work-life balance	14 %	40 %
Poor communication	0 %	40 %
Increased working hours	14 %	20 %

Source: Authors.

5 Discusssion

Based on the results, we are now able to answer the research questions that have been set.

First research question: What are the advantages and disadvantages of working from home during the Covid-19 pandemic and what are the least common ones?

The most common advantage of working from home during a Covid-19 pandemic is time flexibility, which reaches 20% representation. I think that thanks to the flexibility of time, they can schedule their working hours so that they have more time for their family and household chores, for example. As can be seen from Figure 1, the second most common advantage with a 17% share is the need for commuting and the associated savings. It is a matter of saving time and then, above all, of saving money. The following very common benefits of working from home during a Covid-19 pandemic are health, i.e., a reduced risk of Covid-19 virus infection and better concentration. Both benefits have been mentioned in up to five professional articles and amount to 13%. In addition to the reduced risk of infection with the virus, there is also the advantage of a reduced risk of spreading the virus. Better concentration can then be caused, for example, by less frequent disturbances, which in my opinion can be caused in the workplace by, for example, co-workers or customers. Other frequently mentioned benefits are the comfort of home and more time spent with the family. The comfort of the home could positively affect the concentration of employees in the home office, and more time spent with the family can be ensured by the aforementioned time flexibility.

Productivity, for example, is one of the least common benefits of working from home during a Covid-19 pandemic. This may be the least common advantage, for example, because productivity at work during the Covid-19 pandemic may not differ significantly from their productivity in the office, as mentioned by Awada et al. [14]. The other least common benefits are also digitalization, better communication, and physical health. These least common benefits are included in the results and Figure 1 as "other" and have therefore always been mentioned in only one of the sources. They reach only 3%. The advantages, which were mentioned in two sources and gain 5%, are better mental health, independence, and the possibility of preparing one's meal. Even these benefits can still be described as less common.

The most common disadvantage of working from home during the Covid-19 pandemic is social isolation. Social isolation has been reported by many sources and is as high as 19%, as shown in Figure 2. I think that social isolation can trigger mental problems and thus negatively affect the health and productivity of workers from home. The second most common disadvantage with 16% representation is the imbalance between personal and professional life. This can be caused mainly by poor scheduling of working hours. Homeworkers, for example, work at a time when they can take care of their family or vice versa. Therefore, as Okuyan and Begen [22] have pointed out, it is appropriate to avoid this disadvantage by more frequent breaks and better working time arrangements. Another often mentioned disadvantage is the lack of equipment, which represents 13%. Lack of equipment is a problem that does not facilitate the job. In my opinion, the employer should properly address this situation and provide their employee with suitable conditions, including all the equipment for the room in which the work is performed. The last common disadvantages include the lack of contact with colleagues and weight gain or even obesity caused by reduced activity representing up to 10%.

The least common disadvantages of the home office trend during the Covid-19 pandemic are complications in performing tasks, poor organization of time, or back pain caused by inappropriate equipment. Further, there are too long meetings. These disadvantages are marked as "other" in the results and were always mentioned in only one of the sources used. They reach only 3%, as can be seen from Figure 2.

https://doi.org/10.1051/shsconf/202213501023

These are the least mentioned disadvantages, in my opinion mainly because they are very individual facts. Every employee working from home has different equipment, different lengths of meetings, and different management, ensuring the least possible risk of complications in performing tasks. This individuality may be the result of a minimal mention of these disadvantages. Other uncommon disadvantages are often increased working hours, technical problems, and more stress. Furthermore, poor concentration and communication are also categorized as purely individual matters. The concentration depends mainly on the environment in which they perform their work. Whether it is a shared room like the living room or a private room where no one is disturbed at work. All these disadvantages reach 6%, as can be seen in Figure 2.

Second research question: What are the differences in advantages and disadvantages in terms of location?

Some differences in the advantages and disadvantages of working from home during the Covid-19 pandemic remained significant in terms of location. However, for some advantages and disadvantages, the differences were small or almost non-existent. The biggest differences in terms of location were the advantages of time flexibility and the disadvantages of social isolation. Both benefits were mentioned in up to five sources with respondents from Europe, while from other continents only once or not at all.

As can be seen from Table 1, the advantage of time flexibility is a difference of 71% in terms of location. As can be seen from Table 2, the difference in social isolation is 51%, slightly smaller than the flexibility of time. Nevertheless, it is a disadvantage with the greatest differences in terms of location. The fact that the biggest differences in terms of location are precisely the advantages and disadvantages is very interesting, as the first research question led us to the conclusion that these are the most common advantages and disadvantages of working from home during the Covid-19 pandemic. Therefore, I would expect them to be mentioned frequently by respondents from all over the world, not just or mainly from Europe. There were also significant differences in the benefits of better focus on work and the need for commuting associated with saving time and money. Even these benefits were mentioned mostly only by respondents from Europe. The difference in work focus was 57%. With no need to commute, as can be seen from Table 1, the difference was slightly smaller than with better concentration. Among the last advantages and disadvantages, which generally vary considerably in terms of location, are the advantages of home comfort and independence, as well as the disadvantage of poor communication. Other disadvantages and advantages are no longer so different in terms of location and the differences are slight. It is remarkable that the comfort of home and independence, as positives of working from home and the disadvantage of poor communication, have, on the contrary, been more widespread on other continents than in Europe. The advantage of independence and the disadvantage of poor communication reach up to 40% representation on other continents, while 0% representation in Europe. With the benefit of home comfort, as can be seen from Table 1, the difference is slightly smaller, as it accounts for at least 14% of Europe. The differences in the remaining advantages and disadvantages were only slight. This is the case, for example, with the advantage of a reduced risk of infection with the Covid-19 virus, where the difference was only 11%. I assumed this fact because it is a global pandemic and therefore this advantage is very important for everyone.

6 Conclusion

This thesis aims to analyze the advantages and disadvantages of the home office trend during the Covid-19 pandemic and the differences in terms of location. Using a literature review study, a subsequent synthesis of advantages and disadvantages, and a comparative method, the most common and, conversely, the least frequent advantages and disadvantages of working from home during the Covid-19 pandemic and their differences in terms of location were identified.

It is not possible to determine exactly whether the transition to a home office trend during the Covid-19 pandemic is exclusively positive or negative, as it has its advantages and disadvantages. However, it can be stated that the most common disadvantages of working from home include social isolation and lack of contact with colleagues, which, however, may depend on whether or not a state of emergency is declared and the associated limited movement of persons. The least common disadvantages of the home office trend in the Covid-19 pandemic are, for example, poor communication and lack of concentration. The consequence may be that this is a very individual matter, as some sources, on the contrary, mentioned concentration as a positive. As for the benefits of working from home during the Covid-19 pandemic, the most common is time flexibility, which is appreciated the most by the employees. Further, there is also no need to commute as a time and money saver. Also a reduced risk of Covid-19 infection, which is likely in a pandemic. This benefit, which affects our health, is no different in terms of location, as it is a global pandemic and therefore valuable benefits from employees around the world. While for the other advantages and disadvantages already mentioned, apart from the lack of concentration, considerable inconsistencies, and differences in terms of location were found. Many of the most common advantages and disadvantages were mentioned by many more respondents from Europe than from other parts of the world. Differences in advantages and disadvantages in terms of location may be due to the different development of countries and cultures. Last but not least, the least common benefits are the comfort of home and more time with the family.

The above results inevitably lead to the question of what causes the differences in advantages and disadvantages in terms of location. What measures can lead to minimizing disadvantages? Based on the data and content that has been collected and used for this research, it is not possible to answer these questions. It is therefore recommended to carry out follow-up research, which can build on the data and results of this study and answer the questions for better understanding.

Acknowledgments

This work was supported by the grant VEGA (No. 1/0038/22) Application of competitive digital games for the team cohesion development and social adaptation of Generation Z.

References

1. L. Yang, S. Liu, J. Liu, Z. Zhang, X. Wan, B. Huang, Y. Chen, Y. Zhang. COVID-19: Immunopathogenesis and Immunotherapeutics. *Signal Transduction and Targeted Therapy*. **5**(1) doi:10.1038/s41392-020-00243-2 (2020)
2. J. Grossmann, V. Korbel, D.Münich. *Práce z domova: možnost, nebo nutnost?* Praha: Národohospodářský ústav AVČR. ISBN 978-80-7344-572-0 (2021)
3. R. Kyzlinková, S. Veverková, H. Vychová. Práce z domova - popis stavu před pandemií a možné konsekvence do nových poměrů organizace práce. *Fórum sociální politiky*. **48**(5), (2020)
4. M. Barriga, R. Holger, R. C. Aguirre, D. Coello-Montecel, P. O. Pacheco, M. I. Paredes-Aguirre. The Influence of Work–Family Conflict on Burnout during the COVID-19 Pandemic: The Effect of Teleworking Overload. *International Journal of Environmental Research and Public Health*. **18**(19) doi:10.3390/ijerph181910302 (2021)
5. P. Kalina. Change Management: COVID-19 and Beyond. *Journal of Human Resource Management*. **23**(1), 38-40 (2020)
6. D. Lazarević, M. Dobrodolac, L. Švadlenka, B. Stanivuković. A model for business performance improvement: A case of the postal company," *J. Bus. Econ. Manag.*, **21**(2), 564–592 doi: 10.3846/jbem.2020.12193 (2020)
7. J. Jaros, V. Melichar, L. Svadlenka. Impact of the Financial Crisis on Capital Markets and Global Economic Performance-Web of Science Core Collection. *Transport Means - Proceedings of the International Conference*. 431–434 (2014)
8. Ľ. Čábyová, P. Krajčovič. *The role of SoLoMo marketing and media in the communication of eco-innovations*. 1st ed. Budapešť: Wolters Kluwer. ISBN 978-963-295-952-8 (2020)
9. J. G. Matúšová. Logo as the Greatest Symbol of Brand. *European Journal of Media, Art & Photography*. **9**(2), 126-133 (2021)
10. K. Stachová, J. Blštáková, Z. Stacho. *Riadenie ľudí v digitálnej a post-pandemickej dobe*. 1st ed. Praha: Wolters Kluwer. ISBN 978-80-8078-606-9 (2021)
11. M. Hitka, Z. Balazova, V. Grazulis, P. Lejskova. Differences in Employee Motivation in Selected Countries of CEE (Slovakia, Lithuania and the Czech Republic). *Inzinerine Ekonomika-Engineering Economics*. **29**(5), 536-547 doi: 10.5755/j01.ee.29.5.13953 (2018)
12. M. Hitka, Z. Rózsa, M. Potkány, L. Ližbetinová. Factors forming employee motivation influenced by regional and age-related differences. *Journal of Business Economics and Management*. **20**(4), 674-693 doi: 10.3846/jbem.2019.6586 (2019)
13. L. Krejčí, Š. Doleželová. Systematizace ne-řidičských činností řidičů autonomních nákladních vozidel. *Perner's Contacts*, **16**(1). https://doi.org/10.46585/pc.2021.1.1662 (2021).
14. M. Awada, G. Lucas, B. Becerik-Gerber, S. Roll. Working from home during the COVID-19 pandemic: Impact on office worker productivity and work experience. *Work*. **69**(4), 1171-1189 doi:10.3233/WOR-210301 (2021)
15. C. Ipsen, M. van Veldhoven, K. Kirchner, J. P. Hansen. Six Key Advantages and Disadvantages of Working from Home in Europe during COVID-19. *International Journal of Environmental Research and Public Health*. **18**(4) doi:10.3390/ijerph18041826 (2021)
16. E. Bolisani, E. Scarso, C. Ipsen, K. Kirchner, J.P. Hansen. Working from home during COVID-19 pandemic: lessons learned and issues. *Management & Marketing. Challenges for the Knowledge Society*. **15**(s1), 458-476 doi:10.2478/mmcks-2020-0027 (2020)
17. G. E. Puglisi, S. Di Blasio, L. Shtrepi, A. Astolfi. Remote Working in the COVID-19 Pandemic: Results From a Questionnaire on the Perceived Noise Annoyance. *Frontiers in Built Environment*. **7** doi:10.3389/fbuil.2021.688484 (2021)
18. Z. Chen. Influence of Working From Home During the COVID-19 Crisis and HR Practitioner Response. *Frontiers in Psychology*. **12** doi:10.3389/fpsyg.2021.710517 (2021)

19. P. Mehta. Work from home—Work engagement amid COVID-19 lockdown and employee happiness. *Journal of Public Affairs*. doi:10.1002/pa.2709 (2021)

20. MPSV ČR. *Práce z domova: Zkušenosti zaměstnanců a vedoucích pracovníků v souvislosti s nárůstem využívání práce z domova během pandemie koronaviru* [online]. Available at: https://www.esfcr.cz/documents/21802/14776072/Vyhodnocen%C3%AD+dotazn%C3%ADku+Pr%C3%A1ce+z+domova.pdf/99450e64-d6a5-453a-82ca-a5ddd1dd6b23

21. M. R. Pfeifer. Human Resources during COVID-19: A Monthly Survey on Mental Health and Working Attitudes of Czech Employees and Managers during the Year 2020. *Sustainability*. **13**(18) doi:10.3390/su131810055 (2021)

22. O. B. Okuyan, M. A. Begen. Working from home during the COVID-19 pandemic, its effects on health, and recommendations: The pandemic and beyond. *Perspectives in Psychiatric Care*. doi:10.1111/ppc.12847 (2021)

23. M. Rieth, V. Hagemann. The Impact of Telework and Closure of Educational and Childcare Facilities on Working People During COVID-19. *Zeitschrift für Arbeits- und Organisationspsychologie A&O*. **65**(4), 202-214 doi:10.1026/0932-4089/a000370 (2021)

24. D. Irawanto, K. Novianti, K. Roz. Work from Home: Measuring Satisfaction between Work–Life Balance and Work Stress during the COVID-19 Pandemic in Indonesia. *Economies*. **9**(3) doi:10.3390/economies9030096 (2021)

25. A. López Peláez, A. Erro-Garcés, F. J. P. García, D. Kiriakou. Working in the 21st Century. The Coronavirus Crisis: A Driver of Digitalisation, Teleworking, and Innovation, with Unintended Social Consequences. *Information*. **12**(9) doi:10.3390/info12090377 (2021)

26. M. S. Khan, H. Umer. Benefits and Challenges of COVID-19 Induced Telework Across Gender: Policy Lessons for Sustainable Future Telework. *SSRN Electronic Journal* doi:10.2139/ssrn.3848640 (2020)

27. R. Farooq, A. Sultana. The potential impact of the COVID-19 pandemic on work from home and employee productivity. *Measuring Business Excellence*. doi:10.1108/MBE-12-2020-0173 (2021)

28. M. Guler, A.K. Guler, M. G. Gulec, E. Ozdoglar. Working From Home During a Pandemic. *Journal of Occupational & Environmental Medicine*. **63**(9), 731-741 doi:10.1097/JOM.0000000000002277 (2021)

29. M. S. Nemteanu, D. C. Dabija, L. Stanca. The Influence of Teleworking on Performance and Employees. *Counterproductive Behaviour*. **23**(58) doi:10.24818/EA/2021/58/601 (2021)

Value Chain Analysis and Attractiveness of the Telecommunications Industry in Mongolia

Tsolmon Jambal[1,*], *Enebish* Jambal[2]

[1]Institute of Technology and Business in České Budějovice, Faculty of Corporate Strategy, Okružní 517/10, 370 01 České Budějovice, Czech Republic

[2]Mongolian University of Science and Technology, School of Business Administration and Humanities, 8th khoroo, Baga toiruu 34, Sukhbaatar district Ulaanbaatar, Mongolia 14191, Mongolia

Abstract. During the 4th Industrial Revolution and the 3rd Wave of Digitalization, the telecommunications sector developed rapidly and played a key role in accelerating economic development and competitiveness. Researchers and experts point out that the sector is an important tool to directly and indirectly support the implementation of the 17 goals of the 2030 Sustainable Development Goals. In this context, a demand it is important for the information and communications technologies (ICT) companies to make strategic changes, modernize their business processes, analyse key technology operations, and evaluate the effectiveness of supply, human resources, financial, and management policies. In order to create a distinct competitive advantage, key activities in the sector, such as procurement, internal operations, and sales channels, will be strategically coordinated, and human resources, technology, finance, and management will be configured to assist. Therefore, industry-wide value chain analysis will explore the challenges related to the supply of information and communication raw materials, technology-based cooperation, and legal and regulatory issues. This research aims to assess the value chain of the telecommunication services and define the gaps, which are drawbacks of the development. This paper will be a foundation for further studies on industry profitability and competitiveness by focusing on the areas that are essential to value creation in end-to-end services such as mobile communications, internet, and cable television.
Keywords: value chain; telecommunication; attractiveness of the industry; ICT; IT; Porter's value chain analysis

1 Introduction

Mongolia's information and communication sector, which is accounted for 2.4 percent of GDP, is dominated by the private sector, which has consistently implemented policies to promote privatization, market liberalization, and competition, increasing the sector's efficiency and revenue. The information and communication sector consists of a range of industries and services – Internet service provision, telecommunications equipment and services, information technology (IT) equipment and services, media and broadcasting, libraries and documentation centers, commercial information providers, network-based information services and other related information and communication activities [1]. Mongolia's telecommunications sector is a unique sector with different technological activities, such as fixed and mobile communications, internet, cable television, and postal services. For the telecommunications industry, assessing a value chain is important to see where the technology is creating value together, to see the coherence of core activities, and to make strategic decisions. According to Porter's study [2], it is clear that the most important role is played by information technology (IT); this has impacts both on differentiation and cost reduction, as it is the lever to create competitive advantage, to create new business, and to change the way firms operate. Few researchers performed Value chain analysis on the example of the PIM industry [3-5]. In this paper we introduce a literature review on historical development of value chain, next, a methodology of value chain assessment, at last, an assessment of value chain of Mongolian telecommunication industry. In this article, we look at the changes and innovations in the value chain theory and methodology, analyze the value chain of the telecommunications industry using M. Porter's generic value chain method, and make connections between technological activities.

1.1 Development of the Value Chain concept

The value chain concept and analysis were first described and initiated by Michael Porter [2]. Changes and innovations in value chain theory and methodology over the course of history have significantly enriched the methodology and techniques of generic value chain analysis from P. Porter's Generic Value Chain Analysis to 1985-1995 [3]. However, research articles since 1996 have developed The Value Network concept, and from 2000, the concepts of Global Value

* Corresponding author: jambal@mail.vstecb.cz

Chain, Added Value Chain, and Sustainable Value Chain have made important contributions to the development of value chain theory. From this it can be concluded that the concept is in transition from the value chain to the value network. The table below shows the development of the value chain concept.

Table 1. Development of the Value Chain concept

Approach to the concept of value chain	Autors
Researchers who have enriched and developed valuation techniques of value chains	Porter and Victor [3], Davenport and Cronin [6], Hergert and Morris [7], Shapiro et al. [8], Porter [9], Rayport and Sviokla [10], Porter [11], Zhang et al. [12], Roper et al. [13]
Researchers who developed The Value Network and the global value chains concept	Porter [14], Achrol [15], Allee [16], Bovel and Martha [17], Anderson [18]
Researchers studying value chain linkages and interrelationships and Collaboration of Sustainable Value Chain	Ensign [19], Dekker [20],
Researchers who have linked value chains to operational management and business process management. The Value Chain can be connected to the concept of business process management	Rainbird [21], Schweizer [22], Horne [23], Bailey and Francis [24]
Researchers who studied The Added Value Chain and Value network	McPhee and Wheeler [25], Peppard and Rylander [26], Fearne and Martinez [27], Guzmán et al. [28], Håkansson and Snehota [29], Möller and Rajala [30], Allee [16], Fearne and Martinez [27]

Source: Authors.

Value System encloses also the concepts of the Supplier Value Chain, Channel Value Chain and Buyer Value Chain. Inside the Value Chain the activities are not interdependent and, starting from the generic model of the Value Chain, it is therefore possible to shape a specific model for a particular firm. Strategy is defined by many authors as management, giving a precise direction or trajectory of activities towards a specific goal, which is formed in the business environment and is influenced by competitive and functional approaches. The literature suggests that strategic management is a set of instructions, decisions and actions that are necessary to achieve competitive advantage [31]. Management has the task of applying all these approaches in managing the overall composition of its activities [32]. As before but also today, strategy is understood as a science to plan and define the direction for action. It is not surprising that strategies were previously used in military actions and military operations. Above all, strategy should be an aid to decision making in the face of unclear or uncertain conditions. Shepherd, Souitaris and Gruber [5] conceptualize strategy as a management tool that can be used to build enterprise value in the long run. Porter points out that it is necessary to be able to properly set a strategy that carries within it the patterns that determine the competitiveness of the firm, such as internal objectives and the ways to achieve them. This is the essence of competitive strategy formulation in the relationship between the firm and its environment [33,34]. In 1939, a new approach to measuring and testing new competitive hypotheses was introduced, which is the positional approach [35,35]. The idea of this hypothesis is based on the concept of, that there is a relationship between performance and market structure. This is because the behaviour of business entities and buyers is based on performance, while the behaviour of business entities is based on market structure [37]. Porter [38] affirms that through this approach, businesses can achieve competitive advantages in the long run if they set and formulate the basic factors affecting the industry structure correctly. He follows this approach and recommends the use of the five forces model. Ongaro and Ferlie [39] also has a clear idea about the definition of strategy, which he summarises as and defines it using the 5Ps. According to him, every strategy needs to start with a plan, which in corporate management terms is referred to as a single, comprehensive and integrated plan designed to ensure the achievement of corporate objectives. Porter attaches importance and significance to complots, where false information is sent to competitors in order to create fear in them and thereby discourage them from taking certain actions [14]. Another point is the position focusing on the firm's view of the environment both outward and inward. The last point will state the character of the enterprise, which reflects the enterprise in such a light as the personalities of the people acting on the enterprise [30]. If the idea of Horne [23] and his need for strategic thinking will be continued, according to Lu, Liu and Falkenberg [40], strategic thinking is not only about thinking about strategy in the context of strategic management, but it is a way to continually look for ways to and how to improve the characteristics of strategic planning. However, he also points out that overanalysing and looking for solutions at any cost can only lead to worse and confusion. Müller, Marius and Buliga et. al. [41] has two perspectives on strategic thinking. The first view is that he views strategy as an intelligent tool that, when set up correctly, can identify a clear plan. The second view is that he views it as an intelligent imagination that is undoubtedly needed to create a clear idea. In the words of Bechtel and Jayaram [42], the most important condition for the elaboration of strategy is the understanding of the principles of strategic thinking, which includes the principle of variance, which notes that the strategic period in the long term cannot be estimated in advance, so how unexpected changes in the field of international politics can occur. Another point is the principle of permanence, which points out that strategy making is a never-ending process, and there is a need to constantly monitor the evolution of undeveloped strategies and the evolution of factors that may influence them. The principle of variation and the creative approach, whereby strategy-making should bear in mind that we are in a constantly changing environment that affects economic, political and social developments. It is therefore necessary to develop strategies in several versions. Cézanne, Lorenz and Saglietto [43] notes that a strategy developed yesterday may not be valid today or tomorrow. It should be kept in mind that a business should constantly come up with new approaches and adapt its products to meet the needs of the market, try to reduce costs, come up with new ideas, innovate and motivate its employees to perform better and better. It is essential to take into account the possibility of risk when drawing up strategies and everyone should be aware that risk can occur not only in our lives but in almost every corporate environment. If we focus on the definition of risk, we need to keep in mind what risks are and what they can cause. Risks can be viewed as a set of methods, procedures and predictions where the goal of management is to learn to anticipate them, estimate them, know how to deal with them and above all prevent them [44]. Risks can be measured according to several methods, one of which is the risk mapping method, which indicates areas with different levels of risk, based on a process indicating different types of hazards and vulnerability [45]. Currently, the business environment is subject to variability, caused by economic and political circumstances. It sheds new light on the issue of business risk management [46]. Different methods are used for risk assessment and principles, which include the use of statistical methods to assess, for example, the level of financial security of a business using cost theory. Such a method can be referred to as Enterprise Risk Management, which manages risks in an integrated holistic manner [47,48]. The purpose of any enterprise is to create a quality product with a special value for a specific market. In order to navigate the business environment clearly, all decisions should be based on the observed results of the situation of the internal and external environment. Therefore, in order to achieve to make the right decision, the right steps need to be taken, and it is this analysis of the business environment that is the crucial step for the strategic manager to take. One of the main elements distinguishing the work of a strategic manager from that of a business project manager is the confrontation of the intention with the environment [49]. The reasons why the business environment must be analysed are simple. It is for the reason of being able to identify and determine the position of the business in the market in which it operates, to be able to respond effectively to changes, to estimate its potential, to be able to predict the behaviour of customers and competitors and to be able to identify risks.

2 Methodology

A theoretical framework of "Value Chains and Competitive Advantage" and a basic methodological model on the concept of value chain have been developed by M. Porter in his publications such as "Competitive Advantage of Creating and Sustaining Superior Performance [2] and Cases in Competitive Strategy [33].

2.1 Porter's value chain analysis

Every company is a collection of activities that are performed to design, produce, market, deliver and support its products. All these activities cab be represented using a value chain, shown in Figure 1. The value chain displays total value, and consists of value activities and margin. Value activities are the physically and technologically distinct activities a firm performs. Margin is the difference between total value and the collective cost of performing the value activities. M. Porter's Generic value chain consists of nine parts of activities that are connected to each other in their own unique ways. It consists of two groups: main functions of the value chain and supporting functions. Value activities can be divided into two broad types, primary activities and support activities. Primary activities are the activities involved in the physical creation of the product and its sale and transfer to the buyer (Figure 1).

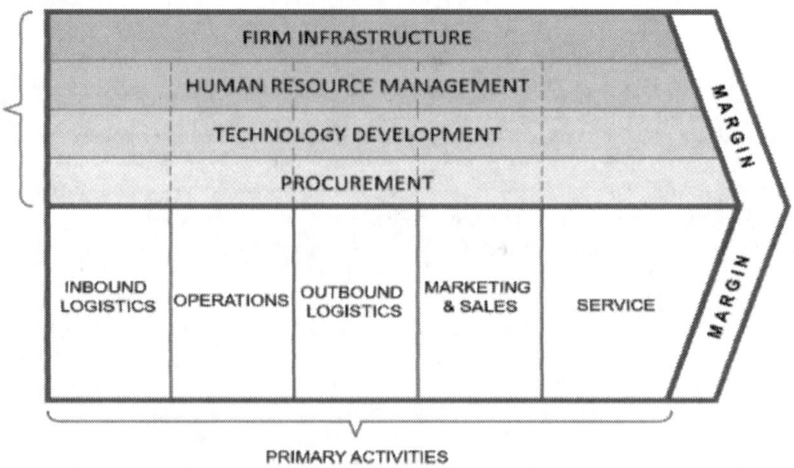

Figure 1. The generic value chain model adapted from Porter

Source: Authors based on [2].

Primary activities can be divided into five generic categories: Inbound logistics, operations, Outbound logistics, Marketing and Sales and Service. Each category is vital to competitive advantage depending on the industry. Support activities involved in competition of any industry are divided into four generic categories: Procurement, Technology development, Human resource management, Firm Infrastructure. Primary and Support activities have different roles in competitive advantage.

Table 2. Methodology of the Value chain analysis

Methodology of the Value chain analysis	
Starting with the generic chain, individual value activities are identified into industry. Each generic category can be divided into discrete activities. Important Steps in Value Chain Analysis Methodology:	
1	The value is assessed by starting from the general chain, in other words, outline independent value operations are performed toward delivering final services to an end-user.
2	Determine whether operations completed by themselves or outsourced from another organization independently or in collaboration.
3	Independent technological and economic activities must be separated.
4	Mapping the Value Chain. Describe the sequence of activities along the flow of the process.
5	Let's see how it affects the difference.
6	Activities are classified according to their contribution to the cost.
7	Formulate research questions [4].

Source: Authors.

2.2 Analysis of the attractiveness of the industry

The analysis of the attractiveness of the industry is evaluated using a grid with 15 factors and the forces that affect them are assigned a score from 1 to 10. A higher score (10) is assigned when the conditions in the industry are maximized in terms of available resources. Otherwise, the point score is low, where the company is unable to meet its obligations in the industry. The optimal point score range is from 75 to 120 points. A score lower than 75 points signals the need for acute repositioning of the company within the industry, for example by entering a new segment [50]. Place the figure as close as possible after the point where it is first referenced in the text. If there is a large number of figures and tables it might be necessary to place some before their text citation.

3 Results and Discussion

3.1 Telecommunication and Information technology sector in Mongolia

Total revenue of the sector reached 1,274.1 billion MNT in the quarter of 2019, which is 14% higher than that of the same period of the previous year. Revenues from mobile services, which account for more than 50 percent of this revenue, increased by 0.7 percent from the same period of last year, while revenues from IP TV and Internet wholesale services decreased by 1.4 to 2.3 percent [51]. A total of 6,675 people work in the sector, of which 827 are engineers and technicians, accounting for 12.3% of the total workforce [52].

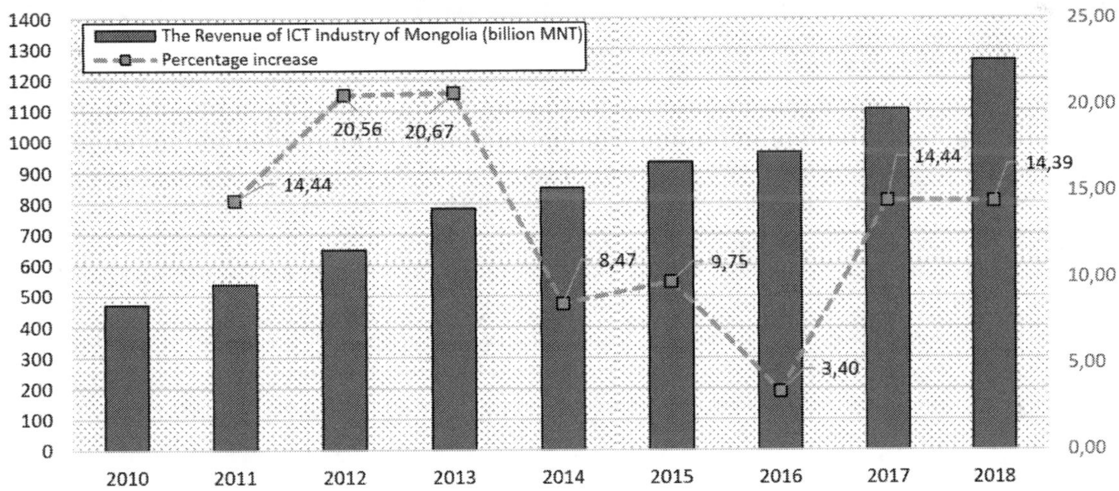

Figure 2. The Revenue of ICT industry of Mongolia (billion MNT)

Source: Authors based on [29].

Mobile and landline telephone services: The number of users of a new type of landline services, the Internet-based triple play service (IPTV), increased by 5.8%, contributing to a total increase of 1.1% in landline services.

Internet services: The total number of Internet users increased by 10.1% to 2.65 million compared to the same period of the previous year, which is 91.3 percent of the population. This is due to 3G and 4G mobile networks, provided by 67 companies licensed to provide Internet services in Mongolia. 2.45 million or 93% of customers are mobile users and 231.1 thousand are landline users.

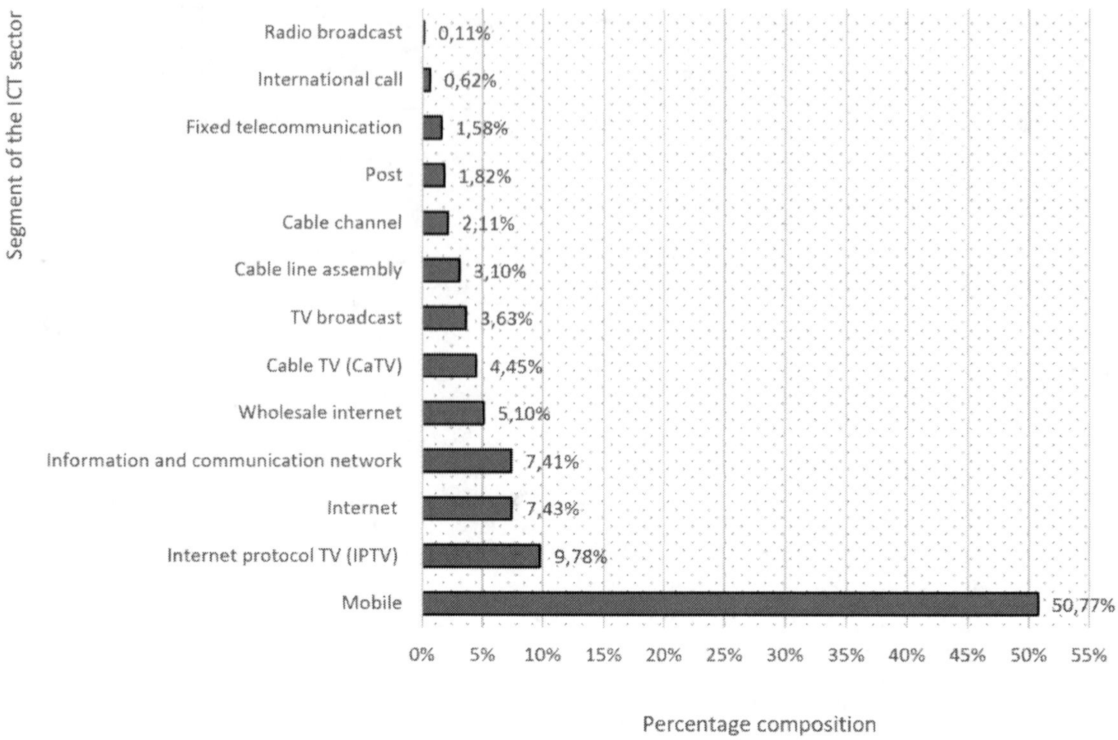

Figure 3. Composition of individual segments of the telecommunications and information technology sector

Source: Authors.

Software production and information technology: There are more than 3,000 companies operating in the field of software production and information technology, of which 95.1% are companies operating in the field of communication and information technology. Including companies with foreign investment, this figure is 97%. Cable TV service: According to a report by NSO-MN in 2019 the number of multi-channel TV service users increased by 6.8% compared to the same period of 2018 and reached 1,047,000. 57% of the total consumption is in Ulaanbaatar and 43% in rural areas [51]. Transmission network: Mongolia is connected to more than 195 countries via Intelsat, APSTAR-5 satellite, fiber optic cable and radio relay lines. As of 2019 there are 37,646 km of fiber-optic cable in Mongolia and 6,294 km in urban areas. Postal service: In our country, revenues from express mail, parcels, packages and logistics services have increased in recent years and as of 2019, accounted for 1.6 percent of the total revenue of the sector [51].

3.2 Value chain of the Mongolian telecommunication and Information technology sector

An overview of the value chain of telecommunications industry reveals that the industry is competing with the same technology in networking and cabling, which is part of the core business. However, in Outbound Logistics, Marketing and Sales and Service parts the industry has differences and variety of companies are competing for a market share. This is shown in Fig 4.

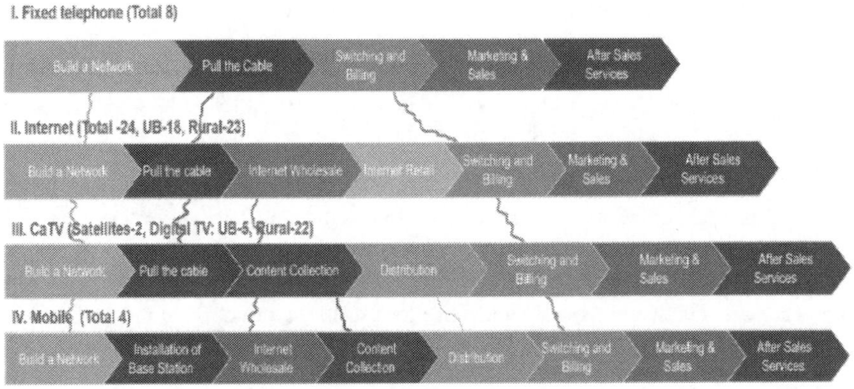

Figure 4. Value chains in the information and communication sector

Source: Authors.

Among others we want to emphasis postal services industry of Mongolia. The focus of the postal industry is the continuous improvement of incoming and outgoing processing as part of the development of the postal network and the maintenance of the network's continuity. There is also a need to pay attention to the development of transportation and logistics, the security of postal items, the timing of inspections and the quality of delivery.

Figure 5. Postal service value chain

Source: Authors.

Note: VAS - Value added service.

According to Figure 5, in the case of postal services, it is preferable to implement a different strategy for the main operations, transport and logistics operations, and distribution services to end users.

3.3 Analysis of the attractiveness of the telecommunications and information industry in Mongolia

The analysis of the attractiveness of the sector was carried out in the Mongolian telecommunications and information sector. The analysis included an assessment of individual factors and strength according to a score from 1 to 10 points based on a subjective assessment (see Table 2).

Table 3. Font styles for a reference

Factor		Potency	Score
1	Growth potential	Growing demand	7
2	Market diversity	Number of segments	8
3	Profitability	Rising / stagnant / declining	7
5	Expansion	Competitors, inflation	8
5	Concentration	Number of dominant enterprises	5
6	Sales	Cyclic, continuous	7
7	Specialization	Focus, differentiation, uniqueness	6
8	Brand	Value, quality, substitution	7
9	Distribution	Channels, need for support	5
10	Pricing policy	Experience effect, elasticity	6
11	Cost position	Competitive, high / low cost	5
12	Services	Guarantee, reliability	8
13	Technology	Leadership, uniqueness	5
14	Integration	Vertical, horizontal, control	8
15	Possibility of entry and exit	Industry barriers	6
Total			**98**

Source: Authors.

Growth potential: In the last decade, ICT revenues in Mongolia have grown by an average of 13.26% year on year. As a result, the use of telecommunications products and services such as mobile and fixed telephones, the Internet, cable television and postal services can be expected to continue to grow.

Market diversity: Telecommunication products and services are widely used in other industries. In this way, they reach many segments and create new opportunities for innovation.

Profitability: Companies operating in the industry are currently attractive in terms of overall corporate profits.

Expansion: The direct impact of inflation is unpredictable due to the high demand for telecommunications services, but there is considerable competition in this sector. New application-based software companies are particularly attractive. The attractiveness of the industry is assessed as above average.

Concentration: The concentration in the telecommunications sector represents 4 dominant mobile operators, which represent more than 50 percent of the sector's total revenue. Small, weak companies have difficulty entering the market, and even though they are strongly attracted. State regulation, obtaining a license and high investment form a major barrier to entry in this market.

Sales: The use of telecommunications products and services in the telecommunications industry is on the rise due to the current situation of COVID-19.

Specialization: In this area you can find many different groups of products and services, which are very different from each other. But the products and services in a given group are not so different.

Brand: There are a number of dominant companies with established brands (Mobicom, Skytel, Unitel, Gmobile) on the market, whose products and services are popular with consumers, but do not guarantee higher quality. Software small companies provide services through their already established 3G and 4G networks or channels.

Distribution: Distribution channels are created through our own established networks and online form. Recently, the post office has become more of a delivery service due to Covid.

Pricing policy: Due to the high level of competition in the market, companies often have a low pricing policy. In recent years, this has become increasingly common on the Internet. Products and services are delivered as complete as possible. Low-cost companies make the sector more attractive with lower prices, but they also face higher costs compared to strong competition.

Cost position: Dominant companies have a cost advantage over small companies due to their high capital potential. Due to the rapid pace of technological innovation, the costs of technological innovation and new investment are high. In recent years, the fact that software development companies have entered the industry at relatively low cost has been rated as moderately attractive.

Services: As already mentioned, there are many types of products and services in this sector. Mobile communication services dominate. The industry tends to be competitive in terms of services, and focuses on after-sales service by providing multiple sets of packages.

Technology: Due to the rapid pace of technological change, significant investment is needed to ensure the professional development of employees in the telecommunications sector in Mongolia.

Integration: Vertical and horizontal integration is considered. Horizontal integration takes place by merging small companies with dominant ones. In the telecommunications sector, start-ups have been cooperating with large companies in recent years. This is beneficial for both parties when introducing new products and services.

Possibility of entry and exit: Barriers to this sector are mainly high competition from large companies (enterprises) and increased demands on initial capital. For these reasons, firms (businesses) may not be interested in entering the industry.

4 Conclusion

Value chain analysis in the Mongolian ICT sector is important in defining strategies for gaining or strengthening the competitive advantage of industries in this sector. As industries in the industry often localize high technology from the outside, they make a difference in the field of Outbound logistics, Marketing and Sales and Service operations. In the case of postal services, the quality and performance received by end-users are affected by the difficulties in the basic operations of transport and logistics of distribution to end users. Internet, telephone and how is this linked to activities in different sectors. The telecommunications and information sector in Mongolia scored 97 points in the analysis of the attractiveness of the sector. The total average value is 6.53, which can be described as an average satisfactory value. Unsatisfactory values were achieved in the areas of concentration factors, distribution, cost position and technology. On the contrary, diversity, expansion, services and integration seem to be very attractive among the researched factors. With regard to the future in terms of the attractiveness of the industry, the possibility of new unforeseen small start-ups can enter the industry. Furthermore, value chain analysis would be very effective in tracing product flows, showing the value adding stages, identifying key actors and the relationships with other actors in the chain. The results of this study contribute to the research and practice of value chain management by improving predictive and reactive decisions to take advantage of VC model visualization and analysis of the attractiveness of the telecommunications and information industry in Mongolia.

References

1. J. Onjala, D. McCormick. Methodology for Value Chain Analysis in ICT Industry Frameworks for the Study of Africa. *Asian Drivers*. (2007)
2. M. E. Porter. *Competitive Advantage, Creating and Sustaining Superior Performance, First free*. New York City: A Division of Simon & Schuster Inc. ISBN 0-684-84146-0 (1998)
3. M. E. Porter, E. M. Victor. How information gives you competitive advantage. *Harvard Business Review*. **63**, 149-160 (1985)
4. J. Peppard, A. Rylander. From value chain to value network: insight from mobile operators. *European Management Journal*. **24**, 128-141 doi.org/10.1016/j.emj.2006.03.003 (2006)
5. D. A. Shepherd, V. Souitaris, M. Gruber, MCreating New Ventures: A Review and Research Agenda. *Journal of Management*. **47**(1), 11–42 doi.org/10.1177/0149206319900537 (2021)
6. L. Davenport, B. Cronin. Strategic information management: forging the value chain. *International Journal of Information Management*. **8**, 25–34 doi.org/10.1016/0268-4012(88)90005-9 (1988)
7. M. Hergert, D. Morris. Accounting data for value chain analysis. *Strategic Management Journal*. **10**(2), 175-188 doi.org/10.1002/smj.4250100207 (1989)
8. J. M. Shapiro, V. M. Singhal, S. N. Wagner. Optimizing the value chain. *Interfaces*. **23**, 102-117 doi.org/10.1287/inte.23.2.102 (1993)
9. M. E. Porter. Towards a dynamic theory of strategy. *Strategic Management Journal*. **12**, 95-117 doi.org/10.1002/smj.4250121008 (1991)
10. J. F. Rayport, J. J. Sviokla. Exploiting the virtual value chain. *Harvard Business Review*. **73**, 14-24 (1995)
11. M. E. Porter. Strategy and the internet. *Harvard Business Review*. doi.org/10.1109/ICSSSM.2007.4280087 (2001)
12. Q. Zhang, M. A. Vonderembse, J. S. Lim. Value chain flexibility: a dichotomy of competence and capability. *International Journal of Production Research*. **40**, 561–583 doi.org/10.1080/00207540110091695 (2002)
13. S. Roper, J. Du, J. H. Love. Modelling the innovation value chain. *Research Policy*. **37**, 961-977 doi.org/10.1016/j.respol.2008.04.005 (2008)
14. M. E. Porter ME What is strategy? *Harvard Business Review*. **74**, 61-78 (1996)
15. R. S. Achrol. Changes in the theory of interorganizational relations in marketing: toward a network paradigm. *Journal of the Academy of Marketing Science*. **25**, 56-71 doi.org/10.1177/0092070397251006 (1997)
16. V. Allee. Reconfiguring the value network. *Journal of Business Strategy*. **21**(4), 36-39 doi.org/10.1108/eb040103 (2000)
17. D. Bovel, J. Martha. From supply chain to value net. *Journal of Business Strategy*. **21**, 24-28 doi.org/10. 1108/eb040101 (2000)
18. S. C. Anderson. The globally competitive firm: functional integration value logistics global marketing and business college strategic support. *Competitiveness Review*. **10**, 33-45 doi.org/10.1108/ eb046397 (2000)
19. P. C. Ensign. Value chain analysis and competitive advantage: assessing strategic linkages and interrelationships. *Journal of General Management*. **27**(1), 18-42 doi.org/10.1177/030630700102700102 (2001)
20. H. C. Dekker. Value chain analysis in interfirm relationships: a field study. *Management Accounting Research*. **14**, 1-23 doi.org/10.1016/S1044-5005(02)00067-7 (2003)
21. M. Rainbird. A framework for operations management: the value chain. International Journal of Physical Distribution & Logistics Management. **34**, 337–345 doi.org/10.1108/09600030410533628 (2004)
22. L. Schweizer. Concept and evolution of business models. Journal of General Management. **31**, 37–56 doi.org/10. 1177/030630700503100203 (2005)
23. J. R. Horne. Are you seeing the big picture in selecting your improvement projects? *TQM Journal*. **26**, 658-666 doi.org/10.1108/TQM-08-2013-0094 (2014)

24. K. Bailey, M. Francis. Managing information flows for improved value chain performance. *International Journal of Production Economics*. **111**(1), 2-12 doi.org/10.1016/j.ijpe.2006.11.017 (2008)

25. W. McPhee, D. Wheeler. Making the case for the added-value chain. *Strategy & Leadership*. **34**(4), 39-46 doi.org/10.1108/10878570610676873 (2006)

26. J. Peppard, A. Rylander. From value chain to value network: insight from mobile operators. *European Management Journal*. **24**, 128-141 doi.org/10.1016/j.emj.2006.03.003 (2006)

27. A. Fearne, M. G. Martinez. Dimensions of sustainable value chains: implications for value chain analysis. Supply Chain Management. **17**(6), 575–581 doi.org/10.1108/13598541211269193 (2012)

28. J. Guzmán, P. Moreno, P. Tejada. The tourism SMEs in the global value chains: the case of Andalusia. *Service Business*. **2**, 187-202 doi.org/10.1007/s11628-008-0034-6 (2008)

29. H. Håkansson, I. Snehota. No business is an island: the network concept of business strategy. *Scandinavian Journal of Management*. **22**, 256–270 doi.org/10.1016/j.scaman.2006.10.005 (2006)

30. K. Möller, A. Rajala. Rise of strategic nets—new modes of value creation. *Industrial Marketing Management*. **36**, 895-908 doi.org/10.1016/j.indmarman.2007.05.016 (2007)

31. R. E. Hoskisson, F. Chirico, J. Zyung, E. Gambeta. Managerial Risk Taking: A Multitheoretical Review and Future Research Agenda. *Journal of Management*. **43**(1), 137–169 doi.org/10.1177/0149206316671583 (2017)

32. T. Saebi, L. B. Lien, N. J. Foss. What Drives Business Model Adaptation? The Impact of Opportunities, Threats and Strategic Orientation. *Long Range Planning*. **50**(5), 567-581 doi.org/10.1016/j.lrp.2016.06.006 (2017)

33. M. E. Porter. *Cases in Competitive Strategy*. New York City: A Division of Simon & Schuster Inc. ISBN 0-684-84148-7 (1982)

34. E. G. Carayannis, E. Grigoroudis, M. Del Giudice, M. R. Della Peruta, S. Sindakis. An exploration of contemporary organizational artifacts and routines in a sustainable excellence context. *Journal of Knowledge Management*. **21**(1), pp. 35-56 doi.org/10.1108/JKM-10-2015-0366 (2017)

35. M. E. Porter, E. M. Victor. How information gives you competitive advantage. *Harvard Business Review*. **63**, 149-160 (1985)

36. C. Lomberg, D. Urbig, C. Stöckmann, L. D. Marino, P. H. Dickson. Entrepreneurial Orientation: The Dimensions' Shared Effects in Explaining Firm Performance. *Entrepreneurship Theory and Practice*. **41**(6), 973-998 doi.org/10.1111/etap.12237 (2017)

37. P. M. Kreiser, B. S. Anderson, D. F. Kuratko, L. D. Marino. Entrepreneurial Orientation and Environmental Hostility: A Threat Rigidity Perspective. *Entrepreneurship Theory and Practice*. **44**(6), 1174-1198 doi.org/10.1177/1042258719891389 (2020)

38. M. E. Porter. *Competitive advantage, creating and sustaining superior performance*. New York: The Free Press. ISBN 0-02-925090-0 (1985)

39. E. Ongaro, E. Ferlie. Strategic Management in Public Organizations: Profiling the Public Entrepreneur as Strategist. *The American Review of Public Administration*. **50**(4–5), 360-374 doi.org/10.1177/0275074020909514 (2020)

40. H. Lu, X. Liu, L. Falkenberg. Investigating the Impact of Corporate Social Responsibility (CSR) on Risk Management Practices. *Business & Society*. **61**(2), 496-534 doi.org/10.1177/000765032092898 (2020)

41. J. M. Müller, O. Buliga, K.-I. Voigt 2018. Fortune favors the prepared: How SMEs approach business model innovations in Industry 4.0. *Technological Forecasting and Social Change*. **132**, 2-17 doi.org/10.1016/j.techfore.2017.12.019 (2018)

42. C. Bechtel, J. Jayaram. Supply Chain Management: A Strategic Perspective. *New Zealand Journal of Crop and Horticultural Science*. **47**(4), 15-34 doi.org/10.1108/09574099710805565 (2019)

43. C. Cézanne, E. Lorenz, L. Saglietto. Exploring the economic and social impacts of Industry 4.0. *Revue d'économie industrielle*. **1**(1), 11-35 doi.org/10.4000/rei.8643 (2020)

44. O. F. Bustinza, E. Gomes, F. Vendrell-Herrero, T. Baines. Product–service innovation and performance: the role of collaborative partnerships and R&D intensity. *R&D Management*. **49**(1), 33-45 doi.org/10.1111/radm.12269 (2019)

45. W. Viriyasitavat, L. D. Xu, Z. M. Bi, A. Sapsomboon. Blockchain-based business process management (BPM) framework for service composition in industry 4.0. *Journal of Intelligent Manufacturing*. **31**(2), 1737–1748 doi.org/10.1007/s10845-018-1422-y (2020)

46. A. Prosvetova, E. Dozhdeva, V. Manyaeva, O. Belyakova. Risk Management In The Process Of The Financial Activity Of The Corporation. *Global Challenges and Prospects of the Modern Economic Development*. **57**, 1169-1175 doi.org/10.15405/epsbs.2019.03.118 (2019)
47. L. Hahnenstein, G. Köchling, P. N. Posch. Do firms hedge in order to avoid financial distress costs? New empirical evidence using bank data. *Journal of Business Finance & Accounting*. **48**(3-4), 718-741 (2021)
48. M. Altuntas, T. R. Berry-Stölzle, R. E. Hoyt. Enterprise Risk Management Adoption and Managerial Incentives. *Journal of Insurance Issues*. **43**(2), 1–42 (2020)
49. M. J. Sousa, Á. Rocha. Skills for disruptive digital business. *Journal of Business Research*. **94**, 257-263 doi.org/10.1016/j.jbusres.2017.12.051 (2019)
50. J. Tichá, I. Hron. *Strategické řízení*. Praha: Česká zemědělská univerzita v Praze. ISBN 978-80-213-0922-7 (2016)
51. CRC. *CRC statistik of Mongolia 2019* [online]. Available at: https://crc.gov.mn/ (2019)
52. Communications and Information Technology Authority (CITA). *CITA-Human Resource Research Report*. (2018)

Tax Reform as an Instrument for the Stability of the National Economy

Ján Dobrovič[1,*], *Yaroslava* Kostiuk[1,2], *Petra* Plachtová[1]

[1]Institute of Technology and Business in České Budějovice, Faculty of Corporate Strategy, Okružní 517/10, 370 01 České Budějovice, Czech Republic
[2]University of Zilina, Faculty of Operation and Economics of Transport and Communications, Univerzitna 1, 010 26 Zilina, Slovakia

Abstract. This article presents the results of the survey on the needs and expectations of people affected by the tax reform so that to assess the tax reform in Slovak Republic along with the overall satisfaction with the tax & customs system and also to identify deficiencies in the area of tax and customs administration. We have started from the existing functional organizational structures and tax administration systems not only in Slovakia, but also in Hungary, Poland, Czech Republic and Slovenia, while the basic prerequisite for the study was to increase the efficiency of the tax system as a whole. Based on the trend analysis, we assume that the tax and customs reform will make a significant contribution to increasing the system efficiency and, ultimately, to a more positive perception of taxes.
Keywords: taxes; tax reform; trend analyses; efficiency; Slovakia

1 Introduction

Tax administration in Slovakia is governed by the principles established after 1989. Despite the efforts made in recent years to reduce the number of employees, the system of financial administration in Slovakia can be considered inefficient in terms of increasing the competitiveness of the state. The general tendency of the present is to complain about inflexible, reluctant bureaucratic apparatus, which, on the one hand, leads to the reluctance of citizens to meet the tax liabilities and, on the other, state measures are getting only stricter [1]. This situation ultimately leads to the aforementioned reduction in competitiveness of the state and reduction of its credibility in the eyes of own [2].

The aim of this contribution is to overview the financial management system in Slovakia, to find critical points that affect this unfavorable situation described above and to propose an optimal solution (in the context of the European Union). Slovakia's financial management needs to be seen in a wider context here. We thus stem from the existing functional organizational structures and tax administration systems not only in Slovakia, but also in Hungary, Poland, Czech Republic and Slovenia, assuming the need to change of the organizational structure, its optimization as well as the introduction of marketing principles in the field of customer orientation to achieve a positive image of tax administration in the eyes of the public. For this purpose, we conducted a survey, the results of which reflect the assessment of the current situation. We anticipate that the tax and customs reform will make a significant contribution to increasing the efficiency of the system and, ultimately, to the positive perception of taxes, perceived today as socially unpopular obligations [1,3].

Slovakia is the eighth most attractive European country in terms of tax systems according to the survey conducted by KPMG International. Its results reflect the views of more than 400 tax specialists in international companies operating in Europe. The evaluation criteria were: attractiveness, administrative difficulty, consistency, long-term stability, the scope of legislation, rates and relations with tax authorities. According to the survey results, at European level, the least attractive area is the scope of tax legislation. The ranking of the states was determined on the basis of absolute attractiveness, which was calculated as the difference in percentage of the respondents, according to which the key aspects of their domestic tax systems were attractive, and the percentage of dissatisfied respondents [1]. In this KPMG chart, which was based on the survey of European business representatives' views on the attractiveness of domestic tax regimes, the best position was reached by Cyprus, followed by Ireland and Switzerland. Positive assessments of these three countries can be explained by the uniform interpretation of tax legislation, minimum changes in tax laws and relatively low tax rates [4].

* Corresponding author: 24574@mail.vstecb.cz

2 Methodology

2.1 Tax and Tax system

Profit reveals the efficiency of a company as well as the positive and negative aspects of its operations [5, 6]. Under International Financial Reporting Standards, profit and profitability are often used to evaluate the performance of companies and to determine changes in economic resources [7]. In contrast, Taxes are the most important non-credit public income. There are many comprehensible definitions, but we will mention those that describe taxes from two different perspectives. From a formal legal point of view, the taxes can be understood as mandatory, statutory payments that are paid by the taxpayer of the state at the specified amount and within the specified time limits to the relevant public budget. Such a definition is related to the features of tax laws and the constitution, but for the purposes of economic tax theory it is not sufficient. That is why it should be broadened by economic considerations. From the economic-financial point of view, the tax represents the fiscal relationship between the taxpayer and the state that is used by the government to set its goals [8]. From this point of view, taxes are financial relations, characterized by non-refundability and unilateralism. But at the same time, these statements suggest the need to define taxes also in terms of philosophy, which will create a space for the examination of qualitative features [9].

Musgrave [10] classifies taxes as following: taxes levied on the goods market and taxes levied on the (production) factors market. For example, on the (production) factors market it is the payroll tax paid by the employer and on the goods market, it is a value added tax or excise tax [11]. The current business environment for SMEs is very demanding and the enterprises are less able to gain access to external resources [12]. Bilan [13] points to the interconnection of the whole system, while referring to the so-called butterfly effect which means that a small change at one place may lead to large changes in the other place.

Taxes and tax laws (more or less) directly relate to each person and his/her finances. They determine when, to whom and for what we have to pay tax, they also determine what is tax-exempt respectively not taxed. Their continuous development is evidence that the tax system cannot be closed in any country. A good tax system should bring the state treasury sufficient revenues, it should not be an administrative burden and must be sufficiently transparent. Its role is to spread the tax burden homogeneously. An optimal tax system puts emphasis on two aspects: justice and efficiency.

The main task in creating the tax system is to create a system that would allow to collect enough funds needed to fulfill the essential functions of the state and the regions, while leaving sufficient financial resources to economic entities for their life and development. The importance of taxes lies mainly in the fact that they represent the main source of state revenues. They represented the most important component of the public budget system - state budget together with tax revenues is generally considered to be an important tool for government and state policy, i.e., an instrument used by the government to implement its economic goals and tasks [1]. Current time places high demands on managers and other employees, forcing them to think about how best to optimize processes [14], not excluding the tax area.

The tax system corresponds to the economic and political organization of the state - its government and, last but not least, to the currently governing political stream [15]. Slovakia is in a stage of development when it seeks to reach the economic level of the European Union countries by creating an attractive business environment [16]. In addition, a good business environment creates a favorable environment for the development of small and medium-sized enterprises (SMEs), which are important part of the national economic system [17]. In the context of the above-mentioned Zubaľová [18] notes that it is not possible to remove all distortions from all areas burdened with taxes, i.e. to achieve the "ideal state". If we consider the absence of an "ideal state" in the interests of optimizing the tax system, we must strive to maximize the social function of benefit and to find a balance between efficiency and justice.

Another important element that can play a role in simplification of the tax system is electronization. For example, Belas and Demjan [19] examining bank cutsomers' satisfaction found that the most significant factors of customer satisfaction are the availability of banking products and services through electronic banking. The "Doing Business 2016" which evaluates the business environment in national economies showed that Slovakia was in 29th place in a better position than the Czech Republic (36th place), which Belás and Sopková [20, p.147] consider to be the result "the simplification of the business establishment in the country and the introduction of an electronic system of filing tax declarations and the process of paying taxes itself".

3 Results and Discussion

In accordance with above-mentioned philosophical-theoretical starting points, we examined the effectiveness feedback of the tax administration of the Slovak Republic. The reform of the tax administration of the Slovak Republic presupposes the acceleration of the management and decision-making processes and the improvement of the transfer of information, including the feedback within the individual tax offices (TOs). Removal of one administrative management interlink will cause that eight tax offices will be managed as executive organizational units of the tax administration. From the perspective of Tax Directorate of the Slovak Republic (TD SR) this form is considerably simpler and more flexible than managing the current number of 102 tax offices through 8 offices acting as administrative centers. Similarly, the management of approximately 8 branches of the tax office (BTO) in the self-governing region assumes lesser requirements concerning directors of eight TOs. The advantage is also the saving on staffing, especially by decreasing the number of managers in Offices of the Tax Directorate of the Slovak Republic (OTD SR) according to the current organizational structure of OTD SR. By creating a smaller number of BTOs with increased number of employees, the transition of some of them from small TOs to TOs/BTOs will ensure the size "unification" of the individual TOs/BTOs with the parallel impact on creating the same organizational structure of all TOs/BTOs. This will allow the application of the same procedures and management system, as well as performance comparison (benchmarking), which (due to the differences in organizational structures) have so far not been possible to compare. In these intentions and on the basis of these assumptions, we have examined the acceptance of the reform by taxpayers, which we consider to be extremely important for its flawless functioning.

The presented research can be methodically classified as "face to face" category. In order to get as much relevant data as possible to identify the need, acceptance or rejection of the reform a simple questionnaire was distributed. The following criteria are considered to be relevant:

- Willingness of the tax office employees
- Competence of the tax office employees
- Tax collection – process/method and simplicity
- Collection of tariffs/customs – process/method and simplicity
- Collection of levies - process/method and simplicity
- Administrative difficulty of the tax and customs agenda
- Time saving
- Availability
- Electronicisation/computerisation - great user comfort
- Transaction costs
- Postage costs
- Labour costs

Selected key criteria were generated according to the expected benefit of the reform (if implemented). This is a qualitative and quantitative research based on the facts of the reform (mentioned in the introductory part of the contribution). The aim of the survey is to identify the needs and expectations of the people most affected by the reform. In addition, our research will spontaneously reveal the need for reform, its (in)effectiveness, through the method of generating criteria. The analysis of individual segments is based on own experience and the perception of the benefits of tax and customs reform from the citizen's point of view. Through a weighted average, we will define a qualitative mark for individual reform factors and and then we quantify them given the number of relevant answers received (i.e. completed questionnaires).

- Time horizon of the survey: 25/01/19 - 15/12/20
- Representative sample: 1 500
- Number of questionnaires issued: 3500
- Number of (completed) questionnaires collected: 2187

For the purpose of the survey, respondents were divided into four age-related categories (intervals), with the following age categories being represented: 18-25 years: 384 respondents, 26-35 years: 369 respondents, 36-45 years: 359 respondents, and 46-60 years 388 respondents. The average number of respondents in the four (age) categories is 375 respondents. Calculated standard deviation is of value of 13.44. In terms of employment, the composition of respondents participating in the research in the following categories was as follows: officials: 373 respondents, entrepreneurs: 383 respondents, students: 360 respondents, not mentioned (unspecified) job: 384 respondents. Questionnaire survey was carried out in four districts, the number of respondents in each category is as follows: Prešov: 374 respondents, Bratislava-city: 385 respondents, Košice: 389 respondents, Banská Bystrica: 352 respondents. From the above qualitative analysis, it can be concluded that the different categories of respondents in terms of age, gender, employment and geographical distribution are homogeneous with minimum variations in the number of groups in each group.

To analyze the questionnaire (evaluation) questions, factor analysis was chosen as the most appropriate method for processing the given data type. Factor analysis is a multidimensional statistical method that focuses on generating new variables and aims to reduce the amount of data (data reduction) with minimal loss of information. It is based on the assumption that the dependence between the observed variables is due to the effect of a smaller number of non-measurable variables in the background, which are referred to as common factors. These common factors are defined as a linear combination of original variables.

The main objective of factor analysis is to know and use the structure of common factors that are considered to be the hidden (underlying) causes of mutually correlated variables, concurrently to explain and elucidate the observed dependencies in the best and most simple way. Thus, in the final solution, each variable should be correlated with a minimum number of factors, an at the same time the number of factors should be as small as possible. To assess the suitability of factor analysis as a tool, a Kaiser-Mayer-Olkin (KMO) test and Bartlett's test of sphericity was conducted (Table 1).

Table 1. KMO test, Bartlett's test

Kaiser-Meyer-Olkin Measure of Sampling Adequacy.		.895
Bartlett's Test of Sphericity	Approx. Chi-Square	91.564
	df	66
	Sig.	.001

Source: Authors.

Table 2. Table of eigenvalues

	Eigenvalues - Extraction: Principal components			
Value	Eigenvalue	% Total variance	Cumulative Eigenvalue	Cumulative %
1	1,185683	9,880695	1,185683	9,88070
2	1,129148	9,409563	2,314831	19,29026
3	1,073402	8,945018	3,388233	28,23528
4	1,054944	8,791202	4,443177	37,02648
5	1,035592	8,629930	5,478769	45,65641
6	1,026562	8,554679	6,505330	54,21109

Source: Authors.

Table 3. Table of factor loadings

	Factor Loadings (Varimax normalized) Extraction: Principal components (Marked loadings are >,600000)					
Variable	Factor 1	Factor 2	Factor 3	Factor 4	Factor 5	Factor 6
Willingness	0,247480	0,407177	-0,204138	0,247481	-0,119360	0,033054
Competence	-0,055020	**0,681540**	0,006961	-0,072000	0,265825	-0,143970
Tax collection	0,208022	0,193628	0,113738	**0,679960**	0,040969	0,043309
Collection o Tariffs	0,132591	0,065282	-0,075041	-0,362360	0,549920	-0,087184
Collection of levie	**0,681574**	-0,003556	-0,035835	-0,123172	-0,073222	-0,030383
Administrative difficulty	0,059084	-0,045698	-0,032242	-0,025110	0,030893	**0,857102**
Time saving	-0,019193	0,122648	**0,774129**	0,022727	-0,152556	-0,235661
Availability	**0,631722**	-0,009950	0,037099	0,165632	0,066763	0,065606
Electronisation	-0,246514	0,479615	0,041579	-0,040427	-0,365937	0,364061
Transaction costs	0,164648	0,278856	0,194761	-0,570120	-0,033999	0,152900
Cost of postage	-0,015977	0,224791	**-0,620793**	0,032046	-0,174028	-0,250511
Labor costs	-0,152111	0,061086	0,068049	0,266025	**0,711466**	0,122212
Expl.Var	1,103872	1,051341	1,092940	1,102301	1,094582	1,060295
Prp.Totl	0,091989	0,087612	0,091078	0,091858	0,091215	0,088358

Source: Authors.

The analysis of the factor model itself is based on the analysis of factor loadings of the individual factors listed in Table 3. It is evident from the table that Factor 1 strongly correlates with the question of the collection of levies (0.68157) and availability (0.63172). Considering the above-mentioned fact and the plus sign of these significant correlations, the first common factor explaining the variability of respondents' replies can be attributed to the name "availability and effectiveness of levies collection". From the point of view of respondents' responses an increasing the efficiency of levies collection and improving the availability of contact points can contribute to increasing the efficiency of tax administration.

The second factor strongly correlates with competence (0.68154) and is therefore called "competence of the tax office employees"; an increasing effectiveness of tax administration is achieved through increasing the competence of its employees. The third common factor strongly correlates with time savings (0.77413) and postage/shipping costs (-0.62079). This factor can therefore be called "time savings". From the results it can be concluded that increasing the effectiveness of tax administration at this point can be achieved by reducing the time that taxpayers have to spend to meet their obligations and also by reducing the postage costs when sending reports. In the context of these conclusions, it seems necessary to electronize the entire tax system to address both of these issues. The fourth factor is highly correlated (0.67996) with tax collection. Thus, with the increasing efficiency of tax collection, it is possible - according to respondents' opinions - to increase the efficiency of the tax system. The fifth factor is correlated with the labour cost issue (0.71147). Thus, this significant factor can be classified as "labour costs". The last significant factor that affects the variability of respondents' answers correlates with the question of administrative difficulty of the tax and customs agenda (0,8571) and therefore can be called "administrative difficulty".

On the basis of factor analysis, we can conclude that the 12 questions in the questionnaire that deal with the satisfaction of respondents with tax administration can be explained by the six common factors that stand in the background. These factors can be defined as follows:

- Factor 1: levies collection and availability,
- Factor 2: competence of tax administration employees,
- Factor 3: computerization of tax administration,
- Factor 4: effectiveness of tax collection,
- Factor 5: labour costs,
- Factor 6: administrative difficulty of tax administration.

Further analyses (Table 3) showed that the interdependence of Factor 1 (levies collection and availability) and Factor 4 (effectiveness of tax collection) creates only one factor-pure character that is dependent on the Factor 1, namely the "collection of levies". Although the "availability" is strongly correlated with Factor 1, we also observe a small dependence on Factor 4. Similarly, there is dependence of "tax collection" on Factor 4 but also dependence on Factor 1. In addition to these dependencies, the relationship of "transaction costs" and "collection of tariffs/customs" with tax collection effectiveness (Factor 4) can also be observed. In this case, however, a direct proportion is evident. In addition to the obvious dependencies (given in Table 3), the analysis of Factor 1 and Factor 5 (labour costs) also shows the dependence of "electronisation of tax administration" on labour costs; according to the respondents' opinion, electronisation is contributing to the reduction of labour costs. Interesting dependence can be seen between "collection of tariffs/customs" and labour costs as well as between "competence of tax administration staff" and labour costs. In this case, the respondents are of the opinion that increasing labour costs results in increased competence of tax administration staff as well as in higher efficiency of the customs collection. In addition to the above-mentioned facts, i.e. the strong dependence of "availability" and "collection of levies" on Factor 1 and "administrative difficulty" on Factor 6, we also see the dependence of "tax administration electronisation" on the Factor 6. According to the respondents' opinion, electronisation of tax administration will increase its administrative difficulty.

4 Conclusion

To ensure homogeneity we chose 1500 questionnaires to create a representative sample from the total number of completed questionnaires. Given the scale and difficulty of the survey, only partial results and findings are presented. In terms of research carried out, it can be said that there is a dissatisfaction with the overall tax and customs system in Slovakia, with the most critical point being the labour costs related to the tax and customs agenda as well as the costs of processing this agenda. At the same time, it points to the need to simplify the tax and customs system in the form of an agenda electronisation, the transparency of public finances and the general mistrust of the citizens against the tax evasion mechanism and the amount of money spent on tax and customs offices.

Based on the expected benefits of the tax reform mentioned in the document "Concept for the Reform of Tax and Customs Administration Reform, with a View to Unifying the Collection of Taxes, Customs Duties and Insurance Contributions" elaborated by the Ministry of Finance of the Slovak Republic, "unifying the collection of taxes, customs duties and insurance contributions into a single point will significantly reduce the administrative burden vis-a-vis the tax/contribution payers, will reduce bureaucracy and improve the perception of citizens or employers about mandatory payments", i.e. it is assumed that the gradual introduction will most likely to reduce the current dissatisfaction.

In addition to predetermined research objectives, the survey also revealed further weaknesses in the area of tax and customs administration, particularly in the area of citizens' education. From the total number of completed questionnaires, 447 respondents were aged 18-25, but only 291 questionnaires were relevant. This indicates low awareness of young people. Therefore, it is necessary to improve information and to "bring closer" the Slovak tax and customs administration system to students of secondary and higher education. Information campaign about the tax and customs reform in the Slovak Republic could partially improve this issue.

References

1. J. Dobrovič. *Hodnotenie výkonnosti manažmentu efektívnej daňovej správy SR v kontexte regiónov pred reformou daňovej správy SR s návrhom jej zefektívnenia po reforme. Performance Evaluation of Effective Tax Administration Management of the Slovak Republic in the Regional Context before the Tax Administration Reform of the SR with a Proposal for Its Higher Effectivity after the Reform.* Prešov: University of Prešov. (2015)
2. J. Belás, A. Chochoľáková, L. Gabcová. Satisfaction and loyalty of banking customers: a gender approach. *Economics & Sociology*. **8**(1), 176-188 (2015)
3. R. Rajnoha, D. Slivková, J. Dobrovič, Globatization and Transer Pricing in Multinational Corporations in Slovakia and OECD Countries – Analytical Study and Decision-Making Model on the Choice of Optimal Transfer – Pricing Method. Ekonomický časopis, **62**(6), 609-630. (2014)
4. A. Korauš, J. Dobrovič, A. Ključnikov, M. Gombár, Consumer approach to bank payment card security and fraud. Journal of Security & Sustainability Issues, 6(1), 85-102. (2016)
5. D. Tamulevičiené, Methodology of complex analysis of companies' profitability. Entrepreneurship and sustainability issues, **4**(1), 53-63. (2016)
6. A. Korauš, R. Štefko, J. Dobrovič, Acqusition Activity in Financial Sector. 12th International Scientific Conference on European Financial Systems, Brno: Masarykova univerzita v Brne. (2015)
7. A. Ključnikov, J. Belás, L. Smrcka, The role of risk-taking and competitive aggressiveness in management of SMEs. Polish Journal of Management Studies, **14**(1), 129-139. ISSN 2081-7452, (2016)7
8. J. Paulík, F. Kombo, A. Ključnikov. CSR as a driver of satisfaction and loyalty in commercial banks in the Czech Republic. *Journal of International Studies*. **8**(3), 112–127 (2015)
9. A. Tvaronavičiené, G. Žemaitaitiené, T. Bilevičievé. Ecosystem for sustainable entrepreneurship: towards smart public procurement review procedures. *Entrepreneurship and sustainability issues*. **4**(1), 39-52 (2016)
10. R. A. Musgrave, P. B. Musgrave. *Public Finance in Theory and Practice*. 5th Ed. New Yok: McGraw-Hill Education. ISBN 0-07-044127-8 (2004)
11. J. H. Wu, Y. M. Wang. Measuring ERP success: The key-users' viewpoint of the ERP to produce a viable IS in the organization. *Computers in Human behavior*. **23**(3), 1582-1596 (2007)
12. J. Belás, P. Bartoš, A. Ključnikov, J. Doležal. Risk perception differences between micro-, small and medium enterprises. *Journal of International Studies*. **8**(3), 20-30 (2015)
13. Y. Bilan. Sustainable development of a company: Building of new level relationship with the consumers of XXI century. *Amfiteatru Economic*. **15**, 687-701 (2013)
14. F. Chromjaková, R. Rajnoha. Economy of Innovation as a Part of the Increase of the Company Performance. *Journal of Competitiveness*. **1**(1), 66-74 (2009)
15. A. Schultzová. Spolupráca členských štátov Európskej únie pri odhaľovaní daňových podvodov. *Ekonomický časopis*. **53**(3), 308- 313 (2005)
16. J. Belás, A. Chochoľáková, L. Gabčová. Bank Customers' Satisfaction, Customers' Loyalty and Additional Purchases of Banking Products and Services: A Case Study from the Czech Republic. *Economics and Sociology*. **8**(3), 82-94 (2015)
17. A. Ključnikov, M. Sobeková Majková, A. Schwendemann, C. H. Knogler. Do SMEs in Slovakia face real difficulties in obtaining financing? Comparison of the studies from Slovakia and the EU. *Journal of International Studies*. **9**(3), 36-52 (2016)
18. A. Zubaľová. *Daňové teórie*. Bratislava: Vydavateľstvo Ekonóm. ISBN 80-225-1789-5 (2003)
19. J. Belás, V. Demjan, Bank customers satisfaction: case studies from Czech Republic. *Actual Problems of Economics*, **12**(162), 315-323 (2014)
20. J. Belás, G. Sopková. Significant determinants of the competitive environment for SMEs in the context of financial and credit risks. *Journal of International Studies*. **9**(2), 139-149 (2016)

AUTHOR INDEX

EDP Sciences – Web of Conferences
17, Avenue du Hoggar
Parc d'Activité de Courtabouf
BP 112
F-91944 Les Ulis Cedex A
France

ISBN 978-1-7138-4390-0

CURRAN ASSOCIATES INC.
proceedings.com

EOS Annual Meeting (EOSAM 2020)

EPJ Web of Conferences Volume 238 (2020)

Online
7 - 11 September 2020

edp sciences